Library of
Davidson College

Field Guide to the Orchids of Costa Rica and Panama

ALSO BY ROBERT L. DRESSLER

The Orchids: Natural History and Classification
Phylogeny and Classification of the Orchid Family

Field Guide to the Orchids of Costa Rica and Panama

Robert L. Dressler

Comstock Publishing Associates
a division of Cornell University Press
ITHACA AND LONDON

Copyright © 1993 by Cornell University

All rights reserved. Except for brief quotations in a review, this book, or parts thereof, must not be reproduced in any form without permission in writing from the publisher. For information, address Cornell University Press, Sage House, 512 East State Street, Ithaca, New York 14850.

First published 1993 by Cornell University Press.

Printed in the United States of America.

Color plates printed in Hong Kong.

⊗ The paper in this book meets the minimum requirements of the American National Standard for Information Sciences—Permanence of Paper for Printed Library Materials, ANSI Z39.48-1984.

Library of Congress Cataloging-in-Publication Data

Dressler, Robert L., 1927–
    Field guide to the orchids of Costa Rica and Panama / Robert L. Dressler.
      p.   cm.
    Includes bibliographical references (p.) and index.
    ISBN 0-8014-2582-4 (cloth). — ISBN 0-8014-8139-2 (paper)
    1. Orchids—Costa Rica—Identification. 2. Orchids—Panama—Identification. I. Title.
QK495.O64D68   1993
584'.15'097286—dc20   93-19556

Contents

The color plates of the orchids follow page 326.

	Preface	vii
1	The Area: Geography, Climate, Vegetation, and People	1
2	Orchid Structure and Ecology	17
3	Classification and Identification: How to Use This Guide	33
4	*Cattleya* and Its Cousins: Subtribe Laeliinae	45
5	*Oncidium* and Its Relatives: Subtribe Oncidiinae	99
6	The Bizarre Subtribes: Catasetinae and Stanhopeinae	133
7	Mostly Miniatures: Subtribe Pleurothallidinae	156
8	Maxillarieae with Four Pollinia	217
9	Miscellaneous Orchids with Corms or Pseudobulbs	262
10	Miscellaneous Orchids without Pseudobulbs, Mostly Terrestrial	277
	Appendix A. Preparing Orchid Materials for Study or Identification	327

Appendix B. Authors of Orchid Names Used in the
Guide 331
Glossary 355
Illustration Credits 361
Index 363

Preface

My goal in writing this field guide was to make it possible for both botanists and nonbotanists to identify at least the genus of most orchids found in Costa Rica and Panama. I have tried to emphasize features that can be seen in the field with the naked eye or a hand lens. I have also tried to avoid unnecessarily complex terminology in the keys; the Glossary explains the special terms I found useful.

For almost all the genera with fewer than 40 species, users of this guide should be able to identify to species most plants they encounter. The keys for *Epidendrum*, *Maxillaria*, *Oncidium*, and *Pleurothallis* are longer and more complex than I would wish—even though I have divided the keys for three of these genera into several smaller keys—but I hope that they will prove useful. Ideally, separate field guides should be written for each of these larger genera, but such guides will not be available in the near future.

I do not provide keys to the species of *Lepanthes* and *Stelis*. The many species of *Stelis* are not well known, and both genera are difficult to identify in the field without a good dissecting microscope. I do not attempt to key the genus *Kreodanthus*. There are at least three species in Costa Rica and Panama, but two are unnamed and the third probably misnamed; all three are rather scarce. Most orchid genera in Costa Rica and Panama consist of a moderate number of species, and I hope that most will be relatively easy to identify.

The guide is based primarily on published and unpublished drawings and published descriptions, in addition to some measurements and observations of museum specimens; where I was able, I have filled the gaps with my own knowledge and that of colleagues. For very few species were the keys written on the basis of descriptions alone. With the exceptions noted above, I have tried to include all the species known to occur in Panama and Costa Rica; I have also included other species that occur in nearby countries and are likely to appear in our area. Small-flowered terrestrial orchids are especially likely to pass unnoticed, and I have included several that will probably be found in Costa Rica or Panama sooner or later.

The species names published or recorded from Costa Rica or Panama for which I have too little information are marked with an asterisk in Appendix B. They may be the names of distinct species, they may be synonyms of other names, or they may be the correct names for species treated here under other names. In most cases, these taxonomic problems cannot be solved without the study of specimens in European museums.

New orchid species and new records are still being found in Costa Rica and Panama. The bad news is that one cannot hope to identify every orchid found there with confidence. The good news is that anyone may find new species and new records. Indeed, if one searches off the beaten path and pays special attention to terrestrials and smaller-flowered epiphytes, new species and new records can scarcely be avoided.

As this field guide goes to press, work continues on a detailed orchid flora of Costa Rica. Still several years from completion, it will make possible a more complete field guide, though even then we can expect to find additional species in Costa Rica. An orchid flora for Panama already exists, but the orchids had not been well sampled when it was written, and many species now known to occur there are not mentioned.

Several additional species were reported or identified while the manuscript for this field guide was being reviewed. I hope that it will be possible to revise and improve the guide within a few years, and I would appreciate any information on sections of the key that need improvement. I would also welcome news of additional species found in Costa Rica or Panama, preferably with information on where specimens are deposited or with pressed flowers (see Appendix A).

No one person can be an expert on 1300 to 1500 species, and I am deeply grateful for the generous help of John Atwood (*Maxillaria*), James P. Folsom (*Dichaea*), Eric Hágsater (*Epidendrum*), Carlyle Luer (the subtribe Pleurothallidinae), Gerardo Salazar (*Mormodes*), and Dariusz Szlachetko (keys to the genera and species of the Spiranthinae). These studies are continuing, and these friends will continue to add to our knowledge of these groups.

Many others have helped me with information, unpublished drawings, photographs, and sympathy. I am very grateful to Wendy Zomlefer for her patience and her determination to finish the seemingly endless series of drawings for the keys. I am also greatly indebted to Luís Acosta, Bryan R. Adams, Eric A. Christenson, Calaway H. Dodson, Günter Gerlach, Ed Greenwood, Alvaro Herrera, Franco Pupulin, Dora Emilia Mora de Retana, Ileana de Terán, and Mark Whitten. Thanks are due the entire Asociación Costarricense de Orquideología, both for their help and for their remarkable tolerance and restraint when I get in the way while a show is being set up and snatch already arranged plants for my wife, Kerry, to photograph. I am especially grateful to Kerry. She took most of the color photographs used here. Several people kindly supplied slides of species or genera that Kerry and I had not been able to find, or better slides of some that we did have (see Illustration Credits).

ROBERT L. DRESSLER

Micanopy, Florida

1 The Area: Geography, Climate, Vegetation, and People

Costa Rica and Panama are especially attractive to the nature-lover and the orchidophile. They are easily accessible and offer a wide range of habitats and an interesting balance between good roads and remote, relatively undisturbed forests. Their orchid floras may seem modest when compared with the incredible richness of Colombia or Ecuador, but Costa Rica and Panama are much smaller countries. For their size, they are extremely rich in orchids and other plants.

Geography

Physically, Costa Rica and Panama make up the narrowest part of the isthmus between North America and South America. The orientation of this part of the isthmus is much more nearly east and west than is usually realized, and the Panamanian portion is somewhat S-shaped. Directions in Panama are usually given in terms of Atlantic side, Pacific side, toward Colombia, and toward Costa Rica, which causes less confusion than north, south, east and west.

Most of the isthmus is either hilly or quite mountainous. Costa Rica's highest peak, at 3819 meters (about 12,000 feet), is Cerro Chirripó, but there are several other peaks over 3000 meters in height in the central and eastern regions (Map 1). The mountain range becomes lower toward the Nicaraguan border, with a few mountains about 2000 meters in height in Guanacaste.

Western Panama is essentially a continuation of the Costa Rican Cor-

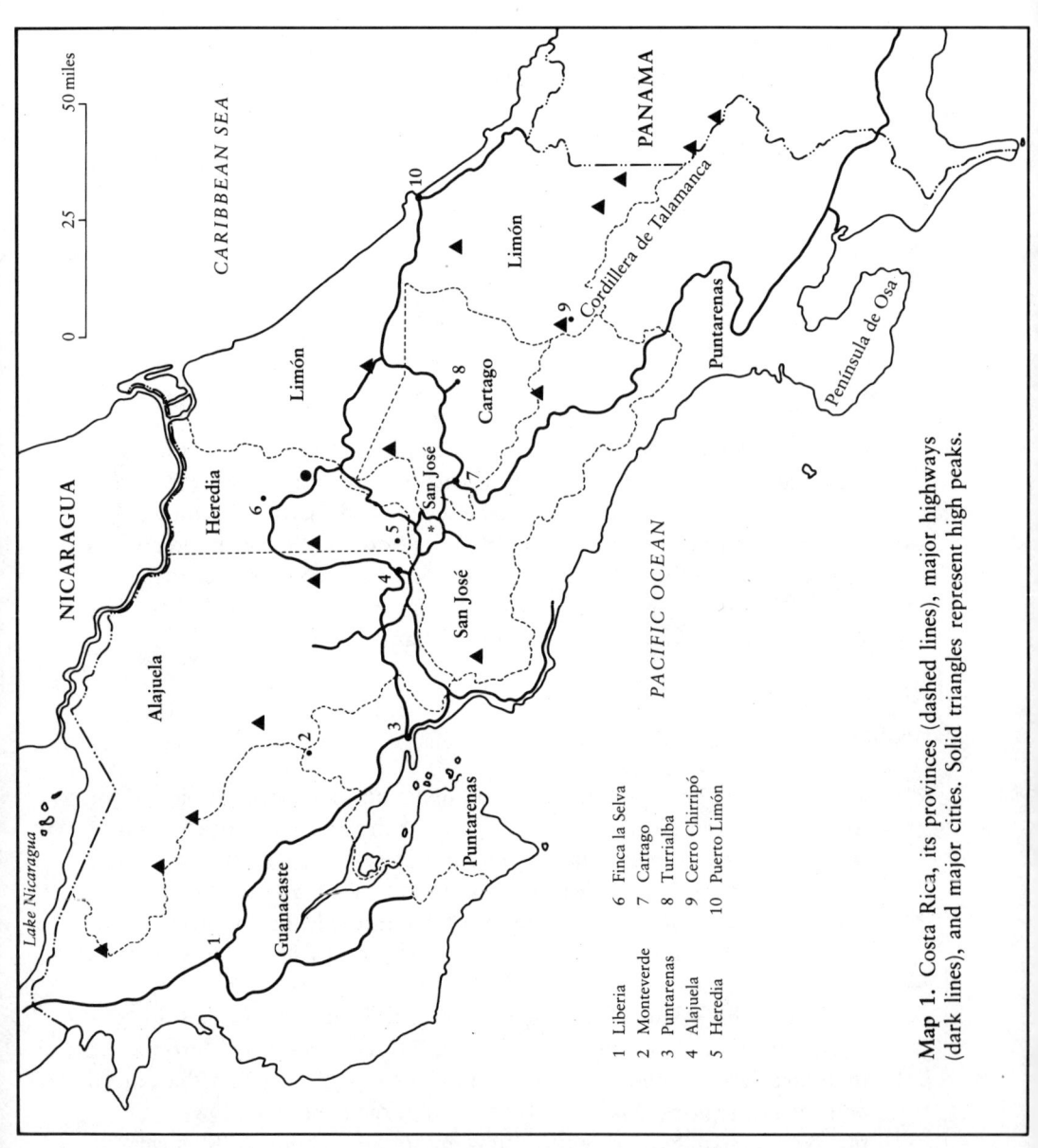

Map 1. Costa Rica, its provinces (dashed lines), major highways (dark lines), and major cities. Solid triangles represent high peaks.

dillera de Talamanca. Volcán Barú is Panama's highest peak, at 3475 meters, and unlike most other mountains in the area its top is relatively dry. There are several other mountains over 3000 meters in Chiriquí and Bocas del Toro. Toward central Panama the mountains decrease in height (Map 2). On the Transisthmian Highway it would be difficult to recognize the continental divide without a highway sign, and the highest point on the highway is less than 200 meters. The principal mountains in Panama and Coclé provinces are all about 800 to 1000 meters. Toward Colombia the mountains become higher and more continuous, with several peaks near 1500 meters. Some of the mountains of eastern Panama are still poorly known except to smugglers and Colombian timber thieves. It is interesting that the relatively low peaks of central Panama are distinctly montane in climate and vegetation. An isolated mountaintop at 1000 meters seems much more montane than a 1000-meter shoulder on a 3000-meter mountain. Thus one finds cool, wet, windswept elfin forest on Cerro Campana, Cerro Gaital, and Cerro Jefe in central Panama.

Central America was probably once an archipelago without continental connection either to the north or the south. Now, the fauna and flora are complex blends of South American, tropical North American, and groups that doubtless go back many millions of years in the Middle American archipelago. On the high peaks one finds many outliers of temperate and alpine North America and a few representatives of the Andean flora.

Climate

The climate of the isthmus is quite variable and closely related to the mountainous topography. In general, though, it is much wetter on the Caribbean side, and there is a marked dry season on the Pacific side. There are a few areas, such as north of El Copé (Coclé Province, Panama), where the dry season winds pour through a low gap in the mountains. In such spots fog and drizzle are nearly constant during the so-called dry season. More rain falls during the rainy season, but there is also much more sunshine. On one high, narrow ridge in eastern Panama I found the north side of the ridge to be quite wet in February. A few steps to the south (as far as one could safely walk) the plants were dry and dusty.

There is some sort of dry season from about mid-December to mid-April throughout the area, but its intensity varies greatly from place to place and year to year. During the dry season the trade winds blow from

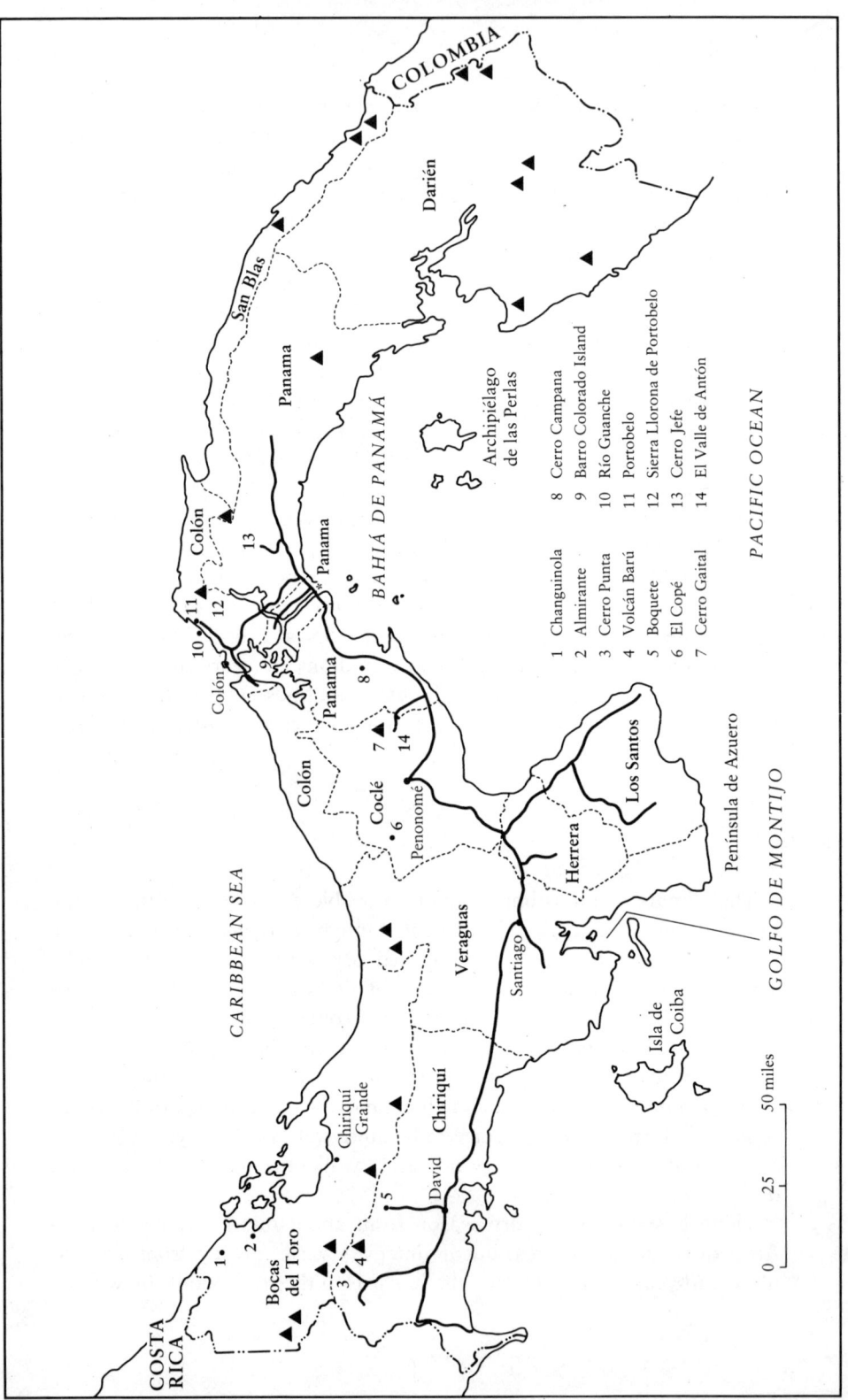

Map 2. Panama, its provinces (dashed lines), major highways (dark lines), and major cities. Solid triangles represent peaks, higher at the two ends of the country and about 800 to 1000 meters in the central area.

the northeast nearly constantly, and the north slopes wring out most of the water that the winds bring to the isthmus. At the beginning of the dry season the cooling breezes are most welcome, and the dirt roads and trails begin to dry out. By the end of the dry season the sky is hazy with dust and smoke, and every breeze deposits black ash from distant fires. The rainy season begins in moderation, but the rainfall increases gradually until October and November, when forests are dripping wet and dirt roads and trails have become quagmires or intermittent streams. For those who wish to get away from civilization, the early dry season and the early rainy season are the best times, though for the wettest areas the late dry season may be preferable.

It is both cooler and moister at higher elevations, and the Caribbean slopes of the mountains are especially wet. On the Caribbean slope of Panama there are wetter areas near Portobelo, especially on the Sierra Llorona de Portobelo, and also along the north coast of Veraguas and the adjacent areas of Colón and Bocas del Toro provinces. The area around Changuinola and Almirante, farther west in Bocas del Toro, receives about the same rainfall as Barro Colorado Island in central Panama. The rainfall in Bocas del Toro is more evenly distributed throughout the year, however, so the vegetation is much lusher than on Barro Colorado, where there is normally a strong dry season. On the Pacific side, western Costa Rica and the western side of the Bay of Panama are especially dry, while Quepos and the Península de Osa are relatively wet. In Panama, Isla de Coiba and the area around the Golfo de Montijo are somewhat wetter than the areas to the east and west. Even in the mountains of the Península de Osa and the Península de Azuero, though, the epiphyte diversity seems lower than in comparable areas in the central mountain range, perhaps because of occasional severe dry seasons.

Vegetation

The Life Zone system of Holdridge is the best known classification of Central American vegetation (Holdridge 1967). Nearly 20 vegetation types are recognized in our area, ranging from tropical dry forest to subalpine rain paramo, with many transitional zones. It is unusual for natural vegetation types to have sharp boundaries; there is usually a gradual transition between different types, with pockets of each type within the hazy boundaries of the other. Still, there is a problem with the Holdridge system (and with some others), especially with regard to epiphytes. Rainfall is considered in delimiting life zones, but average annual rainfall is the factor used. The forests near Almirante and Changuinola and those of Barro Colorado Island are both listed as "tropical moist forest," but as noted above, the seasonal distribution of the rain is quite different in the two areas. For epiphytes, especially, the dry season can be a critical factor.

For our purposes, I characterize the natural vegetation of Costa Rica and Panama using six rather broad categories: dry forest and scrub, seasonally dry forest, moist and wet forests (usually with at least a slight dry season), cloud forests, elfin forest, and subalpine/alpine vegetation. Maps 3 and 4 show the natural vegetation types; that is, what the vegetation would be in each zone if there were no human disturbance. This is, of course, hypothetical in some areas.

Many factors affect the distribution of plant species. Moisture, sunlight, and temperature are the major physical factors, and moisture may override or compensate for temperature, in that plants typically found on high mountains may occur at much lower elevations if there is enough moisture throughout the year. *Miltoniopsis roezlii* and *Oerstedella pseudoschumanniana* are thought of as high-elevation plants (for central Panama), yet I have found *Oerstedella* near sea level near the Río Guanche (Colón Province), and *Miltoniopsis roezlii* grows near sea level in the wet Chocó of Colombia. Once must emphasize that there is a spectrum of vegetation types and that each plant species has its own range of tolerance and requirements, which rarely correspond exactly to a life zone. Any attempt to map or characterize vegetation types in detail will be frustrating, if not misleading.

Dry Forests and Scrub. In these areas there is a strong dry season. The forest canopy may be quite open, or the woody vegetation may be of scattered shrubs or small trees. Dry forest and scrub are typically bone dry in the dry season, and orchid diversity is quite low. Still, orchids that can survive here may be quite numerous, as is the case with *Bletia purpurea, Brassavola nodosa, Catasetum maculatum, Encyclia cordigera, Epidendrum stamfordianum, Oncidium ampliatum,* and *Schomburgkia undulata.* All these plants have fleshy corms or pseudobulbs and can store water to survive a prolonged drought.

Seasonally Dry Forests. Here the trees are taller and the canopy is relatively continuous. During the rainy season the uninitiated might class these forests as "tropical rain forest," whatever that may mean. In the dry season, however, the mud on the trails is dry and cracked, many trees lose their leaves, and many terrestrial plants wither and dry, with only fleshy underground parts remaining to sprout up with the first rains. The diversity of orchids is greater than in the dry forests. On Barro Colorado, for example, more than 90 orchid species have been found, about as many as occur in the state of Florida. Most of the epiphytes in these forests have pseudobulbs or fleshy leaves. *Cattleya skinneri* and *C. patinii* occur in seasonally dry forests, and *Aspasia, Brassia, Encyclia, Maxillaria, Oncidium,* and some of the more resistant pleurothallids can be expected there.

Moist and Wet Forests. The wetter forests are more interesting to the botanist, and especially to the orchidist, because the diversity of tree species and epiphytes is much greater. In the very wet forests a bewildering diversity may grow on a single tree. Naturally, most of the epiphytes grow high in the canopy, and orchid hunters soon learn to carefully examine every fallen tree or branch, where plants can always be taken with a clear conscience. There are, however, some orchids that tolerate more shade and regularly occur on tree trunks, such as *Aspasia, Brassia, Cochleanthes, Kefersteinia,* and *Kegeliella*.

Cloud Forests. Where frequent cloud cover keeps the forest wet much of the time one finds a marvelous diversity in both trees and epiphytes. Epiphytic trees grow on other trees, and it is often hard to determine the relationship between the various types of foliage supported by a single trunk. In cloud forests even cycads (*Zamia pseudoparasitica*) may occur as epiphytes, and there are many orchids, ferns, bromeliads, and woody epiphytes as well. Pleurothallids and other orchids with little storage tissue occur here in great diversity. As in the wet forests, most epiphytes live in the canopy, but many others grow on tree trunks or vines where there is enough light. There are also many terrestrial orchids, some growing in rather deep shade. In the very wettest forests, as on the upper Sierra Llorona de Portobelo, orchid diversity seems to be much lower, but one may still find interesting species in these dripping wet forests. In both cloud forests and elfin forests, a large part of the moisture may condense directly on the foliage. In these areas a dry season road may be quite dry except where a tree overhangs the road and water dripping from the leaves causes a wet spot.

Elfin Forests. On exposed mountain peaks where the soil is rocky or very poor in nutrients (usually because of leaching by rainfall) one finds a variant of cloud forest with small, often gnarled trees and no continuous canopy. Here, the distinction between terrestrial and epiphyte breaks

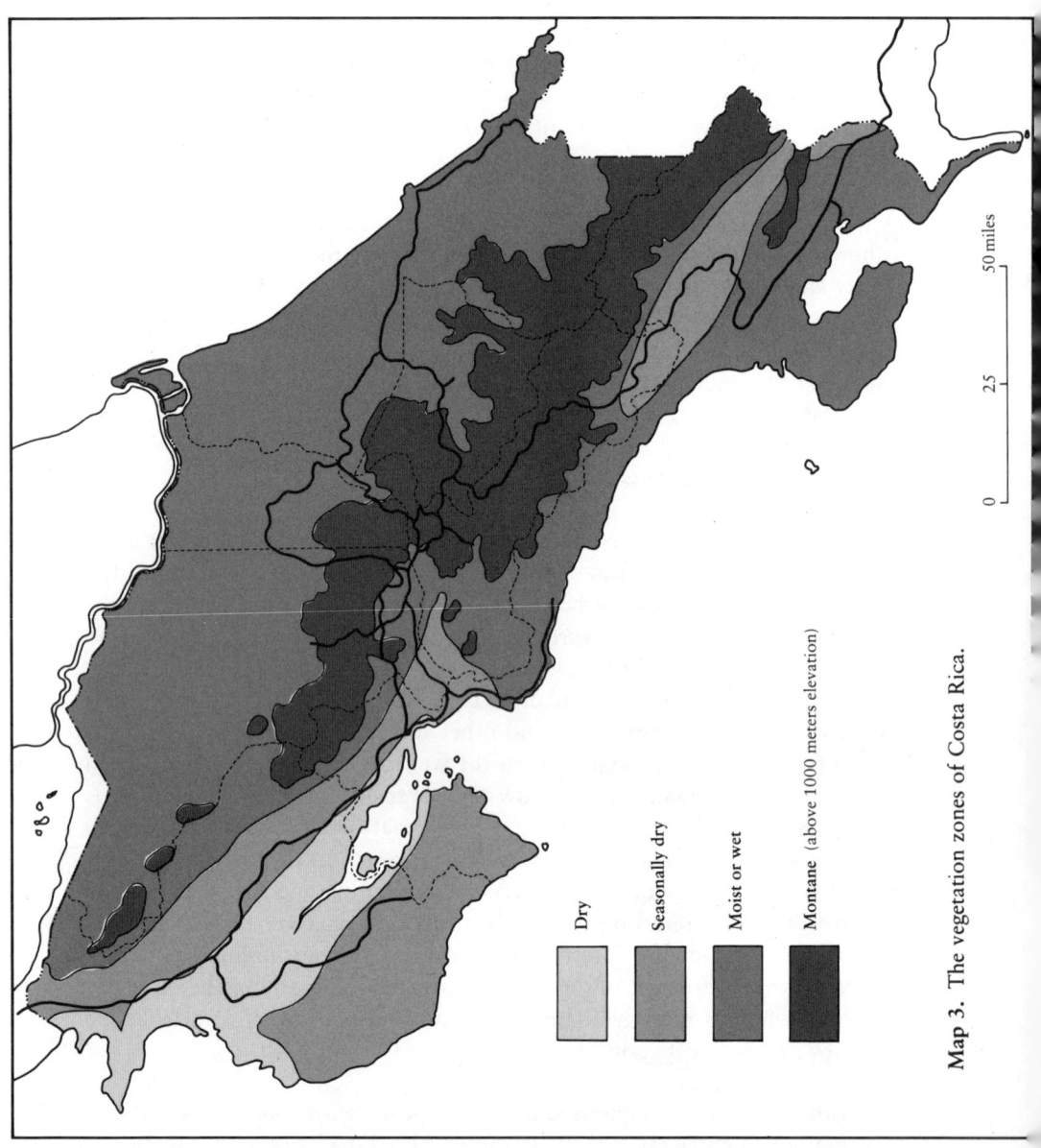

Map 3. The vegetation zones of Costa Rica.

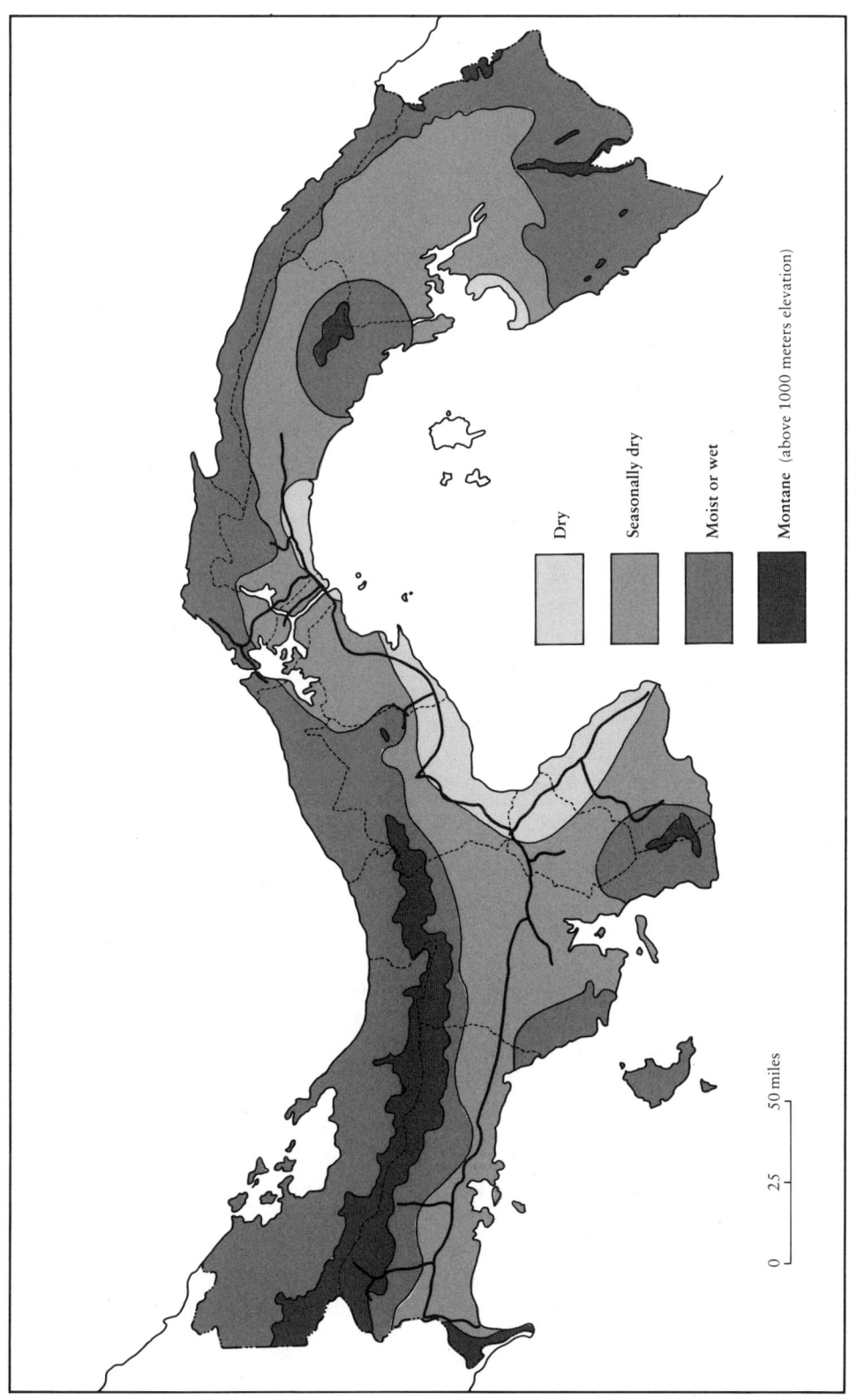

Map 4. The vegetation zones of Panama.

down, as terrestrials grow on tree trunks and fallen epiphytes grow happily in the moss. In the elfin forest one is usually walking on tree roots, and there may be large and treacherous cavities beneath the roots. In the extreme case it is very hard to identify a forest floor. The diversity of orchids in the elfin forest is mind-boggling, and many of the orchids are easy to find without cutting or climbing trees. One of the least popular inhabitants of some elfin forests is saw-grass, or *Scleria*, a climbing sedge with minutely saw-toothed leaves. It should be treated with the greatest respect (or a sharp machete).

Subalpine and Alpine Vegetation. At the highest elevations, lack of soil, low temperatures, and strong winds make life difficult for most plants, and one again finds grassy or scrubby vegetation reminiscent of the driest lowland areas. These mountaintops may receive plenty of rainfall, but they also have maximum drainage and drying winds, and plants growing there may be stressed by drought. The diversity of orchids is lower at the highest elevations, but one may find here such alpine plants as *Brachionidium, Fernandezia, Pachyphyllum,* and *Telipogon*; all are nearly impossible to cultivate at low elevations.

Disturbed and Man-made Vegetation. Those who enjoy nature may complain bitterly about the destruction of natural vegetation, but in fact, much of the world is covered by greatly altered vegetation. Thus it may be appropriate to say a bit about disturbed habitats in Costa Rica and Panama.

PASTURES. At lower elevations, orchidists generally consider pastures to be unnecessary spaces between trees. The typical pasture is covered by pits and pools of noxious mud; the grass may be higher than the average orchidist, and it is certain to be full of ticks during the dry season, to say nothing of unfriendly bulls at any season. At higher elevations, though, especially in Costa Rica, pastures are worth a look. The surface is, if anything, even wetter and more treacherous, but isolated trees left in the pasture may bear many orchids, and one may find interesting things on the trees around the pasture and even on the fence posts.

GRASSLANDS. Grasslands, which are especially common on the Pacific slopes, are often made up of tall grasses of little use to cattle or anything else. The problem is that they are burned nearly every year. The people of the region feel burning is necessary to kill the snakes or to bring the rain, and besides, it has always been done. Reforestation is nearly impossible under these conditions, and the soil is gradually degraded until the area is virtually useless. I took one European ecologist to see the grassy slopes of Cerro Campana, in Panama, and he could not understand why the area was maintained in such sterile grassland. Forests would be much more productive for sustained use, but the tradition is to turn forests into pastures, which then degrade into very poor grasslands.

SECOND-GROWTH FOREST. In wetter areas, pastures and other disturbed habitats will revert to forest if left alone. Forest soils normally carry an assortment of small seeds that cannot germinate in the shade. If a tree falls or a plot is cleared, the secondary vegetation starts growing with the first rains, if not before. Shrubs, vines, and small trees may sprout from the roots. The regeneration is slower if pastures or gardens are maintained for a time, but it follows much the same pattern. The first stages of reforestation involve fast-growing trees and vines (including saw-grass), some of them with harsh or spiny stems and leaves. Young second growth is dense and nearly impenetrable without a machete; even if one can hack through, it is not a pleasant habitat. Eventually the stronger trees shade out some of their competitors and the understory becomes clearer. As the seedlings of these secondary trees cannot grow in the shade of their parents, the forest finally reverts to more natural conditions if there are seed sources nearby.

Even the older second-growth forests are poor orchid habitats, though one may find a few terrestrials or a few common epiphytes on the larger trees. Young secondary forest has no redeeming features whatever. Unfortunately, one must often bash through young secondary forest just to get away from the roadside or to get from one patch of forest to another.

ROAD CUTS. At low elevations, road cuts are usually no more interesting to the orchidist than pastures. In higher or wetter areas, however, road cuts become much more exciting. In high mountain areas old road cuts may carry a great variety of both terrestrials and epiphytes. In some cases the cuts are rocky, and many epiphytes grow just as well on rocks as on trees. Even when the road cut is of clay, it may be steep enough to provide a good habitat for plants that normally grow on trees or rocks.

People and Cultures

Costa Rica and Panama are more different in their histories and cultures than one might expect from geographic proximity. Costa Rica was settled from the north, and the Spanish conquerors were accompanied by Nahuatl-speaking (Aztec) soldiers. Anyone who learned to speak Spanish in Mexico will recognize such words as *ayote, chapulín, chile, hule, petate, zacate, zopilote,* and many other plant and animal names based on Nahuatl, yet these words are virtually unknown in Panama and South America. In Panama *cajeta* is a small box; in Costa Rica, as in Mexico, it is a dessert made of milk (once sold in small boxes). In place of *tú* (thou), the Costa Ricans generally use *vos*, considered archaic in most other areas.

In Costa Rica the Conquest decimated the Indian population, whether through disease or deliberate slaughter, and there are relatively few Indians and cholos (part-Indians) in the population. Blacks are frequent along the Caribbean coast and are becoming more integrated into the national culture. Most Costa Ricans are of European background, and blond hair and blue eyes are common, both in the cities and on remote farms.

Virtually all visitors are charmed by Costa Rica. Not only are the scenery and climate delightful, but the people are friendly, polite, and cultured. Throughout the hemisphere, Costa Ricans are known as "ticos," supposedly because of their immoderate use of the diminutive suffix *-ico*; the alternative forms *-ito* and *-illo* are also frequently heard. In San José one is apt to be treated like an old family friend, even by taxi drivers. Literacy is very high in Costa Rica, and visitors may be quite startled by the culture and knowledge of even a poor farmer. Costa Rica is famous for its working democracy and gets along very well without an army. I do not mean to imply that there is no danger from thieves or pickpockets in Costa Rica. The massive influx of refugees from Nicaragua and El Salvador and the high level of unemployment among Costa Ricans and refugees alike means that there are thieves, though one imagines that tico thieves will be more polite than their refugee counterparts.

Incidentally, there are virtually no house numbers in San José. When my wife and I lived a few months in the San Pedro area our address was 200 meters south of the fig tree and 20 meters east. If that venerable old *higuerón* should fall, the city may have to put up a marker in its place to prevent total confusion. Some years ago I heard of a Spanish-speaking visitor who was told that the house he sought was 70 *varas* (an archaic measure, about 33 inches) south of where Zutano's *pulpería* (a small store selling juices, soft drinks, and usually a bit of everything else) used to be. The visitor did not understand either *vara* or *pulpería*, much less know where Zutano's *pulpería* had been. Now, fortunately, the ticos have adopted the more widely known meter to give directions, but finding one's way can still be a bit confusing to the uninitiated.

With the great diversity of orchids, or *guarias*, in Costa Rica, it is not surprising that ticos are quite orchid-conscious. Many cultivate orchids, and doing so is by no means a rich person's hobby. A farmer or an auto mechanic may have an impressive collection of orchids and know a great deal about them. By and large this is a good thing, but it puts considerable pressure on the orchids in the wild. Most people are conscientious and collect sparingly, but a few *materos* (plant sellers) collect plants for sale, and these too often strip every site they visit, with little thought of leaving anything for seed. Many of their plants die from neglect in the matero's yard. Some conservationists will have nothing to do with them, but I fear that this minor inconvenience does not inhibit the materos very much.

Panama was a cultural backwater of Colombia until outside interests promoted revolution and independence. The Panama Canal brought commerce to Panama City on a large scale, and it is now a large and cosmopolitan city, with all the advantages and disadvantages that that implies. Colón, on the Caribbean coast, has all the disadvantages and few of the advantages. There are large Indian groups in Panama, especially the Chocó, Cuna, and Guaymí. Of these, the Cuna have been the most efficient in protecting their area and taking advantage of foreigners. The Chocó are the most friendly, perhaps so much so that they have a hard time protecting their area and their interests against the pressures from invaders both black and white. The campesinos of the Pacific coast—Chiriquí, Veraguas, Herrera, and Coclé—are mainly of European stock, and they are as friendly as country folk usually are over most of the world. There has long been a considerable black population on the Caribbean coast of Central America. For many years these blacks spoke English and had more contact with the West Indies than with the Spanish-speaking cultures of the Pacific slope. More recently, West Indian laborers were brought in to construct the Panama Canal, and many of them remained in Panama. A few of the older generation still survive and speak West Indian English. The Spanish-speaking population of Darién is largely black and of Colombian origin. Both sides of the Panama-Colombia border are still relatively lawless areas, and the inhabitants move back and forth without benefit of passport.

Orchid Study

Both Costa Rica and Panama are rich in orchid species, and it is interesting to compare botanical knowledge of the two areas. Costa Rica has a long tradition of resident naturalists interested in orchids. For the most part their material was sent to the United States or Europe for study, but it was collected by knowledgeable residents who knew the orchids well enough to recognize novelties when they found them. The important contribution of these resident naturalists is reflected in the many orchids named in honor of A. Alfaro, G. Acosta, P. Biolley, A. and C. Brade, A. M. Brenes, O. Jiménez, C. H. Lankester, H. Pittier, A. Sancho, A. Tonduz, J. Valerio, C. Wercklé, and Amparo de Zeledón. Most collecting in Panama, on the other hand, has been done by short-term visitors. There are, and have been, a number of orchid aficionados, but, with the exception of C. W. Powell, they did not systematically prepare specimens for study. Also, many of the most interesting areas in Panama are still relatively inaccessible. At present, about 950 orchid species have been reported from Panama, and nearly 1200 from Costa Rica. In part, this reflects the greater concentration of high mountains in Costa Rica, but it also reflects the more intensive sampling there. I expect that continued collecting will increase the orchid flora of Panama more than that of Costa Rica, but eastern Costa Rica is still relatively poorly known. Many of the species known from western Panama are likely to be found in eastern Costa Rica as well. Each country probably has more than 1200 orchid species.

SOME GENERAL REFERENCES ON ORCHIDS AND NATURAL HISTORY

Allen, P. H. 1956. *The rain forests of Golfo Dulce*. Gainesville: University of Florida Press.
Atwood, J. T. 1989. *Orchids of Costa Rica, part 1*. Icones Plantarum Tropicarum Fasc. 14. Sarasota Fla.: Marie Selby Botanical Gardens.
Boza, M., and A. Bonilla. 1981. *The national parks of Costa Rica*. Madrid: INCAFO.
Dressler, R. L. 1981. *The orchids: natural history and classification*. Cambridge, Mass.: Harvard University Press (paperback ed. 1990).
Dressler, R. L., and G. E. Pollard. 1974. *The genus* Encyclia *in Mexico*. Mexico: Asociación Mexicana de Orquideología. (Most of the Central American species occur in Mexico.)
Gómez, L. D., ed. 1985. *Vegetación de Costa Rica*. San José: Editorial Universidad Estatal a Distancia.
Herrera, W. 1986. *Clima de Costa Rica*. San José: Editorial Universidad Estatal a Distancia.
Holdridge, L. R. 1967. *Life zone ecology*. San José: Tropical Science Center.
Janzen, D. H., ed. 1983. *Costa Rican natural history*. Chicago: University of Chicago Press.
Leigh, E. G., Jr., A. S. Rand, and D. M. Windsor, eds. 1982. *The ecology of a tropical forest: seasonal rhythms and long-term changes*. Washington, D.C.: Smithsonian Institution Press.

15 References

Mora de Retana, D. E., and J. T. Atwood. 1992. *Orchids of Costa Rica, part 2.* Icones Plantarum Tropicarum Fasc. 15. Sarasota, Fla.: Marie Selby Botanical Gardens. (A continuing series.)

Ridgely, R. S. 1976. *A guide to the birds of Panama.* Princeton: Princeton University Press.

Rodríguez C., R. L., D. E. Mora, M. E. Barahona, and N. H. Williams. 1986. *Géneros de orquídeas de Costa Rica.* San José: Editorial de la Universidad de Costa Rica. (Color paintings of most genera; text bilingual.)

Stiles, F. G., and A. F. Skutch. 1989. *A guide to the birds of Costa Rica.* Ithaca: Cornell University Press.

Terry, R. A. 1956. *A geological reconnaissance of Panamá.* Occasional Paper no. 23. San Francisco: California Academy of Sciences.

Williams, L. O., and P. H. Allen. 1980. *Orchids of Panama.* St. Louis: Missouri Botanical Garden.

2   Orchid Structure and Ecology

This chapter describes structural features of orchids that may be important in identification and briefly discusses a few aspects of orchid ecology, including pollination. These subjects are treated in greater detail elsewhere (Dressler 1981, for example).

General Structure

Growth Habit. Though there is much diversity in the ways orchids grow, they show two basic growth habits. The great majority of monocotyledons, including most orchids, have a modular, or sympodial, habit. Each module has limited growth. In the temperate zone a module is usually one year's shoot, including a section of rhizome, the stem and leaves, and an inflorescence. In the wet tropics some orchids may produce several modules during a single year. Typically, the new shoot arises from the base of an older shoot, and their bases together form part of the rhizome, but new shoots may arise from any part of an older shoot. Many species of *Epidendrum* have new shoots that arise well above the base of the old stem, so that they eventually have a branched, bushlike form. Such plants usually form an occasional new shoot from the base, and it, too, branches to form a small bush. In other species of the same genus, each new shoot arises well above the base of the older one, and these become scrambling, vinelike plants. This is especially typical of cloud forest plants growing in thick moss. New shoots may even arise from the tips of the old shoots, as in *Scaphyglottis*, where each shoot from the base eventually forms a shrublike branched chain of shoots, with the upper shoots progressively smaller than the basal shoot.

The other principal growth habit shown by orchids is quite different, in that a shoot may continue to grow indefinitely at the tip. This monopodial habit is typical of *Vanda* and its Old World allies but is also found in some New World orchids. The inflorescence of monopodial plants is always lateral; a terminal inflorescence would end the stem. Monopodial plants typically lack a distinct rhizome and produce roots along the stem. They may branch, but each branch has the potential for indefinite growth. In cultivation, the branches may be cut off and planted when they have roots.

Roots. Orchid roots are never thin and fibrous, like those of a lawn grass, but they are otherwise quite diverse. They are always somewhat fleshy, but they vary from relatively thin to nearly spherical. Epiphytes generally have a layer of dead, spongy cells on their outer surface, the velamen, that functions in holding and absorbing water and nutrients and may also be protective. Terrestrials may also have a velamen, but it is usually less obvious. Terrestrials, and some epiphytes, tend to have fleshy roots that serve as storage organs. Many terrestrials with thin leaves lose their leaves during the winter or dry season, and a rhizome or thick roots survive to grow again the next year. Some terrestrials with thinner roots also have thickened root-tubers, or tuberoids, among their roots. *Habenaria* and its relatives usually have a pair of globose structures that are mainly root but include some stem tissue. One of these tuberoids survives during the dormant season, and a new one is formed by the new shoot, the older one shriveling and disappearing before the dormant season.

Stems. Orchid stems are quite as variable as the roots; they may be short or long and slender or quite thick. The long, slender stems may be either soft and herbaceous or quite tough and woody, with hard, fibrous tissues. Stems normally have either leaves or bracts scattered along their length. The point at which a leaf is attached is termed a node and is quite obvious in orchid stems because the base of the leaf or bract usually clasps the stem, causing a definite line around the stem. Each node normally has a bud just above the leaf, and branches normally arise at nodes. The space between two nodes is termed an internode. As noted above, the rhizome is a compound structure made up of the bases of several successive shoots. There may be no definite rhizome in plants that sprawl or climb. In descriptions, the unqualified term *stem* usually refers to all the stem except the rhizome and the inflorescence.

Corms and Pseudobulbs. The botanical term *corm* refers to a short, squat stem, usually of several internodes, like that of *Gladiolus*. The corm is usually produced underground, but the typical corm of *Bletia*, for example, may be exposed on a rock. There is really no sharp distinction between a corm and a pseudobulb. The term *pseudobulb* refers to any thickened stem, especially one that is not buried in the soil. Pseudobulbs may be flattened, globose, or long and cigar-shaped. A useful feature in identification is whether the pseudobulb is made up of several internodes (has distinct rings around it) or a single internode (rings at base and apex, but none around the thick part).

Leaf Types. Like most monocotyledons, the orchids normally show parallel leaf venation; that is, there are many parallel veins in the leaf and the veins do not branch. A few orchids do not conform to the family description and have reticulate venation. Aside from these few exceptions, one finds three principal types of leaves in the orchids (though some borderline cases fit poorly into any category). The primitive condition for the orchid family appears to be a pleated, or plicate, leaf, which is just what the term implies. A pleated leaf has several to many major veins and is more or less folded at each major vein. Many terrestrial orchids have soft-herbaceous leaves, with all the veins about the same size. These leaves are thin and relatively flat, or slightly folded along the midline. Conduplicate leaves also have all veins about the same size, or they may have a larger vein along the midvein. These leaves are usually somewhat thick or leathery, and they have a definite fold at the midline.

Leaf Development. This may be useful in identification if one can find young leaves that are not fully expanded. In their development, conduplicate leaves are folded just once, with the two halves flat and pressed together, while both pleated and soft-herbaceous leaves are rolled up during their development. Pleated leaves are somewhat folded at each major vein, but soft-herbaceous leaves are smoothly rolled and not folded at all.

Leaf Arrangement. Leaves may be either spirally arranged or arranged in two opposite ranks, alternating with each other in straight lines up the stem. A few orchids have two or more leaves attached at the same level.

Fleshy Leaves. Many epiphytes, especially, have thick, fleshy leaves that may be either dorsoventrally flattened, laterally flattened, cylindrical, or somewhat three-angled. These are usually modifications of the conduplicate leaf, but *Vanilla* has thick, fleshy leaves that are rolled during their development.

Bracts and Sheaths. The basal part of the leaf in many orchids is wide and tubular and clasps the stem, while the rest of the leaf is flat and spreads out as the leaf blade. A leaflike structure, or bract, that sheaths a stem or pseudobulb (with or without a leaf blade) is called a sheath. Any small scalelike structure at a node, or any leaflike structure that is not really a foliage leaf, is usually called a bract. Every node that lacks a normal foliage leaf has some sort of sheath or bract. The bracts are especially noticeable on the inflorescence, where they may occur on the stalk below the flowers, and each flower has a bract at its base. If the inflorescence is branched, there is a bract at the base of each branch.

Inflorescences. Orchid inflorescences may be terminal, as in most Laeliinae, or lateral. In the orchids of Central America, lateral inflorescences are usually basal, being produced from the rhizome or at least near the base of the shoot; a few, such as *Campylocentrum* and *Lockhartia*, may have lateral inflorescences from the upper part of the stem. Orchid inflorescences normally have several to many flowers on a common stalk, or peduncle, though one-flowered inflorescences do occur, and the flowers may be fascicled, or densely clustered, with or without an obvious common stalk. Flowers, like leaves, may be spirally arranged on the stem, or they may be two-ranked along the stem. Also, whether they are spiral or two-ranked, the flowers may be secund—that is, they all twist to one side and face the same direction.

Botanists sometimes refer to the inflorescences of *Lycaste* or *Maxillaria* as being one-flowered, or solitary, which is a bit confusing, since the plants usually bear several or many solitary flowers. It may be clearer to say that each flower is borne on a separate stalk, or peduncle, as far as one can see. In fact, several of these so-called one-flowered inflorescences are borne on a common stalk, but the stalk is so short that it is hidden by the sheaths at the base of the pseudobulb.

In some cases, many flowers open at about the same time. In many other cases, the flowers are produced successively, or one at a time, and the rachis (the inflorescence stem above the peduncle) may be very short. When such an inflorescence has its first flower, it may appear to be one-flowered, but after a number of flowers have been formed the rachis is much more obvious. In *Pleurothallis*, especially, one may find flowers that are densely clustered at the end of a common stalk, and this type of inflorescence also may produce one flower at a time, in which case there is a cluster of bracts at the end of the common stalk. In all pleurothallids, the pedicels do not fall off with the flowers, so these persistent stalklets are a distinctive feature of older inflorescences.

Flower Structure

Ovary. In most plants, the ovary, the part that develops into the fruit, is obvious in the center of the flower. The orchids, however, have what is called an inferior ovary. This is not a value judgment; it indicates that the other flower parts are attached above the ovary. Further, the orchid ovary often remains slender until after pollination, so the ovary is only slightly thicker than the pedicel when the flower opens and it is not conspicuous.

Sepals and Petals. The outer, green sepals of most dicot flowers and some monocot flowers are very different from the inner, white or brightly colored petals. The six perianth parts of most lilylike plants are so similar that botanists usually use the term *tepals*, implying that they are neither sepal nor petal, but in the orchids the three outer segments are a bit different from the three inner segments, and it is customary to call them sepals. Two of the three inner parts are similar and are called petals, but the third of the inner parts is usually larger and more complex than the other petals (see Fig. 1). This distinctive third part, called the lip, is a characteristic feature of the orchid flower. Often lobed, the lip may surround the column and may have keels or a thick, fleshy callus somewhere in the center (see Fig. 2).

Spur. Flowers often have a saclike or tubular nectary, or spur, that helps lure pollinators to the flower. In orchids the spur is usually at the base of the lip, but it may also be formed by the lateral sepals, usually with an extension of the lip into the spur (see Fig. 2).

Column. Most flowers have several to many stamens surrounding the ovary or style. In orchids these parts are united into a single structure, the column—or, if you wish to impress people with your erudition, the gynostemium. Orchids never have more than three stamens, and most have only a single fertile stamen. The sterile stamens and the basal portion of the fertile stamen are united with the style to make up the column. The anther, or pollen-bearing part of the stamen, is near the tip of the column, except in the lady's slippers, which have a fertile anther on each

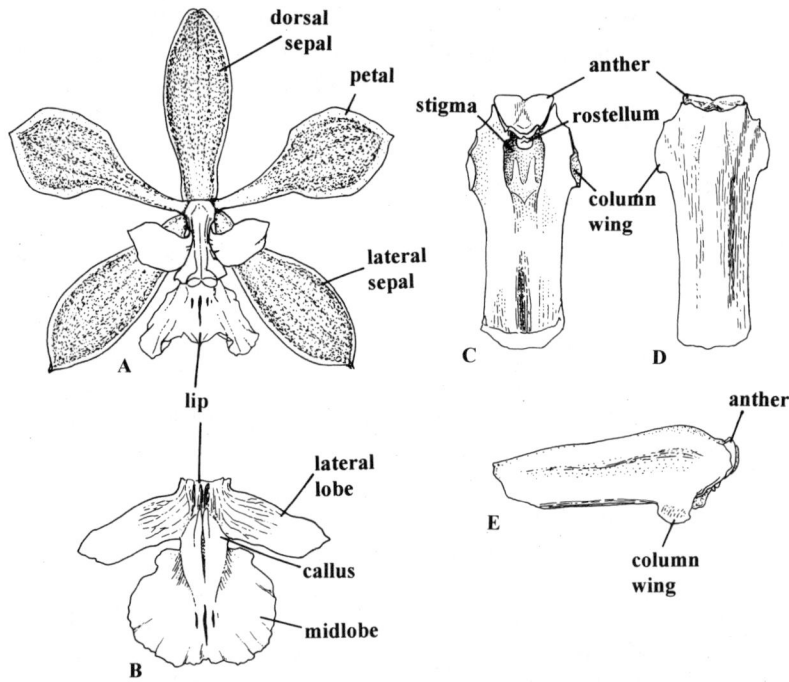

Figure 1. Structure of an orchid flower. A, an *Encyclia* flower, front view. B, the lip, flattened out. C–E, ventral, dorsal, and lateral views of the column.

side of the column. The anther may be dorsal, apical, or ventral with respect to the column.

There are often "wings" spreading or pointing forward from the sides of the column, usually near the anther and stigma. These probably represent the two lateral sterile stamens, and they are often useful in identification (see Fig. 1). Also, the tip of the column may form a hood over the anther.

Stigma and Rostellum. The stigma is on the underside of the column, beneath or a bit behind the anther. This is where the pollen must be deposited if the flower is to produce seeds. In most orchids a part of the stigma serves to attach the pollinia to a pollinator. In some cases this rostellum is essentially a partition between the fertile stigma and the anther, and the edge of the partition deposits glue on anything that touches it, including potential pollinators. The pollinia then stick to this glue on the pollinator. In the most highly evolved orchids the rostellum forms a

Flower Structure

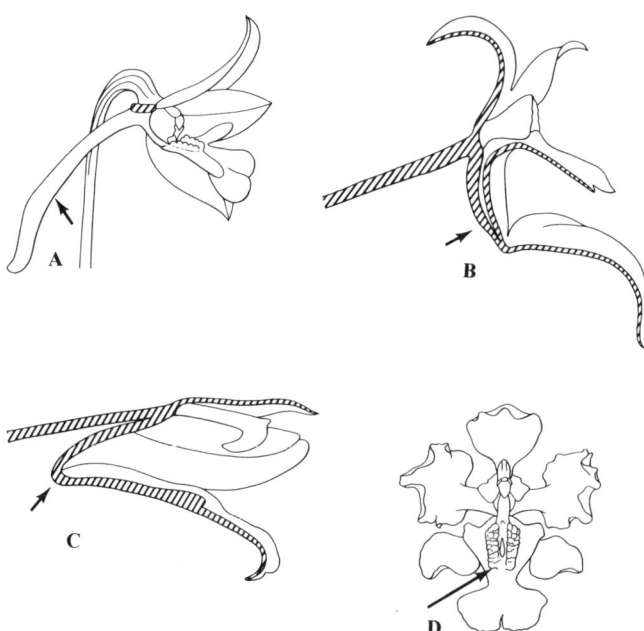

Figure 2. Various structures of orchid flowers. A, side view of a flower with a prominent spur projecting back from the base of the lip (arrow). B, section of a *Systeloglossum* flower showing the prominent column foot (arrow). The column foot and the lip together form a nectary. C, section of a *Maxillaria* flower with a prominent column foot projecting back at the base. This gives the flower a very distinct "chin" (arrow). D, an *Oncidium* flower, showing the callus (arrow).

clearly defined, sticky pad that is already attached to the pollinia when the flower opens. Thus, when the viscidium is touched by a potential pollinator, it and the attached pollinia are removed as a unit.

Pollinia. An unusual feature of the orchids is the presence of pollinia, bodies made up of many united pollen grains. The number of pollinia and their texture are important in orchid classification. In larger flowers it is easy to see the pollinia by tipping the anther back or by brushing a toothpick or other small object up the ventral surface of the column, but a hand lens or a dissecting microscope may be needed to see the pollinia of very tiny flowers. In most relatively primitive orchid groups the pollen masses are soft and easily crushed or spread out with a little pressure. *Habenaria* and some other orchids have sectile pollinia, with each mass of pollen subdivided into many small packets. The pollinia are firm or even quite hard in most orchids of the subfamily Epidendroideae.

Caudicles and Stipes. Pollinia often have definite stalks, and these are of two quite different sorts. Caudicles are essentially extensions of the pollinia themselves and are usually mealy or elastic. These are produced inside the anther and may become attached to viscidia. In many cases, though, they merely project so that they are easily glued to a pollinator but are not attached to the rostellar glue until it is moved by an insect or other pollinator. Stipes, however, are not produced in the anther; these are usually straps of cellular tissue that separate from the column surface as the flower matures. A stipe is always attached to a viscidium, and there are usually tiny caudicles between the stipe and the pollinia. The caudicles provide a breaking point so that the pollinia may separate from a pollinator and remain in the stigma. The stipe is characteristic of the most highly evolved orchids, the members of the tribes Vandeae, Cymbidieae, and Maxillarieae.

Ecology

Epiphytism. One of the few things that everyone knows about orchids is that they grow in trees. In fact, many species grow in the ground, but a large part of the family is epiphytic, or tree dwelling. This habit is not unique to orchids. Many aroids, bromeliads, peperomias, ferns, and mosses are epiphytes. Growing in trees offers several benefits, especially in wetter areas. It is clear that there is strong competition among plants that grow in the soil. This competition is greatly reduced for plants that can germinate and grow on the bark of a tree. In denser and wetter forests little sunlight reaches the forest floor, which limits the growth of plants that grow there. In wet areas the soil is generally very poor in nutrients and may also be so waterlogged that it is unfavorable for many plants, at least for part of the year. All these problems may be avoided by growing in trees, where there is more light and better drainage. Mineral nutrients are nowhere abundant in tropical forests, but the humus and other debris that accumulate around epiphytes offer some nutrition, and epiphytes avoid competition from tree roots that form a network in the upper soil.

To grow in trees, a plant must be able to withstand some drying. Orchids have several obvious features that help them adapt to their epiphytic habitat. The spongy velamen layer around the roots protects the roots from drying and may also serve to absorb the first, most mineral-rich water when it rains. Pseudobulbs are an obvious adaptation to store water and reduce drying, as are the fleshy leaves of many orchids. In effect, the epiphytic habitat is somewhat analogous to the desert habitat, and, indeed, some cacti share tree trunks and branches with the orchids. The most cactus-like orchids are, of course, those of drier forests. *Catasetum* and *Cyrtopodium* may look somewhat like short but lush corn plants during the rainy season, but during the dry season their leaves drop and they become quite cactus-like, even to the presence of spines. The adaptations to drought are less striking in the cloud forests, but even there a few sunny days with little rain can leave the moss quite dry.

Epiphytes are most diverse and abundant in relatively wet forests. Dry or seasonally dry forests have relatively few species of epiphytes, though the species that can tolerate dry forest conditions may be abundant. The greatest diversity and abundance occur in cloud forests. Diversity is relatively low in very high montane habitats, where it is colder and often drier. The abundance and diversity of orchids are also somewhat lower in exceptionally wet forests such as those of the Sierra Llorona de Portobelo near the north coast of Panama.

Fungal Symbiosis and Saprophytes. Many flowering plants normally have fungi growing in or on their roots and benefit from the relationship. *Mycorrhizae* is the technical term for the combination of root and fungus. The fungus may enhance the flowering plant's ability to absorb nutrients from the soil. Under natural conditions, most orchids cannot germinate without a symbiotic fungus. The orchid seed is so small that it cannot contain much food for the germinating plantlet, and this is probably one reason that orchid seedlings require a fungal partnership. In cultivation the seeds are supplied with mineral nutrients and sugar (under aseptic conditions), and they grow quite well without a fungus. In nature, however, the seeds apparently cannot grow without the proper fungus to digest organic matter and pass carbohydrates on to the orchid embryo, allowing the plantlet to continue its growth.

Botanists have long argued over whether each orchid requires a special type of fungus or whether many different fungi can be mycorrhizal partners for any given orchid. As usual, neither extreme viewpoint is quite correct, but recent studies show that each orchid grows best with only a few fungi, and in most cases an orchid genus normally occurs with a particular genus of fungus in nature. The mycorrhizal fungus is critical for germination, and some terrestrial orchids continue to have mycorrhizae throughout their lives. The root-tubers of *Habenaria* and its relatives apparently do not have fungi during the resting season but acquire new fungi each growing season.

Saprophytic flowering plants lack chlorophyll and obtain their energy from organic material in the soil; they are always dependent on mycorrhizal fungi. In southern Central America only a few orchids are saprophytic throughout their life cycles, especially *Wullschlaegelia* and *Uleiorchis*. Several other orchids, however, such as *Eulophia, Govenia, Oeceoclades,* and the genera related to *Catasetum,* are clearly saprophytic for a time as seedlings. These orchids grow in soil or rotten wood and form an irregular, coralloid body. These plants may grow for a while before they are seen, and the first visible shoot may be quite large. Saprophytes get all their energy from their fungal associates, and the orchids may actually be parasitic on the fungi.

Pollination. Flowers are organs of reproduction, primarily sexual reproduction. When flowers receive pollen from other flowers, the resulting seeds combine genetic material from both parents. The features that distinguish orchids from one another and from other flowers are largely those related to pollination, so studying pollination helps one to understand the diversity of the orchid family. Five main classes of pollinators are important for the orchids (see Table 1). Each class differs in size, shape, and flight, and each class perceives and reacts to a somewhat different spectrum of colors and odors. With these characteristics in mind, one can often make an educated guess as to the pollinator of a flower without ever seeing an insect or bird near the flower. Such guesses should be distrusted, but they are frequently correct.

REWARDS. Pollinators are normally looking for something when they visit a flower, and, all other things being equal, flowers that offer some

Table 1. Pollination syndromes in orchids

Pollinators	Flower characteristics		
	Odor	Color	Form
Bees	Sweet, day	Diverse, ultraviolet, no pure red	Gullet
Butterflies	Agreeable, day	Vivid red, yellow	Tubular
Moths	Sweet, strong	White, cream, pale green	Tubular
Nectar flies	Sweet or disagreeable	Green, yellow, brown, purple	Cupped, shallow
Birds	None	Vivid red, yellow, cerise	Tubular or narrow cup

sort of reward will be the most successful in attracting pollinators. The two most common rewards are nectar and pollen. The pollen of most orchids, however, is compacted into more or less solid pollinia that cannot be eaten by bees. Grasshoppers or caterpillars may eat the pollen, along with the rest of the flower, but pollinia do not function as a reward for pollinators. Nectar is the reward offered by most orchids, but a surprising array of other rewards is offered.

Oil. A few groups of bees gather oil to feed their larvae, and some flowers have oil glands rather than nectaries. These oil glands probably represent modified nectaries—that is, their ancestors probably offered increasing amounts of oils in their nectars as they adapted to different pollinators. While orchid nectar is usually somewhat hidden in a spur, orchids with oil may offer it in shallow, open glands. Few other insects are likely to "rob" such flowers. In our area, oil glands occur in *Ornithocephalus*, *Sigmatostalix*, and some species of *Oncidium*, especially those with long-beaked anthers. Observers should beware of small brown bees on *Ornithocephalus* or *Sigmatostalix*; they are not the stingless bees (*Trigona*) they appear to be; they do sting.

Pseudopollen. While orchid pollen is not available as a reward, some flowers produce a pollen-like substance that may be gathered by bees and appears to be quite nutritious. In Central America, *Polystachya* has mealy pseudopollen on the lip and is pollinated by small bees that gather this food.

Perfume. Perfume is usually used as an advertisement rather than a reward, but male euglossine bees gather floral perfumes and store them in special organs on their hind legs. A number of orchids offer perfume as a reward and are normally pollinated only by male euglossine bees, of which more in a few paragraphs.

MIMICRY AND DECEIT. Many orchids offer little or no reward. Possibly they are merely stingy with their energy, but there may be other benefits associated with deceit, such as a greater percentage of cross-pollination or pollination over greater distances.

In the simplest case, a flower with little or no reward looks like some other flower that does offer a reward, or a scarce flower may benefit by looking like a much commoner flower. Many species of *Oncidium* offer no reward and may be mimicking either their more generous cousins that offer oil or the flowers of malpighiaceous vines that have oil glands and resemble the *Oncidium* flowers. It has been suggested that the flowers of *Epidendrum radicans* mimic milkweed flowers. Both have the same colors and attract the same butterflies, but the *Epidendrum*, offering no reward, attracts far fewer pollinators.

Many orchids show so-called generalized food flower mimicry; that is, they do not mimic any particular flower, but they do look promising to the uninitiated pollinator. This is, essentially, false advertisement. Such flowers depend on naive pollinators, and in most cases the pollinators soon learn that the flowers offer no reward. In a study of *Cochleanthes lipscombiae*, James Ackerman found that the bees carrying *Cochleanthes* pollinia were always relatively young. That is, older, experienced bees knew enough to avoid *Cochleanthes*. There is the possibility, though, that *Cochleanthes lipscombiae* mimics the flowers of a legume vine that grows in the same area. Other orchids that appear to be generalized food flower mimics are *Cleistes*, *Cattleya*, and *Phragmipedium*.

Mimicry, however, is not limited to mimicking other flowers. Some orchids mimic rotten meat and attract carrion flies. I know of none in our area, but some South American species of *Masdevallia* and Asiatic species of *Bulbophyllum* have unattractive colors and revolting odors. Other orchids, such as the European bee orchid, mimic female insects and attract male insects of the appropriate species. Several Australian orchids are pollinated by pseudocopulation, and in the Andes, *Trichoceros* (*la mosca*) mimics female tachinid flies and is pollinated by male tachinids. The related genus *Telipogon* is probably pollinated by the same bristly flies.

POLLINATION BY EUGLOSSINE BEES. As mentioned above, male euglossine bees gather perfumes. Botanists still do not know just how the bees use the perfumes, but it is easy to see how the orchids use the bees. The euglossine bees are tropical American relatives of the bumblebees; three genera are important in orchid pollination (Table 2). Two small genera of nest parasites, *Aglae* (one species) and *Exaerete* (five species), look more like metallic wasps than bees and play a minor role in orchid pollination. The euglossine bees are all rather long-tongued, and some incredibly so. The females visit nectar flowers, including those with long, slender spurs, and pollinate some orchids. The males visit the same food flowers and are strongly attracted to certain perfumes. The males have brushes on their front feet, which they use to pick up droplets of perfume from the surfaces of flowers. A male bee may repeatedly brush on the surface of a flower and then hover near the flowers while transferring the perfume to its inflated hind legs, which have saclike structures for storing

Table 2. Characteristics of the principal genera of euglossine bees

Genus	Size of bees	Surface appearance	Period of activity	Approximate no. species
Euglossa	Small to medium	Shiny, metallic	All year	110
Eulaema	Medium to large	Hairy	All year	14
Eufriesea	Medium to large	Hairy	Seasonal	60

the perfumes. Since each bee returns repeatedly to the same flowers, the orchids have been able to evolve quite bizarre mechanisms that involve the bees slipping through the flower or into a bucket-like lip.

Bees in general, and the euglossine bees especially, have a sort of small shelf at the rear of the thorax, the scutellum, that seems designed as a place to attach pollinia. The viscidium can be stuck under the edge of the scutellum, where it is not easily brushed off, and the pollinia then project backward so that they will pollinate another flower when the bee backs out of it. Many orchids place their pollinia under the scutellum, but the pollinia may be attached almost anywhere on the bee (see Fig. 3). The structure of the pollinia, stipe, and viscidium vary a good deal, and one can usually identify the orchids at least to genus on the basis of pollinia alone.

A number of tropical American orchids are pollinated by euglossine bees, including all the *Catasetum* and *Stanhopea* groups, the *Chondrorhyncha* complex, *Dichaea*, *Lockhartia*, *Lycaste*, *Macradenia*, *Macroclinium*, *Notylia*, *Trichocentrum*, and *Trichopilia*.

FALL-IN AND FALL-THROUGH FLOWERS. I noted above that the euglossine bees may brush many times on a single inflorescence or flower. An awkward position on a slippery surface greatly increases the probability that a bee will slip at least once during its visit to a given plant. In *Coryanthes*, the bucket orchid, any bee that slips—or hits a drop of watery liquid from the glands at the base of the column—is almost sure to fall into the cuplike lip where the watery liquid accumulates. Smaller bees cannot fly if their wings are wet, and there is no room in the lip for larger bees (which visit different species of *Coryanthes*) to spread their wings. The sides of the lip are quite smooth and slippery, but there is a virtual stairway on the surface of the lip that leads out beneath the column. This, then, is the only choice open to a bee that falls into the lip. In crawling out beneath the column the bee passes beneath the stigma and then under the anther, where the viscidium of the pollinia is firmly placed

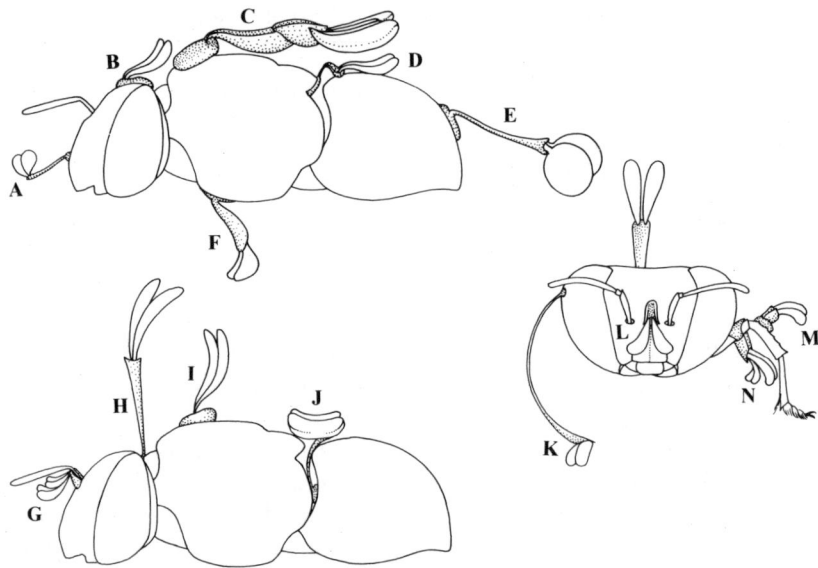

Figure 3. Outlines of euglossine bees showing where the pollinia of different orchid genera are attached. A, *Notylia*. B, *Peristeria elata*. C, *Catasetum*. D, *Acineta*. E, *Cycnoches*. F, *Dressleria*. G, *Kefersteinia*. H, *Kegeliella*. I, *Peristeria* (most species). J, *Coryanthes*. K, *Macradenia*. L, *Trichocentrum*. M, *Clowesia warscewiczii*. N, *Chaubardiella*.

between the bee's thorax and abdomen. In spite of this ordeal, some bees return to other *Coryanthes* flowers and pollinate them.

In *Stanhopea* the lip, column, and column wings form a sort of vertical tube through which bees can fall. When a bee slips and falls, the pollinia are neatly deposited behind its thorax. In most species of *Gongora*, the bee must stand upside down at the base of the slippery lip to reach the perfume. If (or when) the bee slips, the column forms a steep slide that guides the falling bee toward the stigma and anther.

TRAP FLOWERS. The flower of *Coryanthes* may be considered a sort of passive trap flower, but other orchids have more elaborate traps in which part of the flower is hinged and movable. In *Peristeria* part of the lip is hinged, and when the bee moves far enough into the flower its own weight causes the midlobe to tip over and push the bee against the column. The lip is actively motile in the tiny flowers of *Acostaea*; when a small fly touches the right trigger, the lip flips up against the column, trapping the insect for a short time. If pollination does not occur, the lip resets and waits for another opportunity.

SLINGSHOTS. In *Catasetum*, *Mormodes*, and *Cycnoches*, the stipe that connects the pollinia to the sticky viscidium is elastic and is stretched over a knob. When a bee touches the proper trigger (different in each genus) the viscidium is released and is thrown from the column with considerable force. The pollinia of *Catasetum* hit the pollinator with such force that the bee may be reluctant to approach another male flower. The female flowers, however, are quite different in appearance. Small bees of the genus *Euglossa* pollinate some species of *Catasetum*, and they may not be able to fly away until the anther cap falls off the pollinia and the pollinia have dried a bit. Even then, the bees may be able to gain altitude only slowly.

SYNCHRONOUS FLOWERING. Several orchids show a very interesting behavior. They have short-lived flowers but coordinate their flowering so that all plants in a population flower on the same days. This may seem a bit improbable, but it can be explained. In the Malaysian orchids that have been studied the flower buds develop to a certain stage and then stop. When the proper environmental cue is given, the buds resume growth, and all flower on the same day several days after the cue is received. The cue is a sudden drop in temperature, usually associated with a rainstorm. The virtue of synchronous, or gregarious, flowering seems to be that the many flowers open at once are a much more effective stimulus for the pollinators, which might ignore the flowers if only a few opened at a time.

Several Central American genera show synchronous flowering: *Palmorchis*, *Psilochilus*, *Sobralia*, and *Triphora*. In all cases the flowers are short-lived, lasting only a few hours in *Palmorchis*, and a day or two in *Sobralia*.

HOW TO STUDY ORCHID POLLINATION. Botanists are still studying orchid pollination, and knowledge of who pollinates what is incomplete. The best way to observe orchid pollination is to live right in the orchid habitat and maintain a garden of orchids where one may occasionally find the pollinators in action. This method, though, is not always practical.

Orchids that offer enticing perfumes or other rewards may be visited several times during their short lifespan, at least in good weather. An *Encyclia* or an *Oncidium* that offers no reward may be in bloom for several weeks, but such a plant may be visited only once or twice during that time. When living on Barro Colorado Island in Panama, I heard pollinators coming to *Oncidium panamense* a few times. In each case, by the time I had located the bee, it had realized that the flowers had no reward and left without my seeing more than a blur of movement.

The euglossine-pollinated orchids, though, are usually visited several times each morning if the weather is good. In fact, one sometimes hears the bees buzzing before seeing the plant. If one is lucky enough to find the bees in action, it is best to catch a bee or two and to press a flower or two, so that both species can be identified (see Appendix A). The bees that brush on the flower surface are always males and cannot sting. After the bees have landed on the flower a couple of times they lose much of their wariness and one can often catch them by hand, but it is much easier with a small net. Insect specimens are normally pinned, but they can be dried in tissue paper or preserved in alcohol. Butterflies are much more delicate and should always be dried (folded in wax paper).

Always take careful notes as to time and place, and label the specimens. Photographs are always helpful but are usually not enough to identify the insects. Hummingbirds are much more difficult to collect, but they can be identified by the use of a field guide. Some orchids have all the earmarks of hummingbird-pollinated flowers, but there are relatively few published observations.

3 Classification and Identification: How to Use This Guide

People have always named plants and animals, and some groups of plants are easily recognized by everyone. The idea that all plants in a particular group might be related to each other in some way developed long before Darwin. Now biologists recognize that all the members of a "natural" group have descended from a single ancestral species and that they share a number of features inherited from that common ancestor. There are, however, still "groups" that are unnatural—that is, they share only a few or poorly correlated features—and their classification will surely change when it is carefully studied.

Classification is based not on isolated features but on the correlations between different features. Further, modern biologists try to distinguish between features inherited from a relatively recent common ancestor, primitive features inherited from a remote ancestor, and parallelisms, or features that have evolved independently in different plants, especially in those that grow under similar conditions or are pollinated by the same type of insect or bird. Though classification may seem arbitrary to the uninitiated, there *are* correlations between the features used in classification. There are complicating factors; all plants that have a given feature are not necessarily relatives, nor do all related plants necessarily have the same features.

Classification of Orchids

The following outline lists the subfamilies, tribes, and subtribes of the orchid family and the genera found in Costa Rica and Panama. The classification will change as botanists learn more about orchids and orchid relationships, but the subtribes, especially, may be of interest to some readers. As one learns to know some of the genera, one will begin to recognize groups of related genera and can then go directly to the relevant section of the book.

    Subfamily Cypripedioideae—*Phragmipedium, Selenipedium*

    Subfamily Spiranthoideae
        Tribe Tropidieae—*Corymborkis, Tropidia*
        Tribe Cranichideae
            Subtribe Cranichidinae—*Baskervilla, Cranichis, Ponthieva, Pseudocentrum, Pterichis, Solenocentrum*
            Subtribe Goodyerinae—*Aspidogyne, Erythrodes, Goodyera, Kreodanthus, Ligeophila, Platythelys*
            Subtribe Prescottiinae—*Aa, Gomphichis, Prescottia*
            Subtribe Spiranthinae—*Beloglottis, Brachystele, Coccineorchis, Cyclopogon, Deiregyne, Discyphus, Eltroplectris, Eurystyles, Funkiella, Galeottiella, Hapalorchis, Lankesterella, Lyroglossa, Mesadenella, Mesadenus, Pelexia, Sacoila, Sarcoglottis, Schiedeella, Spiranthes, Stenorrhynchos*

    Subfamily Orchidoideae
        Tribe Orchideae
            Subtribe Habenariinae—*Habenaria*

    Subfamily Epidendroideae, primitive groups (Many of these may eventually become separate subfamilies.)
        Tribe Gastrodieae
            Subtribe Gastrodiinae—*Uleiorchis*
            Subtribe Wullschlaegeliinae—*Wullschlaegelia*
        Tribe Palmorchideae—*Palmorchis*
        Tribe Triphoreae—*Psilochilus, Monophyllorchis, Triphora*
        Tribe Vanilleae
            Subtribe Pogoniinae—*Cleistes*
            Subtribe Vanillinae—*Vanilla*

    Subfamily Epidendroideae, advanced groups
        Tribe Arethuseae
            Subtribe Bletiinae—*Bletia, Calanthe, Spathoglottis*
            Subtribe Chysiinae—*Chysis*
        Tribe Cymbidieae
            Subtribe Catasetinae—*Catasetum, Clowesia, Cycnoches, Dressleria, Mormodes*

Subtribe Cyrtopodiinae—*Cyrtopodium, Galeandra*
Subtribe Eulophiinae—*Eulophia, Oeceoclades*
Subtribe Goveniinae—*Govenia*
Tribe Dendrobieae
Subtribe Bulbophyllinae—*Bulbophyllum*
Tribe Epidendreae
Subtribe Sobraliinae—*Elleanthus, Sobralia*
Subtribe Arpophyllinae—*Arpophyllum*
Subtribe Coeliinae—*Coelia*
Subtribe Laeliinae—*Acrorchis, Barkeria, Brassavola, Cattleya, Caularthron, Dimerandra, Encyclia, Epidendrum, Helleriella, Hexisea, Homalopetalum, Isochilus, Jacquiniella, Laelia, Myrmecophila, Nidema, Oerstedella, Platyglottis, Ponera, Reichenbachanthus, Scaphyglottis, Schomburgkia*
Subtribe Pleurothallidinae—*Acostaea, Barbosella, Brachionidium, Brenesia, Dracula, Dresslerella, Dryadella, Lepanthes, Lepanthopsis, Masdevallia, Myoxanthus, Octomeria, Ophidion, Platystele, Pleurothallis, Restrepia, Restrepiella, Restrepiopsis, Salpistele, Scaphosepalum, Stelis, Trichosalpinx, Trisetella, Zootrophion*
Subtribe Polystachyinae—*Polystachya*
Tribe Malaxideae—*Liparis, Malaxis*
Tribe Maxillarieae
Subtribe Cryptarrheninae—*Cryptarrhena*
Subtribe Zygopetalinae—*Chaubardiella, Chondrorhyncha, Cochleanthes, Dichaea, Galeottia, Huntleya, Kefersteinia, Koellensteinia, Pescatorea, Warrea, Warreopsis,*
Subtribe Lycastinae—*Bifrenaria, Lycaste, Neomoorea, Teuscheria, Xylobium*
Subtribe Maxillariinae—*Chrysocycnis, Cryptocentrum, Maxillaria, Mormolyca, Trigonidium,*
Subtribe Ornithocephalinae—*Ornithocephalus, Phymatidium, Sphyrastylis*
Subtribe Oncidiinae—*Ada, Amparoa, Aspasia, Brassia, Cischweinfia, Comparettia, Fernandezia, Hybochilus, Goniochilus, Ionopsis, Lemboglossum, Leochilus, Leucohyle, Lockhartia, Macradenia, Macroclinium, Mesospinidium, Miltoniopsis, Notylia, Oncidium, Osmoglossum, Otoglossum, Pachyphyllum, Plectrophora, Psychopsis, Psygmorchis, Rodriguezia, Rossioglossum, Scelochilus, Sigmatostalix, Systeloglossum, Ticoglossum, Trichocentrum, Trichopilia, Trizeuxis, Warmingia*
Subtribe Stanhopeinae—*Acineta, Coeliopsis, Coryanthes, Gon-*

gora, *Horichia, Houlletia, Kegeliella, Lacaena, Paphinia, Peristeria, Polycycnis, Sievekingia, Stanhopea, Trevoria*
  Subtribe Telipogoninae—*Stellilabium, Telipogon*
Tribe Vandeae
  Subtribe Angraecinae—*Campylocentrum*
  Uncertain position—*Arundina, Eriopsis* (Maxillarieae, but may deserve its own subtribe)

Using the Field Guide

Botanical Terminology. I have tried to avoid the more obscure and unnecessary botanical terms, but some are the only alternatives to long and wordy descriptions. The terms I do use are defined in the Glossary, and some are illustrated in Chapter 2 and in the Glossary. The terms that describe shape usually represent a rather idealized form (see Fig. 6 in the Glossary), but, of course, the real world tends to produce shapes that aren't quite one thing or the other. It is quite acceptable, however, to hyphenate any combination of these terms to describe a leaf or other structure, as in elliptic-ovate, triangular-ovate, and so on.

Descriptions. Ideally, a flora or field guide would have detailed descriptions of each species so that one could check the plant to be identified against the descriptions. That, however, would inflate this field guide to encyclopedic proportions and would require several more years of work with a generous travel budget. The information supplied here will rarely be enough for the reader to be certain that the correct identification has been reached; if the measurements or color of the plant reached by the key are very unlike those of the plant being identified, it may be that one of us has made an error.

Measurements (**M**). Measurements are given in metric units, the units most used in botany, even though most North Americans are more accustomed to the English system. Inches certainly serve well for measuring large flowers, but many orchids of Central America have very small flowers. I cannot imagine describing the Pleurothallidinae, for example, in fractions of an inch. For general purposes, 25 millimeters (mm) or 2.5 centimeters (cm) are almost an inch, and a yard (3 feet) is only about 3 inches short of being a meter (= 10 decimeters, 100 centimeters, or 1000 millimeters).

As a measurement of flower size, I have usually given the length of the sepals; as the outermost segments, they are easy to measure. In most cases the petals are a bit shorter and often wider than the sepals. For the size of a plant or shoot, I generally do not include the inflorescence, at least in the epiphytes. For the herbaceous terrestrials treated in Chapter 10, however, there may be no sharp line between plant and inflorescence, and the height given usually includes the inflorescence.

Ideally all measurements should give ranges of variation, if not statistical treatment. Since I have collated these measurements from many sources, I have had to settle for what I could find. Where only a single figure was found for some measurement, a qualifying "about" or "approximately" is always implied.

Color (**C**). Flower color reflects my own experience or notes in many cases, and published information or data from herbarium labels in other cases. Remember that two persons may give very different color descriptions of the same flower and that many species show great variation in pattern and intensity of color. There may even be pure white, yellow, or pale green forms in species that normally have other colors masking the base color.

Geographic code (**D**). To save space, I have adopted a code to indicate the general distribution and ecology of the orchids treated here:

No	Northern (often Nicaragua)
CR	Costa Rica
Pma	Panama
SA	South America (usually Colombia or Ecuador)
r	Reported from, but presence not certain
wPma	Western Panama (Chiriquí, Bocas del Toro)
ePma	Eastern Panama (Darién)
Cb	Atlantic (Caribbean) lowlands, up to about 1000 meters
Pc	Pacific lowlands, up to about 1000 meters
Mt	Mountains 800–2500 meters
Pk	High mountains, above 2000 meters
Wt	Exceptionally wet areas, usually at intermediate elevations

Species whose distribution is described as as "No, Pma" are to be expected in Costa Rica, and those listed as "CR, SA" are to be expected in Panama. The ecological zones (Cb, Pc, etc.) refer only to habitats where the species occur in Costa Rica and Panama; in some cases I have too little information to indicate these zones. Please note that these ecological zones overlap. This is quite intentional; few plants respect lines drawn on a map, even when they correspond to climatic differences. Most plants that grow in the lowlands are likely to turn up in the lower mountains, and mountaintops at 800 to 1000 meters elevation may be startlingly montane, with species that otherwise occur at much higher elevations. Similarly, species that seem very montane may occur in much lower sites where it is wet enough. The zonation that I indicate here is approximate, but it should give some idea of where to expect the plants. In a few cases, where the localities I have visited are borderline, I have listed both zones (Pc/Mt, for example).

Other Symbols and Abbreviations. To keep the descriptive data from overwhelming the keys, I have used the following abbreviations and symbols:

>	More than or longer than*
<	Less than or shorter than
C	Color notes; refers to flower color unless otherwise specified
D	Distribution
dS	Dorsal sepal
fasc	Fascicle or fascicled flowers; the dense cluster of flowers is not stalked unless qualified as "stalked fasc"
fls	Flowers
lf	Leaf
lS	Lateral sepal
lvs	Leaves
M	Measurements
P	Petals
pl	Plant or shoot, usually refers to the height or length of a shoot or unit of growth; for plants that branch or produce new shoots on top of old ones, any attempt at measurement is futile
ps	Pseudobulb
S	Sepal or sepals
sub = lvs	Subequal to the leaves
Syn	Synonym
synS	Synsepal; that is, the two lateral sepals united into a single structure

*For larger plants the size may be given as "to > 1 m" or "to > 50 cm." An alternative would be to describe the plants as "to at least 1 m," and so on. Unfortunately, museum specimens of larger plants are usually incomplete and measurements are often lacking.

Color Plates. The color plates illustrate most of the orchid genera of our area, but there are some (mostly small or inconspicuous) genera for which I have no photographs. When possible, the photographs were taken in Costa Rica or Panama, or taken of cultivated plants collected there. For a few genera I have used photographs of plants collected in other areas, or of cultivated plants of unknown origin, but I believe that all are of species found in Costa Rica or Panama.

Names and Synonyms. I have tried to use the correct name for each species, but some genera are being studied and others are in need of study, so the names now in use may prove to be incorrect. I list only the more important synonyms that have been used in recent times. In some cases it is difficult to decide whether a name refers to a valid species or should be a synonym of some other. Such doubts are indicated in a few cases. I have used some names that have not yet been published but should be published before this book is available. If I inadvertently publish any new names that lack proper published descriptions (*nomina nuda*), I apologize to the botanical nomenclators, who understandably disapprove of invalid names. "*Neowilliamsia* (= *Epidendrum*) *cuneata*" means that the species in question has a name as a *Neowilliamsia*, now considered a subgroup of *Epidendrum*, but *N. cuneata* may not have a valid, published name under *Epidendrum*. When I key something as, for example, "*Baskervilla* species," it may be a distinct species for which the name has not yet been published, or it may be a new species only recently recognized as such for which a name has not been chosen, much less published.

How to Identify with a Key. The general key at the end of this chapter is designed to lead you to the proper chapter for identification of most orchids that grow in Costa Rica and Panama. Other keys are given in the following chapters, so that you may continue until an orchid has been identified to genus or species. A key is essentially a series of alternatives. Some plants will stubbornly straddle the fence and more or less fit both alternatives; I have tried to key such plants under both alternatives, but there may be other species that should have been keyed the same way.

The first couplet in the general key is:

1. Terrestrial plants without leaves .... Chapter 10
1. Plants with leaves, *or* epiphytes 2

This choice is an easy one. If the plant you have at hand is growing in the ground and has no leaves, go directly to Chapter 10. If the plant has definite leaves *or* if it grows on a tree, continue through the general key and check the alternatives under couplet 2. At each choice be sure to read the phrases in both alternatives carefully. Then make your choice and continue to the next couplet until the key leads you to a name. After the first choice in each key, each couplet includes a number in parentheses—for example, 12(3). Thus, if neither choice under couplet 12 seems right and you think you may have made an error, you may trace the path back to couplet 3 and recheck previous alternatives. Features that are especially important in identifying a species or distinguishing it from its close allies may be set in **boldface** type in the key or description.

The general key below and the keys to genera in the chapters are accompanied by rather diagrammatic sketches to help you visualize some of the features in the couplets. Sketches do not indicate size differences, and when the first alternative is something quite specific and the second is essentially "anything else," the second alternative may not be illustrated. In such a case the other possibilities are usually illustrated within the next few couplets.

While the keys will help to identify orchids, I strongly recommend that you look through the photographs in the book and become familiar with the more common genera. If you have trouble keying out a plant, does it look like an *Epidendrum*? An *Oncidium*? A *Pleurothallis*? A *Maxillaria*? If it looks like one of these, try the key in the appropriate chapter. Anything that looks much like a *Pleurothallis* is probably treated in Chapter 7, if not under *Pleurothallis* itself. After you have used the guide awhile, you should have a fairly good idea of what some orchid groups look like. The orchids in Chapters 9 and 10 are the leftovers; both are miscellaneous lots of orchids that are only distantly interrelated.

VARIATION. Any discussion of plant classification and identification must say something about variation. We are well aware that each individual of *Homo sapiens* is different from every other (with the relative exception of identical siblings). The point is that such variation is typical of most species. The fact that two plants look rather different in size, shape, color, or some other feature does not mean that they are necessarily different species. Virtually all natural populations are variable. In practice, botanists need to study the distribution and variation of the plants' features to determine whether two plants are representatives of distinct species or merely two variants in a single variable population. Obviously, unraveling the pattern of relationships is much more difficult when only a few plants are available for study. The measurements I cite in the key are largely taken from the literature, and in too many cases these may be based on a single specimen. You will almost certainly find plants that are smaller or larger than the measurement(s) given. Similarly, there is variation in flower color and the size and shape of the flower parts.

EXCEPTIONS. Part of the interest and frustration in biology is that there are exceptions to virtually all rules. Thus, some plants will emphatically not come out to their correct chapters in the general key. For example, nearly all orchids of the subtribe Laeliinae have terminal inflorescences. Still, *Brassavola acaulis* and *Epidendrum rousseouae* have lateral inflorescences, and the inflorescence of *E. stamfordianum* appears to be lateral. Such plants may be difficult to key, but the genera are easily recognized by comparison with the photographs. Similarly, most of our *Malaxis* have definite pseudobulbs or corms, but *M. blephariglottis* and *M. tipuloides* have neither pseudobulbs nor corms. These plants are very distinctive, though, and should key out correctly in either Chapter 9 or Chapter 10.

General Key

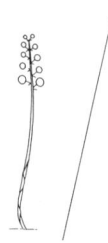

1. Terrestrial plants without leaves . . . . .Chapter 10
1. Plants with leaves, *or* epiphytes 2

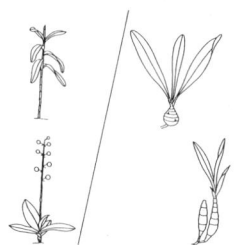

2(1). Without pseudobulbs or corms, stems not thickened. . . 3
2. With pseudobulbs or corms 10

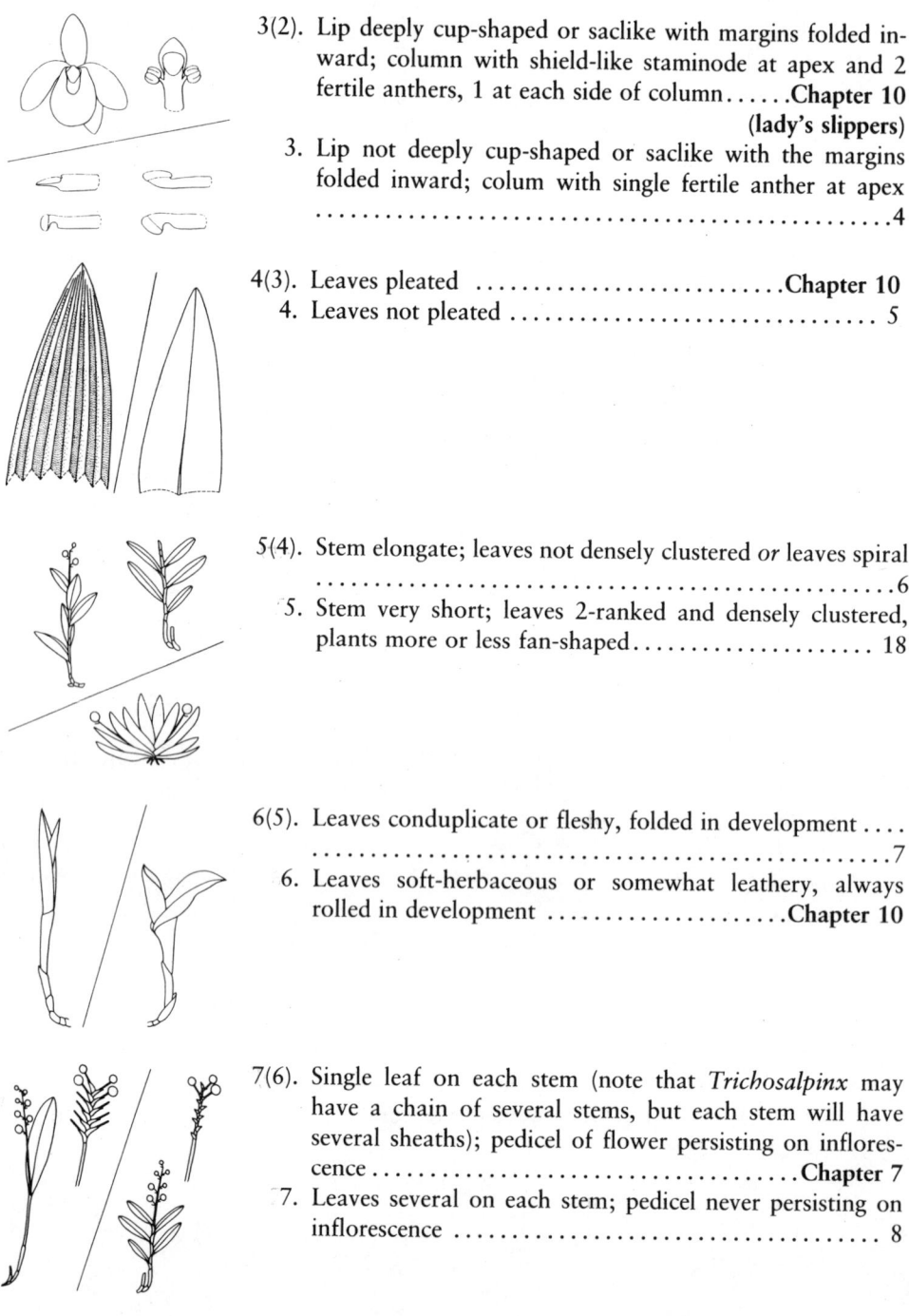

3(2). Lip deeply cup-shaped or saclike with margins folded inward; column with shield-like staminode at apex and 2 fertile anthers, 1 at each side of column......**Chapter 10 (lady's slippers)**
3. Lip not deeply cup-shaped or saclike with the margins folded inward; colum with single fertile anther at apex 4

4(3). Leaves pleated ....Chapter 10
4. Leaves not pleated 5

5(4). Stem elongate; leaves not densely clustered *or* leaves spiral 6
5. Stem very short; leaves 2-ranked and densely clustered, plants more or less fan-shaped.................... 18

6(5). Leaves conduplicate or fleshy, folded in development.... 7
6. Leaves soft-herbaceous or somewhat leathery, always rolled in development ....Chapter 10

7(6). Single leaf on each stem (note that *Trichosalpinx* may have a chain of several stems, but each stem will have several sheaths); pedicel of flower persisting on inflorescence..Chapter 7
7. Leaves several on each stem; pedicel never persisting on inflorescence 8

43 General Key

8(7). Inflorescence lateral, a toothbrush cluster of white or greenish flowers with spurs................ **Chapter 10** (***Campylocentrum***)
  8. Inflorescence usually terminal *or* flowers not as above

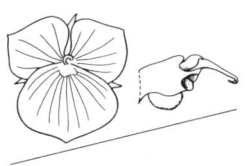

9(8). Petals and lip similar; sepals inconspicuous; column with long beak, pollinia borne on a long stalk (stipe) viscidium hooked ....**Chapter 8** (***Telipogon***)
  9. Petals and lip usually dissimilar; column not beaked *or* the viscidium not hooked; pollinia without a long stalk ....**Chapter 4**

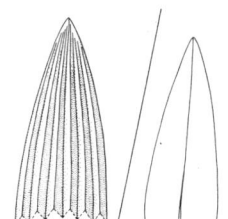

10(2). Leaves pleated 11
  10. Leaves not pleated 16

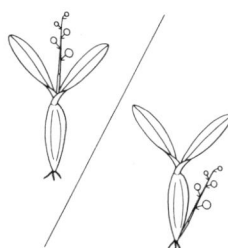

11(10). Inflorescence terminal......................**Chapter 9**
  11. Inflorescence lateral 12

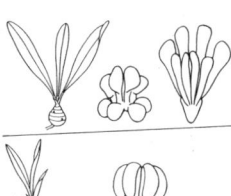

12(11). Pollinia 8; plants with corms ....**Chapter 9**
  12. Pollinia 2 or 4; plants with corms or pseudobulbs.... 13

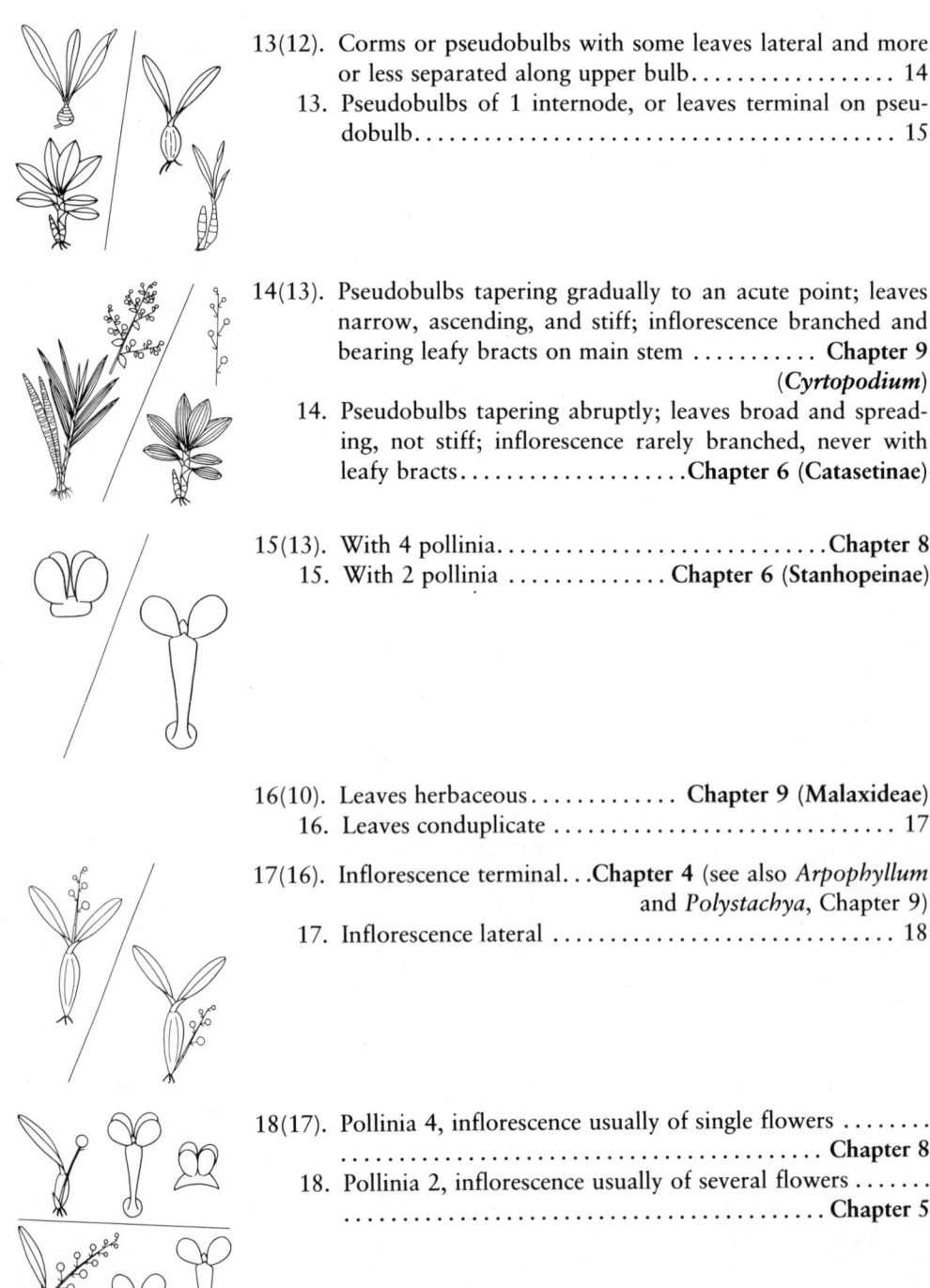

13(12). Corms or pseudobulbs with some leaves lateral and more or less separated along upper bulb.................. 14
13. Pseudobulbs of 1 internode, or leaves terminal on pseudobulb.. 15

14(13). Pseudobulbs tapering gradually to an acute point; leaves narrow, ascending, and stiff; inflorescence branched and bearing leafy bracts on main stem .... **Chapter 9 (*Cyrtopodium*)**
14. Pseudobulbs tapering abruptly; leaves broad and spreading, not stiff; inflorescence rarely branched, never with leafy bracts....................**Chapter 6 (Catasetinae)**

15(13). With 4 pollinia............................**Chapter 8**
15. With 2 pollinia .... **Chapter 6 (Stanhopeinae)**

16(10). Leaves herbaceous............. **Chapter 9 (Malaxideae)**
16. Leaves conduplicate 17

17(16). Inflorescence terminal...**Chapter 4** (see also *Arpophyllum* and *Polystachya*, Chapter 9)
17. Inflorescence lateral 18

18(17). Pollinia 4, inflorescence usually of single flowers .... **Chapter 8**
18. Pollinia 2, inflorescence usually of several flowers .... **Chapter 5**

4 *Cattleya* and Its Cousins: Subtribe Laeliinae

Epiphytic or sometimes terrestrial; stems slender or forming pseudobulbs, usually of several internodes; leaves **conduplicate** or **fleshy;** inflorescence almost always **terminal;** flowers variable, with 2, 4, 6, or 8 firm pollinia, usually with caudicles, rarely with viscidium.

The subtribe Laeliinae—a major group of tropical American orchids—includes several familiar genera, such as the showy cattleyas and the ubiquitous epidendrums. Most orchids with conduplicate leaves and terminal inflorescences are members of the Laeliinae; some Laeliinae have distinct pseudobulbs and others have a slender reed-stem habit.

Key to Genera of Laeliinae

    1. Plants without pseudobulbs, stems slender 2
    1. Plants with definite pseudobulbs, stems thickened 13

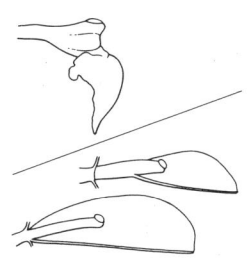

    2(1). Lip solidly united with column up to column apex; pollinia with distinct semiliquid viscidium . . . . . ***Epidendrum***
    2. Lip free from column, or only partially united 3

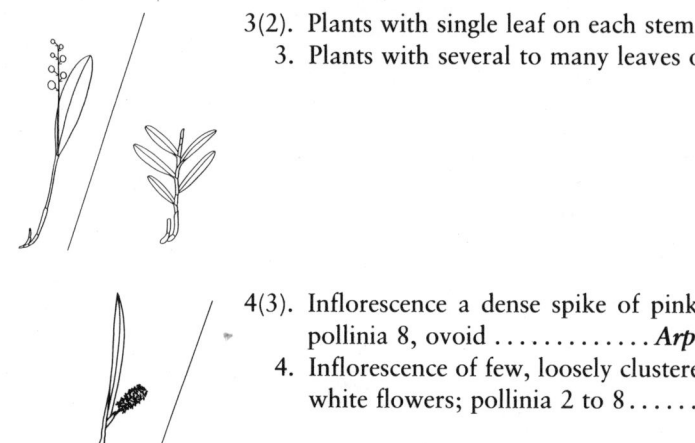

3(2). Plants with single leaf on each stem 4
3. Plants with several to many leaves on each stem 6

4(3). Inflorescence a dense spike of pink or magenta flowers; pollinia 8, ovoid .... *Arpophyllum* (Chapter 9)
4. Inflorescence of few, loosely clustered, white or green and white flowers; pollinia 2 to 8 5

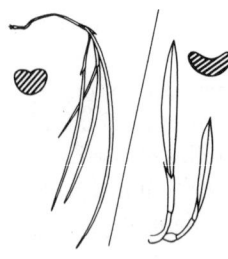

5(4). Leaves narrow and cylindrical; new stems arising from tips of older stems; flowers small, with a fleshy lip ......
.... *Reichenbachanthus*
5. Leaves fleshy, at least somewhat flattened; new stems arising from bases of older stems; flowers large ....
.... *Brassavola*

6(3). Leaves cylindrical or laterally flattened; midlobe of lip fleshy; flaplike stigmatic lobes at each side of rostellum; flowers green, ocher, or bronze .... *Jacquiniella*
6. Leaves dorsoventrally flattened; lip flat or membranous
.... 7

7(6). Column without distinct foot; lip not movable........ 8
7. Column with distinct foot; lip often hinged to column foot; inflorescences both terminal and lateral 11

Key to Genera of Laellinae

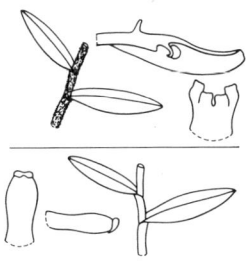

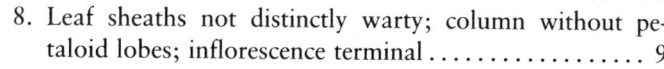

8(7). Leaf sheaths warty; column with petaloid lobes at apex; inflorescences often both terminal and lateral .... ***Oerstedella***

8. Leaf sheaths not distinctly warty; column without petaloid lobes; inflorescence terminal 9

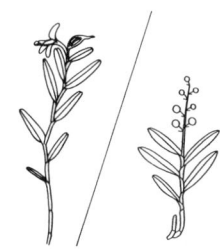

9(8). Leaves fleshy; flowers very delicate, fascicled .... ***Acrorchis roseola***

9. Leaves thin; flowers in a raceme 10

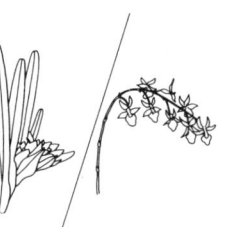

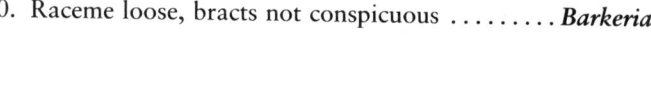

10(9). Raceme dense, 1-sided, with conspicuous bracts......... .... ***Isochilus***

10. Raceme loose, bracts not conspicuous .... ***Barkeria***

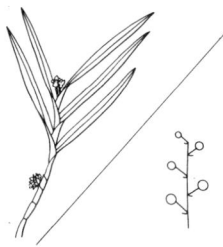

11(7). Inflorescences short, dense; sepals < 10 mm long....... .... ***Ponera striata***

11. Inflorescences lax or producing 1 flower at a time; sepals > 10 mm long..................................... 12

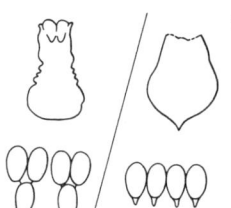

12(11). Pollinia 6; plants to 50 cm tall ..... ***Platyglottis coriacea***

12. Pollinia 4; plants to 3 m tall... ***Helleriella nicaraguensis***

48   *Cattleya* and Its Cousins

Plants with definite pseudobulbs

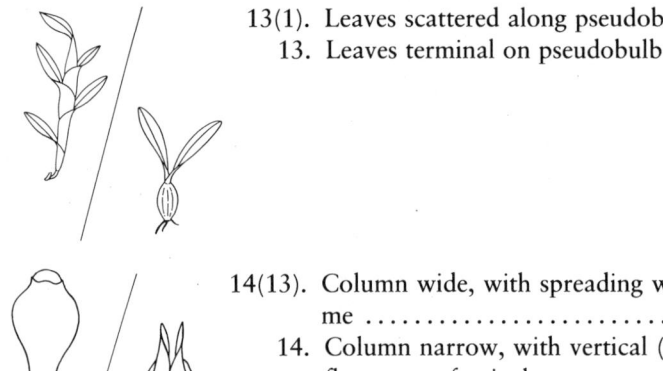

13(1). Leaves scattered along pseudobulb 14
13.   Leaves terminal on pseudobulb 15

14(13). Column wide, with spreading wings; inflorescence a raceme .... ***Barkeria***
14.   Column narrow, with vertical (parallel) lobes at apex; inflorescence fasciculate. .... ***Dimerandra***

15(13). Pseudobulbs arising from tops of old pseudobulbs, forming chains 16
15.   Pseudobulbs arising from bases of old pseudobulbs, not forming chains. 18

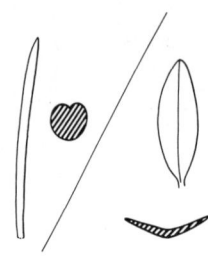

16(15). Leaves cylindrical; lip very fleshy.... ***Reichenbachanthus***
16.   Leaves dorsoventrally flattened 17

17(16). Flowers bright orange-red; pseudobulbs ellipsoid, without distinct stalks .... ***Hexisea***
17.   Flowers green and white or pinkish, *or* pseudobulbs slender or distinctly stalked .... ***Scaphyglottis***

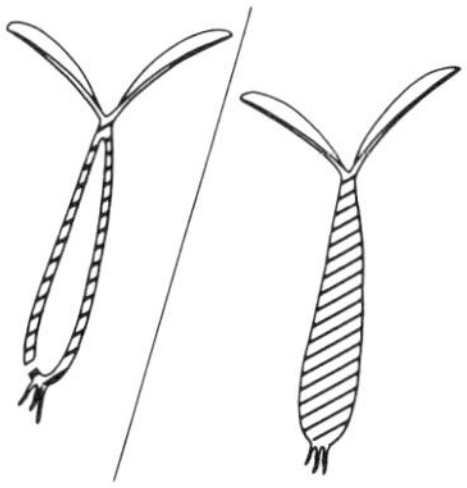

Key to Genera of Laellinae

18(15). Pseudobulbs hollow, usually with ants dwelling within ... 19
18. Pseudobulbs solid 20

19(18). Sheaths of pseudobulb smooth, whitish .... *Caularthron bilamellatum*
19. Pseudobulbs not whitish, surface ridged; inflorescence very long. .... *Myrmecophila tibicinis*

20(18). Stems only slightly thickened, but leaves very fleshy; flowers white; lip enfolding column. .... *Brassavola*
20. Pseudobulbs distinctly thickened; leaves usually flat 21

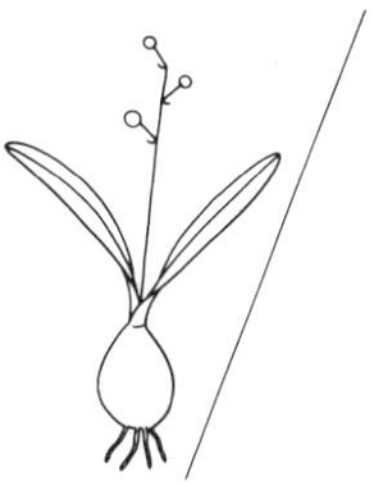

21(20). Pseudobulbs onion-shaped; lip 3-lobed 22
21. Pseudobulbs never onion-shaped 23

22(21). Leaves leathery; plants large; lip never with loose, powdery meal on surface. .... *Encyclia* **subgenus** *Encyclia*
22. Leaves thin; plants small (pseudobulbs to 1.5 cm long); lip uppermost, with loose, mealy "pseudopollen"; flowers yellow or greenish yellow, densely clustered .... *Polystachya* (Chapter 9)

23(21). Pseudobulbs club-shaped, > 12 cm long 24
23. Pseudobulbs various, not club-shaped or < 3 cm long ...
.... 25

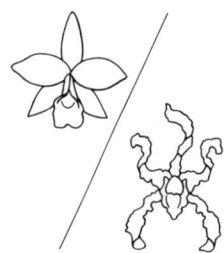

24(23). Inflorescence short; sepals and petals not markedly crisped
.... **Cattleya**
24. Inflorescence several times longer than pseudobulb; sepals and petals markedly crisped .... **Schomburgkia**

25(23). Pseudobulbs stalked, < 3 cm long, each with 1 leaf... 26
25. Pseudobulbs not stalked, or much > 3 cm long, with 1 or several leaves 27

26(25). Plants pendent; leaves fleshy; flowers solitary, membranous .... **Homalopetalum pumilio**
26. Plants erect; leaves thin; flowers racemose, white, fleshy
.... **Nidema** (see also **Scaphyglottis**)

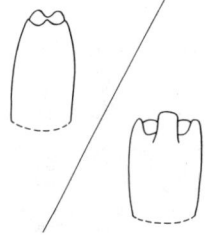

27(25). Pseudobulbs flat; inflorescence much longer than leaf, not branched; column apex with shallow notches between triangular teeth; pollinia 8 .... **Laelia rubescens**
27. Pseudobulbs various; inflorescence usually shorter, sometimes branched; column apex with deep notches between blunt teeth; pollinia 4.... **Encyclia** subgenus **Osmophyta**

Acrorchis. Epiphytic in moss or on rocks at high elevations; reed-stem, stems loosely clustered, slender, with several 2-ranked leaves; leaf sheaths dull red, warty; leaves elliptic or elliptic-ovate, retuse, fleshy, pale beneath; inflorescence terminal, flowers produced from a cluster of bracts, 1 or a few at a time; flowers delicate; lip weakly 3-lobed, with lateral lobes small, midlobe subcircular, callus fleshy, with transverse ridges in front; 1 species known; C: white or pale pink with orange callus; D: CR, Pma (Mt, Pk); M: pl to 14 cm, S 10 mm .... *Acrorchis roseola*
Plate 7(1)

Barkeria. Epiphytic; leaves several, 2-ranked, scattered along more or less thickened stem, often lost during dry season; inflorescence terminal with several to many flowers; lip not clearly lobed; column wide.

Key to *Barkeria*

1. Lip united to column for half column length; C: bright rose-purple, yellow keels on lip; D: No, rCR; M: pl to 20 cm, S 1.5-2.5 cm.... .... *Barkeria skinneri*
1. Lip free from column, or united for less than one-fourth column length 2
2(1). Lip > 2 cm long, base truncate (when spread); C: rose-purple; D: No, CR (Mt); M: pl to 40 cm, S 2.5 cm ...... *Barkeria lindleyana*
Plate 7(2)

   This showy species is in flower for Costa Rica's independence day, so it is called *Quince de septiembre*.
2. Lip about 1 cm long, base of lip tapering; C: white with purple veins on lip; D: No, CR, Pma (Mt, Pc); M: pl to 10 cm, S 12 mm .... *Barkeria obovata*
(Syn. *B. chinensis*)

   The geographically incongruous synonym reflects an old error in labeling, but an earlier name has been found.

Brassavola. Epiphytic or growing on rocks; shoots loosely clustered, stems only weakly thickened, but leaves very fleshy to nearly cylindrical,

1 on each stem; flowers white or greenish white; base of lip enfolding column; pollinia 8.

Key to *Brassavola*

1. Stems nearly as long as leaves; leaves narrow, gradually tapering; ovary and pedicel nearly as long as leaves; lateral lobes of lip fringed, midlobe long and narrow; **C**: white, or S, P pale yellowish green; **D**: No, SA; **M**: pl to 40 cm, S 10 cm.. *Brassavola cucullata*
This species is found in Nicaragua and Venezuela; I have found no records for Costa Rica or Panama.
1. Stems much shorter than leaves; leaves fleshy, tapering abruptly; ovary and pedicel much shorter than leaves; lip not fringed, midlobe much shorter than blade of lip............................... 2

2(1). Leaves very long, hanging downward; inflorescence basal; **C**: white; **D**: CR, Pma (Pc); **M**: pl to 50 cm, S to 10 cm .... *Brassavola acaulis*
Plate 3(5)
2. Leaves < 30 cm long, erect; inflorescence terminal............. 3

3(2). Lip smooth, margins not toothed; **C**: S, P pale yellowish green, lip white with purple spots in throat; **D**: No, CR, Pma, SA (Cb, Pc); **M**: pl to 20 cm, S 6 cm .... *Brassavola nodosa*
Plate 3(6)
3. Lip with raised veins, margins toothed; **C**: S, P greenish white, lip white with yellow-green nerves in throat; **D**: No, rCR; **M**: pl to 20 cm, S 9 cm................................. *Brassavola venosa*
Recent authors distinguish this species by the features given in the key and by its wider leaves. Leaves are generally wider on the Caribbean coast than on the Pacific, but there is a great deal of variation in this feature.

Cattleya. Epiphytic; pseudobulbs club-shaped, with 1 or 2 leaves; inflorescence terminal from a flat spathe; lip enclosing column, more or less trumpet-shaped; pollinia 4, disklike, with straplike caudicles.

Cattleya

Cattleya is the orchid that everyone recognizes as such. Cultivated hybrids are largely descended from South American species, but *C. dowiana* has played an important role in hybridization. Both *C. dowiana* and *C. skinneri* are listed as threatened species (Appendix I of CITES: *Appendices to the Convention for the International Trade in Endangered Species of Wild Fauna and Flora* [U.S. Fish and Wildlife Service]), so they cannot be legally exported, or imported into the United States.

Key to *Cattleya*

1. Pseudobulb with 1 leaf; column > 5 cm long; **C:** S, P yellow, lip purple heavily veined with orange-yellow; **D:** CR, Pma, SA (Cb); **M:** pl to 35 cm, S 6–8 cm .... ***Cattleya dowiana***
Plate 1(1)

    The *guaria de Turrialba* is probably the showiest of all cattleyas but is temperamental in cultivation. Its normal habitat is in tall trees in lowland tropical forest; its survival depends on the preservation of lowland forests.

1. Pseudobulb with 2 leaves; column about 15 mm long or less 2

2(1). Lip only slightly wider than petals: **C:** reddish orange, lip marked with purple within; **D:** No (Pc); **M:** pl 30 cm, S 2–2.8 cm .... *Cattleya aurantiaca*

    This small-flowered *Cattleya* is not definitely known from Costa Rica or Panama, though it occurs in Nicaragua.

2. Lip much wider than petals; flowers rose-purple 3

3(2). Throat of flower paler than lip; flowering March to April; **C:** rose-purple, **white or pale** area in throat; **D:** No, CR (Pc, Mt); **M:** pl to 30 cm, S 5.5 cm .... ***Cattleya skinneri***
Plate 1(3)

    The much admired *guaria morada* is the national flower of Costa Rica.

3. Throat of flower darker than lip; flowering September to October; **C:** rose-purple, **throat dark**; **D:** CR, Pma, SA (Cb, Pc); **M:** to 40 cm, S 3–5 cm .... ***Cattleya patinii***
Plate 1(2)

    Both this species and *C. skinneri* occur in Costa Rica, but only *C. patinii*, which ranges widely in northern South America, is known in Panama. Many plants of *C. patinii* are self-pollinating and thus disappointing in cultivation. Central American plants are often called *C. deckeri*, but that name is based on a Mexican plant, probably a natural hybrid between *C. aurantiaca* and *C. skinneri*.

Caularthron. Epiphytic; pseudobulbs clustered, cigar-shaped, smooth, whitish, hollow, normally occupied by ants, each with 2–3 terminal leaves; inflorescence terminal, subequal to pseudobulb or shorter, with few flowers; sepals and petals broad, spreading (if flower opens), lip 3-lobed, with hollow callus near middle; 1 species in our area; **C:** white or pinkish, callus yellow; **D:** No, CR, Pma, SA (Cb, Pc); **M:** pl to 25 cm, S 12 mm .... *Caularthron bilamellatum*

Abundant in dry forests; the flowers are self-pollinating and rarely open.

Dimerandra. Epiphytic; stem fleshy, slender cigar-shaped, from a narrow stalk; leaves scattered along upper part of pseudobulb; leaves narrowly oblong, leathery, notched; inflorescence terminal, very short, producing 1 flower at a time; flowers very flat, column with 2 wings at apex that are laterally flattened and parallel with each other. Flowers pink or rose-purple with darker lip and white center.

Dimerandra is often common in drier forests, but many are self-pollinating with short-lived flowers.

Key to *Dimerandra*

    1. Petals broadly rhombic, at least two-thirds as wide as long; lip distinctly notched apically; **D:** No, CR, Pma, SA (Pc, Mt); **M:** pl 20 cm, S 13–14 mm .... ***Dimerandra latipetala***
    1. Petals less than two-thirds as wide as long; lip not markedly notched at apex 2
    2(1). Three rows of callus united into transverse plates at apex, with a few separate, smaller plates in front; **D:** Pma, SA (Cb, Mt); **M:** pl 20 cm, S 1.9 cm .... ***Dimerandra elegans***
    Plate 7(3)

2. Three rows of callus free from each other at apex, without separate plates in front; **D:** No, CR, SA (Cb, Pc); **M:** pl 20 cm, S 1.8 cm .... *Dimerandra emarginata*

Encyclia. This genus is made up of 3 distinctive subgenera, and a good case could be made for treating them as different genera. The subgenus *Dinema*, including only *Encyclia polybulbon*, is not known in Costa Rica or Panama.

Encyclia subgenus *Encyclia.* Epiphytic or lithophytic; pseudobulbs conic-ovoid or onion-shaped; leaves straplike, leathery; lip 3-lobed, lateral lobes enfolding column, column with 3 short, triangular teeth at apex, notches between teeth shallow; capsule not winged. Plate 2(1–4).

Key to *Encyclia* subgenus *Encyclia*

1. Lip without distinct lateral lobes, united with column for about three-fifths column length; **C:** S, P greenish yellow shading to brown or purplish brown at tips, lip cream with dull violet markings; **D:** No, rCR, SA; **M:** pl to 25 cm, S 7–10 mm ... *Encyclia luteorosea*
1. Lip with distinct lateral lobes, united with column for less than half column length 2
2(1). Midlobe of lip about 2.5 cm wide; column without trace of wings or teeth on sides; **C:** S, P green to reddish brown, lip white with rose-purple spot in center (Costa Rica) or heavily veined with rose-purple throughout (Panama); **D:** No, CR, Pma, SA (Pc); **M:** pl to 50 cm, S 2.5–3.5 cm .... *Encyclia cordigera*
Plate 2(1, 2)

Known as the *semana santa* in Panama because it flowers during Holy Week, this is one of the loveliest orchids of Central America. It grows in seasonally dry lowlands and is a tough, easily cultivated plant. The flowers are variable and include some very fine forms. The Panamanian form may be called var. *rosea*; the name *randii*, sometimes used, refers to a different, Brazilian species. This species is still often called *Epidendrum atropurpureum*, but that name belongs to a West Indian plant now in the genus *Psychilus*.

2. Midlobe of lip < 1.8 cm wide; column usually with a distinct tooth or wing at each side................................... 3
3(2). Column with prominent rounded or squarish wings............ 4
3. Column with narrow or triangular teeth..................... 7
4(3). Ovary and pedicel rough and warty or almost spiny; lip acute; column wings bluntly triangular; **C**: green, lip yellowish with red markings; **D**: No, CR (Mt); **M**: pl 35–40 cm, S 2.5–3.1 cm....... .... *Encyclia tuerckheimii*
4. Ovary and pedicel relatively smooth; lip obtuse or notched 5
5(4). Column distinctly bowed (downward) in middle; wings squarish; **C**: S, P shiny, brown to dark brown-red, lip white with purple veins; **D**: No, CR?; **M**: pl to 35 cm, S 1.2–1.7 cm... *Encyclia guatemalensis*
5. Column straight, not as above............................. 6
6(5). Midlobe of lip about 12 mm wide, veins warty, lateral lobes obovate, much wider above base; **C**: S, P basally pale green, then dark red-brown, lip pale yellow or greenish yellow, veins lined with dark red; **D**: No, rCR; **M**: pl 40–50 cm, S 2.3–3 cm ....*Encyclia alata*
Plate 2(4)
6. Midlobe of lip much less than 12 mm wide, veins smooth; lateral lobes not wider above; **C**: S, P green to brownish yellow, lip with rose-purple streaks; **D**: No, CR, Pma (Cb, Pc); **M**: pl 25 cm, S 11 mm .... *Encyclia amanda*
Common in central Panama.
7(3). Lateral lobes of lip much wider at tips than base.............. 8
7. Lateral lobes of lip not expanded above 9
8(7). Column without lateral teeth (wings); sepals and petals abruptly wider in upper halves; **C**: S, P green suffused with reddish brown, lip pink or cream; **D**: No, CR; **M**: pl 30–35 cm, S 1.8–2.4 cm .... *Encyclia selligera*
8. Column with small but definite lateral teeth (wings); sepals and petals gradually wider above; **C**: S, P olive green with purple spots, lip **violet** with white margins; **D**: No, CR, Pma (Mt); **M**: pl 60–80 cm; S 11–12 mm .... *Encyclia mooreana*
Plate 2(3)
9(7). Ovary and veins on midlobe of lip with small warts; lateral lobes of lip not spreading; **C**: S, P green, lip cream or yellowish; **D**: No, CR, Pma (Pc); **M**: pl 30–35 cm, S 1.3–2 cm .... *Encyclia gravida*
Many forms of this species are self-pollinating, and some set seed without opening. Even the flowers that open are relatively drab.

9. Veins on midlobe of lip not warty, lateral lobes of lip spreading, midlobe marked with violet veins; **C:** S, P cream to pale greenish yellow, midlobe cream or yellowish cream with violet veins, lateral lobes basally suffused with **brown**; **D:** No, CR, Pma (Mt); **M:** pl 35–45 cm, S 1.3–2 cm . . . . *Encyclia ceratistes*

Encyclia subgenus *Osmophyta*. Epiphytic or lithophytic, pseudobulbs various, often flattened; lip various; column with three fleshy, blunt teeth separated by deep, rounded sinuses; capsule often winged. Plates 2(5, 6) and 3 (1–3).

If one considers the variation of this group in South America or Florida, it may seem reasonable to divide it into several genera. If one looks carefully at the group in Mexico and Central America, however, such a division seems quite impossible because of intermediates between the apparent groups. Various names have been used for parts of this group, but the oldest generic name, *Prosthechea*, has been used for only 1 species.

The flowers of the cockleshell orchids of the *E. fragrans* complex are so similar that they are almost always confused, but the species are easily distinguished by details of the plant and flower.

Key to *Encyclia* subgenus *Osmophyta*

1. Lip distinctly 3-lobed. 2
1. Lip undivided 10
2(1). Lateral lobes of lip each larger than midlobe 3
2. Midlobe larger than lateral lobes. 5
3(2). Sepals, petals, and midlobe obtuse; midtooth of column not longer than lateral teeth; **C:** S, P yellowish brown or brown, lip white with red dots; **D:** No, CR (Mt); **M:** pl to 25 cm, S 4–6 mm . . . .*Encyclia ochracea*
3. Sepals, petals, and midlobe of lip acute; midtooth of column long and finger-like. 4
4(3). Pseudobulbs narrow, usually 5–10 cm long; flowers 4–6; sepals > 8 mm long; **C:** S pale brownish green, P and lip cream; **D:** No, CR, Pma (Mt); **M:** pl to 20 cm, S 8–13 mm . . . . *Encyclia pseudopygmaea*
4. Pseudobulbs ellipsoid, to 3 cm long; flowers usually 1 or 2; sepals to 6 mm long; **C:** S, P cream or pale green, lip white with 1–3 purple spots or streaks; **D:** No, CR, Pma, SA (Mt); **M:** pl to 8 cm, S 5–6 mm . . . . *Encyclia pygmaea*

5(2). Midlobe of lip distinctly warty............................... 6
5. Midlobe of lip veined or smooth, not warty................... 7
6(5). Pseudobulbs each with long "neck"; midlobe of lip deeply notched; lateral lobes acute; C: S, P reddish to blackish brown, lip greenish cream to cream-yellow; D: No, CR, Pma (Mt); M: pl to 50 cm, S 10–13 mm.................................*Encyclia varicosa*
6. Pseudobulbs without long necks; midlobe not deeply notched, lateral lobes rounded; C: S, P pale green suffused with brown or mahogany brown, lip cream or yellowish, lateral lobes with red-brown veins; D: No, CR, Pma, SA (Mt); M: pl to 25 cm, S 4–6 mm .... *Encyclia livida*
7(5). Midlobe of lip obtuse or rounded; midtooth of column fleshy; plants whitish 8
7. Midlobe tapering to acute point; midtooth of column fringed; plants not whitish... 9
8(7). Sepals 1.6–2 cm long; midlobe of lip longer than wide, with wide sinuses; pseudobulbs flattened, sharp-edged; inflorescence often branched; C: S, P brown to dark purple, lip cream; D: No, CR, wPma (Mt); M: pl to 20 cm, S 1.6–2 cm .... *Encyclia campylostalix*
8. Sepals about 6 mm long; midlobe of lip wider than long, with narrow sinuses; edges of pseudobulbs rounded; inflorescence not branched; C: S, P green, lip cream; D: Pma (Mt); M: pl 13 cm, S 6 mm .... *Encyclia fortunae*
9(7). Sepals and petals spotted; C: S, P greenish cream or pale green with red-brown spots, midlobe of lip rose-purple; D: No, CR, Pma (Mt); M: pl 50 cm, S 2.4–4.4 cm .... *Encyclia prismatocarpa*
Plate 2(5)
9. Sepals and petals without spots; C: S, P pale greenish yellow, may have violet specks, lip cream with 1 or 2 violet streaks; D: CR, Pma (Cb, Mt); M: pl 45 cm, S 2.4–4.4 cm ....*Encyclia ionocentra*
These plants occur at lower elevations than *E. prismatocarpa* and range much farther east in Panama; the whole complex needs careful study.
10(1). Narrow stalk of lip as long as column, midtooth of column fringed, lip acute; C: S, P pale green to olive-tan, lip basally cream then violet-purple; D: No, CR, Pma (Mt); M: pl to 60 cm, S 4.2–5.5 cm .... *Encyclia brassavolae*
Plate 2(6)
10. Not with above combination of features 11

11(10). Lip not conspicuous, about twice as wide as column, not much longer than column; **C**: greenish or yellowish, S, P usually spotted; **D**: No, CR, Pma, SA (Mt); **M**: pl to 70 cm, S 10–13 mm........ .... *Encyclia vespa*
                                                   Plate 3(3)
   11. Lip conspicuous, usually concave and white or greenish marked with colored veins, at least at base 12

12(11). Pseudobulbs each with only 1 leaf........................ 13
   12. Pseudobulbs each with 2 to several leaves................... 16

13(12). Lip flat or convex; column nearly as wide as long............. 14
   13. Lip definitely concave 15

14(13). Lip curved, elliptic, acuminate, much longer than wide; **C**: S, P white flushed with pink, lip with violet lines; **D**: Pma (Mt); **M**: pl 30 cm, S 2.8–3.7 cm.. *Encyclia sima*
           Very fragrant.
   14. Lip flat, subtriangular or 5-angled, wider than long; **C**: pinkish red or purple with greenish yellow margins; **D**: No, CR, Pma (Mt); **M**: pl 35 cm, S 1.3–2 cm .... *Encyclia spondiada*

15(13). Inflorescence from new growth, before growth of pseudobulb; spathe generally thin; pseudobulbs generally ovoid; **C**: cream, lip with purple lines; **D**: Pma, SA (Cb, Pc); **M**: pl 35 cm, S 2–3 cm ... . *Encyclia (fragrans* subsp.) *aemula*
           The flowers of *E. (fragrans* subsp.) *aemula* are quite similar to those of *E. fragrans*, and these species have long been confused. They flower at different seasons and usually occur in different habitats. This plant lacks a valid name as a species of *Encyclia*, which should soon be remedied.
   15. Inflorescence from mature pseudobulb, with a thick, leathery sheath; pseudobulbs generally narrower, ellipsoid or narrowly ovoid; **C**: S, P white, often spotted with purple-violet (in Panama), lip with purple-violet lines; **D**: No, CR, Pma, SA (Mt); **M**: pl to 20 cm, S 2–3.5 cm .... *Encyclia fragrans*
                                                 Plate 3(2)
           The plants of Panama and Ecuador with spots on sepals and petals have been known as *E. chimborazoensis*, but some Panamanian plants lack the spots and can scarcely be distinguished from *E. fragrans*. *E. fragrans* flowers during the dry season (December to April).

16(12). Plants creeping; pseudobulbs widely spaced on rhizome, distance between pseudobulbs nearly equal to length of pseudobulb 17
16. Pseudobulbs clustered, distance between pseudobulbs much shorter than length of pseudobulb. 19

17(16). Leaves usually 3, up to 2.5 cm wide; peduncle about 5 cm long, exposed; inflorescence from new growth; lip clearly concave, with a thick, fleshy callus; leaf tip obliquely notched; C: white or cream, lip with purple streaks; D: No, CR (Mt); M: pl 15 cm, S 1.9–2.4 cm .... *Encyclia vagans*
17. Leaves usually 2, < 12 mm wide; peduncle short, covered by spathe; inflorescence from mature pseudobulb; lip nearly flat, callus flat; leaf tip acute or narrowly obtuse. 18

18(17). Base of lip squarish; column with a distinct triangular lateral expansion on each side in middle; C: white or cream, P with 1 and lip with several violet streaks; D: No, CR, Pma (Mt); M: pl 15 cm, S 13–15 mm. .... *Encyclia abbreviata*
18. Base of lip clearly cordate, with distinct retrorse lobules; column very slightly widened in middle; C: cream, lip with several violet lines; D: No, CR (Mt); M: pl 25 cm, S 1.6–2.6 cm .... *Encyclia neurosa*

19(16). Lip acuminate or narrowly acute; flowers normally 2, back-to-back, on short stalk (< 2.5 cm); C: white or cream, lip with maroon-red lines; D: No, CR, SA; M: pl to 40 cm, S 4–4.5 cm.............. .... *Encyclia baculus*
(Syn. *Epidendrum pentotis*)
19. Lip broadly acute, obtuse or notched 20

20(19). Lip distinctly notched, wider than long; column midtooth thin, toothed or fringed; C: cream or greenish white, lip with purple lines; D: No, CR; M: pl to 30 cm, S 1.5–2 cm .... *Encyclia radiata*
20. Lip acute or obtuse but not notched. 21

21(20). Callus of lip smooth, lip more or less tinged with purple between lines; sepals and petals long-acuminate, twisted; C: S, P green, lip basally white with deep purple veins, remainder deep purple flushed with yellow-green; D: No, CR, Pma (Mt); M: pl to 35 cm, S 3–4 cm .... *Encyclia cochleata*
In Costa Rica and Panama this well-known species usually has rather small flowers.
21. Callus of lip fuzzy, lip never tinged purple between lines; plant whitish green 22

22(21). Sepals and lip about 2 cm long; lip obtuse; flowers opening widely; C: cream or white, lip with purple lines; D: CR, wPma (Mt); M: pl 25 cm, S 2–2.6 cm . . . .*Encyclia ionophlebia*
Plate 3(1)
22. Sepals and lip up to 1.7 cm long; lip acute; C: cream or greenish white, lip with purple lines; D: No, CR, Pma, SA (Pc, Mt); M: pl 25 cm, S 1.4–1.7 mm . . . .*Encyclia chacaoensis*
Flowers often self-pollinating.

Epidendrum. Usually slender-stemmed with many 2-ranked leaves (reed-stem), but may have pseudobulbs or few leaves; inflorescence usually terminal; narrow base of lip **united with** full length of the **column**, with an opening at column apex; anther usually dorsal; pollinia 2 or usually 4, with a semiliquid viscidium; the apex of the stigma (rostellum) is slit by the removal of the viscidium. Plates 4(1–6), 5(1–6), 6(1, 2).

General Key to *Epidendrum*

1. Plants with pseudobulbs, part or all of stem above rhizome distinctly thickened . . . . **Key 1**
1. Plants without pseudobulbs, stems uniformly slender above rhizome 2
2(1). Inflorescences all lateral, without leaflike bracts . . . . **Key 3**
2. Inflorescences terminal or both terminal and lateral 3
3(2). Sides of column aperture forming a narrow slit beneath anther; caudicles hard and bony; rachis of inflorescence markedly flattened; new stems arising above bases of older stems (plant thus becoming bushy) . . . . ***Epidendrum* (*Neowilliamsia*)**
Key 12
3. Column aperture not forming a narrow slit; caudicles soft and mealy 4
4(3). Floral bracts and bracts of peduncle laterally flattened, keeled, spreading, prominent, similar 5
4. Floral bracts not laterally flattened and keeled, bracts of peduncle inflated, cylindrical, or unlike floral bracts 6
5(4). Shoots arising from bases of older shoots; flowers green, fleshy . . . . **Key 9**

5. Shoots arising above bases of older shoots; flowers not fleshy, usually not green .... *Epidendrum* (*Neowilliamsia*)
Key 12

6(4). Stems branched, *or* new shoots arising above bases of older shoots 7
6. Stems not branched, new shoots arising from base 9

7(6). Stems branched, branches irregular or unequal, plants thus rather shrubby... Key 10
7. New shoots arising above bases of older shoots but not otherwise branching; shoots all about the same length, or upper shoots shorter 8

8(7). Dwarf plants with narrow, fleshy leaves; shoots somewhat separated on rhizome; flowers short-stalked on a zigzag rachis; pollinia 2................................. *Epidendrum* (*Epidanthus*)
Key 11
8. Plants without above combination of features .... Key 8

9(6). Inflorescence with inflated sheath or clustered bracts unlike bracts above sheath .... Key 2
9. Inflorescence without distinct sheaths, basal bracts of peduncle similar to upper bracts 10

10(9). Midlobe of lip deeply divided, lobes narrow and spreading, often wider than across lateral lobes, midlobe not stalked; inflorescence often branched and flowering repeatedly; pollinia of "birdwing" type, with bladelike edges above ....
.... *Epidendrum paniculatum* complex
Key 5
10. Midlobe not deeply divided, or with a distinct stalk; not with above combination of features................................. 11

11(10). Inflorescence with distinct peduncle, this usually longer than upper leaves.. 12
11. Peduncle short or inconspicuous 13

12(11). Dwarf plants with narrow, fleshy leaves; shoots somewhat separated on rhizome; flowers short-stalked on a zigzag rachis; pollinia 2 .... *Epidendrum* (*Epidanthus*)
Key 11
12. Plants without above combination of features .... Key 4

13(11). Flowers solitary or umbellate, both peduncle and rachis very short or concealed; leaves fleshy; flowers green or greenish .... Key 7
13. Flowers usually several, racemose, with a definite rachis .... Key 6

Keys to *Epidendrum*

Key 1
Plants with definite pseudobulbs.

1. Pseudobulb basal with slender, leafy stem on top, grooved; C: white with yellow callus; D: CR, Pma, SA (Mt); M: pl to 1 m, S 4–5 mm .... *Epidendrum blepharistes*
(Syn. *E. dolabrilobum*)
This species usually lives on ant nests. It is not recorded from Panama, but I have found dead, fallen plants that appear to be of this species in central and western Panama.
1. Pseudobulb thickened throughout or at least in upper part, never with a leafy stem above.................................... 2

2(1). Pseudobulbs oblong, without slender basal stalk; inflorescence lateral; C: green; D: CR, Pma, SA (Cb, Pc); M: pl 12 cm, S 10–12 mm .... *Epidendrum rousseauae*
(Syn. *E. laterale*)
Plate 5(6)
A very un-*Epidendrum*-looking plant with typical *Epidendrum* flowers.
2. Pseudobulbs elongate or distinctly stalked; inflorescence terminal or pseudolateral (on rudimentary pseudobulbs in *E. stamfordianum*) 3

3(2). Some leaves 2-ranked on upper part of slender pseudobulb...... 4
3. Leaves terminal on relatively short pseudobulb 5

4(3). Flowers produced 1 at a time from closely spaced, overlapping bracts; C: pale green; D: Pma (Cb, Mt); M: pl to 35 cm, S 10–12 mm .... *Epidendrum ellipsophyllum*
4. Flowers racemose, all flowering at once; bracts prominent but not overlapping; C: probably green; D: CR (Mt); M: pl 15 cm, S 11–12 mm .... *Epidendrum bracteosum*

5(3). Plant pendent, with long, fleshy leaves; C: S, P greenish yellow, lip white; D: No, CR, wPma (Mt); M: pl to 1 m, S 7–7.5 cm .... *Epidendrum parkinsonianum*
Plate 4(2)
5. Plant erect 6

6(5). Lip not markedly 3-lobed 7
6. Lip distinctly and deeply 3- or 4-lobed 9

7(6). Leaves fleshy, subcylindrical; flowers sparsely fuzzy externally, sepals 4–5 mm long; C: yellow-green, S bronzy; D: No, CR, Pma (Cb, Mt, Pc); M: pl to 15 cm, S 4.5–5 mm ....
.... *Epidendrum stangeanum*
> This species seems to be a close relative of the plants that some authors still call *Lanium*. It has 2 deeply notched pollinia, and those who would put every plant with a slight difference in pollinia into a different genus would have to name a new genus for this one.

7. Leaves flat, leathery; flowers smooth externally, sepals at least 8 mm long... 8

8(7). Stems slender; flowering from mature growth; C: greenish white; D: No, CR (Cb, Mt, Pc); M: pl 15 cm, S 7–8 mm ....
.... *Epidendrum octomerioides*
8. With distinct pseudobulbs; flowering from young growth, before pseudobulb develops; C: green; D: Pma (Mt); M: pl 25 cm, S 1.8 mm .... *Epidendrum volutum*

9(6). Inflorescence usually appearing lateral; leaves 2–3; midlobe of lip wider than long, fringed; C: S, P greenish or pale yellow with purple spots, midlobe orange; D: No, CR, Pma, SA (Cb, Pc); M: pl to 50 cm, S 1.8–2.2 mm .... *Epidendrum stamfordianum*
9. Inflorescence clearly terminal, though often on a young shoot; midlobe of lip much longer than wide 10

10(9). Lateral lobes of lip deeply fringed; inflorescence usually on a mature pseudobulb; C: white, or S, P greenish; D: No, CR, Pma, SA (Cb, Pc); M: pl to 30 cm, S 5.5–6.5 cm .... *Epidendrum ciliare*
Plate 6(1)
> A common and rather showy species. The plants are variable, and unscrupulous vendors sell some forms as *Cattleya* when there are no flowers.

10. Lateral lobes not fringed; inflorescence on young shoot, often before new pseudobulb is formed 11

11(10). Midlobe of lip narrowly elliptical; lip subequal to sepals and petals; C: white; D: No, CR, Pma (Mt); M: pl 30 cm, S 4.5–5 cm ....
.... *Epidendrum oerstedii*

11. Midlobe more than half as wide as long; lip much shorter than sepals and petals; **C:** S, P pale green, lip white; **D:** CR, SA (Mt); **M:** pl 30 cm, S 1.9–2 cm .... ***Epidendrum glumibracteum***
Similar to the South American *E. purpurascens*, but apparently distinct.

Key 2

Stems slender, inflorescence from definite sheath, bracts of sheath unlike either foliage leaves or floral bracts (flowers normally several, not 1–2); sheath bracts sharply different from bracts of peduncle or floral bracts.

1. Sheath of several overlapping bracts......................... 2
1. Sheath of 1 or 2 bracts 4
2(1). Lateral lobes of lip smooth or slightly toothed, not deeply divided; **C:** rose-purple, lip with white spot; **D:** CR (Mt); **M:** pl to 2 m, S 1.5–1.9 cm .... ***Epidendrum pfavii***
Plate 4(6)
2. Lateral lobes of lip fringed or deeply divided 3
3(2). Plant small (< 50 cm tall); lateral lobes of lip with hairlike divisions; with wide gap between lateral lobes and midlobe; **C:** S, P green with dark carmine spots, lip white; **D:** No, CR, Pma, SA (Mt, Pc); **M:** pl to 40 cm, S 10–12 mm ....***Epidendrum criniferum***
3. Plant large (to 1.3 m tall); lateral lobes with finger-like divisions, without gap between lateral lobes and midlobe; **C:** S, P greenish yellow with red spots and streaks, lip white, spotted rose; **D:** No, CR, Pma, SA (Cb, Mt); **M:** S 1.8–2.2 cm....................
.... ***Epidendrum raniferum***
4(1). Spathe bracts narrow, not inflated; inflorescence arched; **C:** S, P greenish, lip orange with white margins; **D:** CR, Pma (Mt); **M:** pl to 35 cm, S 2.5–4 mm.................. ***Epidendrum arcuiflorum***
4. Spathe bracts much inflated 5
5(4). Erect, robust plant (> 30 cm tall); midlobe of lip narrowly lanceolate; **C:** S, P pale green, lip white; **D:** No, CR, Pma, SA (Mt); **M:** pl to 1 m, S 3.5–4.2 cm..................... ***Epidendrum lacustre***
(Syn. *E. obesum*)
5. Smaller plants (< 20 cm tall); midlobe of lip rounded 6
6(5). Lower sheaths of stem inflated and spreading; leaves obovate, obtuse; plant erect; **C:** probably green; **D:** CR (Mt); **M:** pl 15 cm, S 13 mm .... ***Epidendrum trianthum***

6. Lower sheaths of stem not inflated or spreading; leaves ovate, oblique, acute; leaves and bracts barred with red; plant pendent; C: pale green; D: CR, Pma (Mt); M: pl to 8 cm, S 3–4 mm .... *Epidendrum obliquifolium*

Key 3
Stems slender, inflorescences lateral only (without subtending leaves or leaflike bracts).

1. Column hood tubelike, longer than rest of column; C: pale green; D: No, CR, Pma (Mt); M: pl to 80 cm, S 11–13 mm .... **Epidendrum phragmites**
   No good specimens have been collected in Panama, but I have found fragmentary material that appears to be this species.
1. Column hood short, scarcely surpassing anther 2
2(1). Leaves broad and notched; base of inflorescence with conspicuous overlapping bracts; C: lvs dark red beneath, fls reddish green; D: CR, SA (Mt); M: pl to 40 cm, S 13–15 mm..................... .... **Epidendrum albertii**
2. Leaves narrowly elliptic, acute; bracts at base of inflorescence small, inconspicuous; C: S, P creamy green, lip white; D: Pma (Mt); M: pl to 45 cm, S 9 mm.................... **Epidendrum brachybotrys**

Key 4
Stems slender, inflorescence with distinct peduncle, usually longer than upper leaf; bracts sheathing, not strongly flattened. If there are conspicuous bracts at base of inflorescence, these are not sharply distinct from sheathing bracts of peduncle.

1. Peduncle somewhat flattened; flowers produced 1 by 1 from a dense cluster of bracts 2
1. Peduncle not markedly flattened; flowers usually several to many, not from a dense cluster of bracts........................... 4
2(1). Lip as wide as long, apex rounded 3
2. Lip longer than wide, tapering to apex; C: S green speckled with purple, P pale green, lip whitish green with white calli; D: CR (Pc); M: S 12 mm .... *Epidendrum adnatum* or C: green; D: CR (Cb); M: to 30 cm, S 13 mm................ .... **Epidendrum lankesteri**

or C: green; D: Pma (Cb, Mt, Pc); M: pl 20 cm, S 12 mm ....
.... ***Epidendrum panamense***
These 3 species are very similar but may be distinct; there appear to be several similar species in the complex.

3(2). Lip with several fleshy ridges radiating from base; leaves obtuse, notched, purplish; C: plant purplish, fls green or purplish green; D: CR, Pma (Mt); M: pl 30 cm, S 15 mm ....
.... ***Epidendrum phyllocharis***
3. Lip without fleshy ridges; leaves acute, green; C: S, P pale green, lip cream; D: Pma (Mt); M: pl 30 cm, S 9–11 mm ....
.... ***Epidendrum allenii***

4(1). Flowers produced several at a time in dense cluster with buds in center, thus flattened or umbel-like 5
4. Flowers in narrow raceme, simultaneous or 1–2 at a time, cluster not flattened or umbel-like 9

5(4). New shoots produced above bases of older shoots; plant **sprawling** and rooting from stem; column **arched**; C: S, P orange-red or red, lip yellow; D: No, CR, Pma, SA (Mt); M: pl to 1 m, S 1.5–1.8 cm
.... ***Epidendrum radicans***
Plate 4(5)
This species is locally abundant along mountain roadsides. The flowers are similar to those of the South American *E. ibaguense*, which has a straight column, but the plants are quite different.
5. New shoots arising at base of older shoots; stems clustered, erect; roots basal; column straight 6

6(5). Lip uppermost, fringed on sides, usually not clearly lobed; C: magenta-pink, lip darker; D: No, CR, Pma, SA (Cb, Pc); M: pl 1 m, S 10–13 mm .... ***Epidendrum imatophyllum***
This attractive species grows on arboreal ant nests and is rarely cultivated with any success.
6. Lip lowermost, may be toothed but not fringed 7

7(6). Lip margins toothed, usually 3-lobed; C: red with yellow callus; D: No, CR, Pma, SA (Cb, Mt, Pc); M: pl 1 m, S 15–16 mm ....
.... ***Epidendrum baumannianum***
(Syn. *E. hawkesii*)
Normally epiphytic and sometimes growing on ant nests,

this may be a geographic subspecies of the well-known *E. incisum* (better known as *E. schomburgkii*).
7. Lip margins not toothed 8

8(7). Lip as long as wide, 3-lobed, midlobe notched, thus appearing 4-lobed; **C:** yellowish brown to reddish brown; **D:** No, CR, Pma (Cb, Pc); **M:** pl to 60 cm, S 6.5–8 mm ....
.... ***Epidendrum galeottianum***
Plate 5(5)
This has been widely known as *E. anceps*, but that name is based on a plant of the West Indies with much wider petals.
8. Lip much wider than long, forward margin shallowly 3-lobed; **C:** pale green; **D:** CR, Pma, SA (Cb); **M:** pl 50 cm, S 8–10 mm ......
.... ***Epidendrum smaragdinum***
(Syn. *E. pachyrhachis*)
Sometimes grows in ant nests.

9(4). Lip subcircular, shallowly notched; plant robust, 1–2 m tall; **C:** S, P green, lip bright red-orange, tip of column may be purplish; **D:** CR, wPma (Pc); **M:** pl to 2.5 m, S 2.8–3.5 cm ....
.... ***Epidendrum pseudepidendrum***
This handsome species is becoming scarce through habitat destruction.
9. Not with above combination of features 10

10(9). Lip distinctly 3-lobed 11
10. Lip not lobed 15

11(10). Midlobe of lip rounded or acute, not notched or 2-lobed 12
11. Midlobe of lip distinctly notched or 2-lobed 13

12(11). With several small flowers at once, sepals < 5 mm long; **C:** S, P gray with pinkish edges, lip white, midlobe dull orange; **D:** Pma (Mt, Pc); **M:** pl 25 cm, S 2.5–4 mm .... ***Epidendrum powellii***
Possibly the same species as *E. arcuiflorum*.
12. One flower at a time, sepals about 3 cm long; **C:** white, S, P may be reddish or greenish; **D:** No, Pma (Cb, Mt); **M:** pl to 50 cm ....
.... ***Epidendrum hellerianum***

13(11). Lip about 2 cm wide; petals obovate, 6–7 mm wide; ovary smooth; C: greenish white; D: No, CR, Pma, SA (Cb); M: pl to 80 cm, S 1.8–2.2 cm............................ *Epidendrum coronatum*
(Syn. *E. moyobambae*)
13. Lip < 1.5 cm wide; petals linear or sublinear, 1–2 mm wide; ovary warty.. 14
14(13). Lip 10–13 mm wide; C: brownish yellow to yellow; D: No, CR, Pma, SA (Mt); M: pl to 1 m, S 12–15 mm......................
.... *Epidendrum polyanthum*
14. Lip about 5 mm wide; C: brownish yellow; D: CR, Pma (Cb, Mt); M: pl to 50 cm, S 4–5 mm................ *Epidendrum myodes*
(Syn. *E. polyanthum* var. *myodes*)
15(10). Inflorescence a narrow, pendent raceme, with many flowers at once; lip fleshy, folded around column so that it cannot be spread without breaking.. 16
15. Inflorescence with few flowers or densely clustered, may be branched; lip not fleshy and folded........................ 18
16(15). Lip margins erose or toothed; inflorescence usually branched; C: purplish brown with white lip; D: Pma (Mt); M: pl to 14 cm, S 4.5–5.5 mm........................... *Epidendrum probiflorum*
16. Lip margins smooth; inflorescence not branched.............. 17
17(16). Leaves ovate to narrowly elliptic, to 7 cm long; C: greenish brown to orange-brown, lip yellow sometimes flushed with purple; D: No, CR, Pma (Mt); M: pl 20 cm, S 5–8 mm...... *Epidendrum carolii*
17. Leaves long-acuminate, to 17 cm long; C: orange-brown to bronzy; D: No, CR, Pma, SA (Mt); M: pl to 30 cm, S 7–8 mm...........
.... *Epidendrum laucheanum*
(Syns. *E. cristobalense*, *E. dolichostachyum*)
Plate 5(3)
Common and variable at higher elevations in Costa Rica, known as *arrocillo*, *rosario*.
18(15). Inflorescence usually branched; dwarf plants; pollinia 2........ 19
18. Inflorescence rarely branched; pollinia 4..................... 20
19(18). Inflorescence slender, lax; leaves several, lanceolate, acute; petals linear; lip as wide as long, flat or slightly convex, papillose; C: pale green; D: CR, Pma, SA (Cb, Mt); M: pl 10 cm, S 3 mm..........
.... *Epidendrum vincentinum*
(Syn. *Epidendropsis vincentina*)
19. Inflorescence stiff, upright; leaves usually 2, at least half as wide as long, obtuse; petals lanceolate, fringed; lip longer than wide, sides

folded upward, not papillose; C: pale green; D: Pma, SA (Mt); M: pl 10 cm, S 8–9 mm .... *Epidendrum flexuosissimum*
Common in some cloud forests.

20(18). Plants with 1–2 leaves per stem and few-flowered inflorescence 21
20. Small plants with several leaves 22
21(20). Sepals and petals acute or acuminate, about 2 cm long; C: olivaceous becoming pink with age; D: CR, Pma (Mt, Pk); M: pl 20 cm, S 2 cm................................. *Epidendrum pallens*
21. Sepals and petals blunt, 12–13 mm long; C: rose-pink or purple; D: CR, Pma (Mt, Pk); M: pl 25 cm, S 12–13 mm ....
.... *Epidendrum paucifolium*
22(20). Inflorescence subequal to leafy stem or longer; flowers produced 1 or 2 at a time; each flower with prominent, rounded nectary at base
.... 23
22. Inflorescence usually shorter than leafy stem, with several to many flowers at once; flowers without prominent nectary 24
23(22). Column forming membranous, toothed hood over anther, hood longer than rest of column; C: S, P translucent pinkish green, lip pale yellow-green; D: No, CR, Pma (Cb, Mt); M: pl 10 cm, S 6–8 mm................................. *Epidendrum macroclinium*
23. Column hood short and fleshy, much shorter than rest of column; C: pale green or pinkish; D: CR (Mt); M: pl 20 cm, S 6–8 mm
.... *Epidendrum physodes*
24(22). Leaves narrow, acute, ascending; C: pale green; D: No, CR, Pma, SA (Mt); M: pl 6 cm, S 3–3.5 mm ...... *Epidendrum miserrimum*
24. Leaves oblong, obtuse, notched, spreading; C: greenish yellow; D: No, CR, Pma (Mt); M: pl 6–10 cm, S 3–4 mm..................
.... *Epidendrum selaginella*

Key 5

Midlobe of lip deeply divided, not stalked, lobes spreading; inflorescence often branched; *Epidendrum paniculatum* complex. The name *E. paniculatum* has been used in Central America but is applicable to a Peruvian species. All of the species keyed here, along with a few others as yet unnamed, have been treated as *E. paniculatum* in most Central American floras.

1. With 1 or a few flowers at a time; leaves ascending; C: S, P pale green, lip purple; D: No, CR, Pma (Cb); M: pl to 40 cm, S 11–12 mm .... *Epidendrum turialvae*

1. Many flowers open at once; leaves spreading.................. 2
2(1). Petals oblanceolate, sepals > 1 cm long; **C:** S, P green, lip white with green border; **D:** No, CR, Pma (Pc); **M:** pl to 70 cm, S 15 mm
.... ***Epidendrum isthmii***
   2. Petals linear or sublinear; sepals about 1 cm long or less........ 3
3(2). Sepals about 5 mm long; plants about 40 cm tall; **C:** brownish pale green, lip white; **D:** CR, Pma (Mt, Wt); **M:** pl to 40 cm, S 5 mm
.... ***Epidendrum subnutans***
   3. Sepals about 8–10 mm long................................. 4
4(3). Callus of lip with prominent purple markings; sepals and petals spreading; plants to nearly 2 m tall, with large, pendent racemes; **C:** S, P green, lip white with purple markings; **D:** CR, Pma (Mt, Pc); **M:** pl to 2 m, S 7–8 mm .... ***Epidendrum piliferum***
                                                                       Plate 6(2)
   4. Lip white; sepals and petals usually reflexed; lateral lobes of lip reflexed; plants to about 25 cm tall; inflorescence < 10 cm long; **C:** S, P green, lip white; **D:** CR (Pc); **M:** pl to 25 cm, S 9 mm........
.... ***Epidendrum resectum***

Key 6
Peduncle short or inconspicuous, stems not branched.

1. Leaves not jointed basally (persisting when dead); small, pendent, fleshy plants... Key 7
1. Leaves jointed basally 2
2(1). Inflorescence of several to many flowers 3
   2. Inflorescence of 1 flower, *or* producing 1 flower at a time 10
3(2). Flower with external spur at base, tip of spur free from ovary; **C:** apricot yellow; **D:** CR; **M:** pl to 35 cm, S 6–8 mm ....
.... ***Epidendrum puteum***
   3. Flowers without external spur 4
4(3). Lip 3-lobed 5
   4. Lip not distinctly 3-lobed 7
5(4). Midlobe of lip pointed; opening between column and lip tilted (facing obliquely downward); **C:** S, P greenish yellow with chestnut markings, lip cream with purplish spots in center; **D:** Pma (Cb); **M:** pl to 40 cm, S 10–12 mm.............. ***Epidendrum dentilobum***
   5. Midlobe of lip notched; opening between column and lip facing forward 6

6(5). Midlobe with distinct isthmus; petals spatulate-oblanceolate, nearly as wide as sepals; C: greenish white; D: No, CR, Pma, SA (Cb); M: pl to 80 cm, S 1.8–2.2 cm .... *Epidendrum coronatum*
(Syn. *E. moyobambae*)
6. Midlobe without isthmus; petals sublinear; C: pale green; D: CR, Pma, SA (Cb); M: pl 50 cm, S 7.5 mm ....
.... *Epidendrum smaragdinum*
(Syn. *E. pachyrhachis*)
Sometimes grows on ant nests.

7(4). Tall plant with short, pendent inflorescence ....
.... *Epidendrum warscewiczii*
Known only from a brief description and a crude drawing showing narrow, bell-shaped flowers hanging from a short inflorescence, from Panama.
7. Plants < 20 cm tall 8

8(7). Each stem with several leaves; inflorescence and sepals fuzzy; C: green flushed with pink; D: ePma, SA; M: pl 5 cm, S 8–9 mm
.... *Epidendrum microphyllum*
(Syn. *Lanium microphyllum*)
8. Leaves 1–2 per stem 9

9(8). Lip with narrow triangular callus in front of column; C: rose-pink or purple; D: CR, Pma (Mt, Pk); M: pl 25 cm, S 12–13 mm
.... *Epidendrum paucifolium*
9. Lip with 2 ovoid calli in front of column; C: greenish white; D: No, CR (Cb, Mt, Pc); M: pl 15 cm, S 7–8 mm ....
.... *Epidendrum octomerioides*

10(2). Plants dwarf, sprawling, each stem with 1 terminal flower (rarely 2), may also have lateral flowers; lip a subequilateral triangle; C: green; D: Pma (Mt); M: pl 5 cm, S 11–12 mm ....
.... *Epidendrum triangulabium*
10. Plants erect, not as above 11

11(10). Petals much wider than sepals, elliptic; C: brown-yellow striped with dull purple; D: No, rCR; M: pl 40 cm, S 3 cm.................
.... *Epidendrum sobralioides*
11. Petals narrow, similar to sepals; lip lobed or not............. 12

12(11). Midlobe of lip linear-lanceolate 13
12. Midlobe as wide as long, lip not or weakly 3-lobed 14

13(12). Rachis of inflorescence exposed; pedicel with ovary shorter than sepals; C: S, P white or pale green, lip white; D: No, CR, Pma, SA (Cb, Pc); M: pl to 80 cm, S 4.5–9 cm .... ***Epidendrum nocturnum***
13. Rachis of inflorescence concealed by leaf bases; pedicel with ovary much longer than sepals; leaves broadly oblong; C: S, P pale green, lip white; D: No, CR, Pma, SA (Mt); M: pl 40 cm, S 7–9 cm .... ***Epidendrum carpophorum***
(Syn. *E. latifolium*)
The flowers of this species are usually self-pollinating and may not open. It usually has very wide leaves, but leaf width varies with growth conditions.

14(12). Inflorescence slender, pendent; lip weakly 3-lobed; C: white, S, P may be reddish or greenish; D: No, Pma (Cb, Mt); M: pl to 50 cm, S 3 cm................................. ***Epidendrum hellerianum***
14. Inflorescence short, erect................................... 15

15(14). Inflorescence bracts wide and overlapping; C: plant purplish, flowers green or purplish green; D: CR, Pma (Mt); M: pl 30 cm, S 15 mm............................. ***Epidendrum phyllocharis***
15. Inflorescence bracts narrow, not overlapping 16

16(15). Lip more or less heart-shaped, widest basally; C: S, P yellow-green, lip white; D: No, CR, Pma (Cb); M: pl to 80 cm, S 2.2–3.6 cm .... ***Epidendrum eburneum***
Plate 4(1)
16. Lip subquadrate, widest near apex; C: pale green; D: CR, Pma (Mt); M: pl to 80 cm, S 13–15 mm........ ***Epidendrum notabile***

Key 7

Inflorescence umbellate or 1-flowered, leaves fleshy, flowers green or greenish, including *Epidendrum difforme* complex. Though the name *E. difforme* has been widely used in our area, it actually belongs to a West Indian species. This group has been treated as a polymorphic complex, but several quite distinct species are found growing together in some areas. Most of the species in this key have been treated as *E. difforme* at one time or another.

1. Sheaths of stem inflated and oblique; inflorescence from a truncate spathe; C: probably green; D: CR (Mt); M: pl 15 cm, S 13 mm ... . ***Epidendrum trianthum***
1. Sheaths of stem not inflated and oblique; inflorescence without prominent spathe 2

2(1). Column hood smooth and fleshy, without teeth or fringe 3
2. Column hood toothed or fringed (*difforme* complex) 7
3(2). Leaves not basally jointed, rotting in place 4
3. Leaves basally jointed, falling off when dead 6
4(3). Sepals about 2 cm long; C: green or bronzy green, about same color as leaves; D: Pma (Cb, Pc); M: pl 5 cm, S 2 cm ....
................................*Epidendrum schlechterianum*
4. Sepals about 1 cm long 5
5. Plant whitish green; flowers usually 2; C: green or reddish green; D: CR; M: pl 8 cm, S 7–8 mm ....*Epidendrum congestoides*
5. Plant dark green or reddish green; flowers usually 4; C: S, P pale green, lip emerald green, column marked with red-purple; D: CR; M: pl 5–6 cm, S 7–10 mm ....*Epidendrum congestum*
6(3). Plants pendent, > 15 cm long; C: entire plant pale gray-green; D: CR, Pma (Mt); M: pl to 60 cm, S 4.3–5 cm ....
....................................*Epidendrum pendens*
Plate 5(2)
6. Plant erect or sprawling, stems < 10 cm long; C: green, lip purplish; D: No, CR, Pma, SA (Pc, Mt); M: pl 10 cm, S 7–14 mm ....
.... *Epidendrum peperomia*
(Syn. *E. porpax*)
7(1). Leaves 2–3 mm wide, fleshy, U-shaped in cross section; C: green; D: CR, Pma (Mt); M: pl 20 cm, S 8–10 mm ....
.... *Epidendrum pudicum*
7. Leaves > 10 mm wide, usually fleshy but flattened 8
8(7). Midlobe surpassing lateral lobes, deeply notched, halves overlapping when flattened; C: green; D: CR, Pma (Pc); M: pl 25 cm, S 3.5 cm
................................*Epidendrum amparoanum*
8. Midlobe notched or not, but halves never overlapping 9
9(8). Lip distinctly and deeply 3-lobed 10
9. Lip not or only shallowly lobed 14
10(9). Lateral lobes of lip surpassing small, notched midlobe; C: yellow-green; D: No, CR, Pma, SA (Mt); M: pl 20 cm, S 2.5–3 cm
................................*Epidendrum barbeyanum*
Plate 5(1)
10. Midlobe of lip always surpassing or subequal to lateral lobes ... 11
11(10). Lip strongly cordate, basal lobes overlapping if flattened, midlobe acute; stem strongly flattened; living leaves with several prominent veins; C: green; D: CR (Mt); M: pl 20 cm, S 2.2–2.4 cm ....
....*Epidendrum storkii*

11. Basal lobes of lip not overlapping; midlobe notched or truncate; living leaves without prominent veins 12

12(11). Midlobe deeply notched; leaf sheaths with winglike keels; leaves broad and very fleshy; C: pale green; D: CR, Pma (Mt); M: pl 20 cm, S 1.8–2.3 cm .... *Epidendrum candelabrum*
12. Midlobe truncate or shallowly notched; leaf sheaths without keels 13

13(12). Base of column markedly thickened, column hood toothed; C: bright green; D: No, CR, Pma (Pc, Mt); M: pl to 45 cm, S 1.3–1.8 cm .... *Epidendrum lagenocolumna*
13. Column not inflated basally, column hood fringed; C: whitish green; D: No, CR, Pma (Mt); M: pl 25 cm, S 13–15 mm .... *Epidendrum firmum* (Syn. *E. majale*)

14(9). Lip margin strongly toothed; petals much wider apically, with narrow bases; C: pale green; D: CR (Cb/Mt); M: pl to 35 cm, S 1.7–1.9 cm .... *Epidendrum odontochilum*
14. Lip margin not toothed 15

15(14). Lip not at all lobed 16
15. Lip shallowly 3- or 4-lobed 21

16(15). Lip ovate, longer than wide, reflexed, forming a 45° angle with column; C: yellowish green; D: Pma (Pc, Mt); M: pl 15 cm, S 1.5–2.6 cm .... *Epidendrum mantis-religiosae*
16. Lip subcircular to kidney-shaped, not markedly longer than wide, not reflexed 17

17(16). Sepals lanceolate, widest near bases and tapering to acute apices; C: translucent green; D: CR, Pma (Mt); M: pl to 15 cm, S 8–15 mm .... *Epidendrum dentiferum*
17. Sepals widest near or above middle, or straplike 18

18(17). Sepals and petals straplike, apices rounded; dwarf plant with 5-angled lip; C: yellow or greenish yellow; D: Pma (Pc); M: pl 10 cm, S 3–3.5 cm .... *Epidendrum hammelii*
18. Sepals widest at or above middle, not at all straplike 19

19(18). Column distinctly arched in middle; stem flattened, sheaths with winglike keels; C: waxy pale green, lip ivory; D: Pma (Mt); M: pl to 25 cm, S 1.6–1.9 cm .... *Epidendrum kerichilum*
19. Column not distinctly arched; stem may be flattened, but sheaths without winglike keels 20

20(19). Petals about as wide as sepals; column shorter than lip blade; basal

calli rounded; C: green; D: CR, Pma (Cb); M: pl to 40 cm, S 1.5–2 cm .... *Epidendrum hunterianum*
20. Petals much narrower than sepals; column longer than lip blade; basal calli narrow and hornlike; C: green; D: CR, Pma (Mt); M: pl to 35 cm, S 1.8–2.7 cm................ *Epidendrum pachyceras*
21(15). Midlobe of lip surpassing lateral lobes 22
21. Midlobe subequal to lateral lobes, lip shallowly 4-lobed in front 23
22(21). Column sharply arched; sepals and petals reflexed, petals straplike; C: pale green; D: Pma (Mt); M: pl to 35 cm, S 10–11 mm........ .... *Epidendrum simulacrum* (Syn. *E. curvicolumna*)
22. Column slightly arched; sepals spreading, petals subparallel with column, oblanceolate; C: waxy pale green, lip ivory; D: Pma (Mt); M: pl 25 cm, S 1.6–1.9 cm .... *Epidendrum kerichilum*
23(21). Column hood fringed, lateral lobes (sides of column opening) spreading and toothed; C: S, P greenish white, lip white; D: CR, Pma (Pc, Mt); M: pl 20 cm, S 13–15 mm...................... .... *Epidendrum trialatum*
23. Column hood toothed but not fringed, lateral lobes not spreading 24
24(23). Stem laterally flattened; inflorescence many-flowered; C: green; D: CR (Cb); M: pl to 25 cm, S 13–16 mm.... *Epidendrum anastasioi*
24. Stem cylindrical; few-flowered; C: pale green; D: Pma (Mt); M: pl 8 cm, S 8.5–9.5 mm..................... *Epidendrum gregorioi*

Key 8

New shoots arising above bases, but not irregularly branched; shoots more or less equal; inflorescences usually pendent (except *Epidendrum macrostachyum*); plants not sprawling (*E. radicans*, with sprawling stems, is in Key 4).

1. Flowers very fleshy, lip U-shaped in cross section with thick, rounded margins; C: brownish yellow; D: CR, Pma, SA (Mt); M: pl to 70 cm, S 10 mm .... *Epidendrum macrostachyum* (Syn. *E. rigidiflorum*)
1. Flowers more or less fleshy, but lip not U-shaped in cross section, margins thin 2
2(1). Erect plants; new shoots arising above middle of old shoots; roots very fleshy; floral bracts inconspicuous 3

2. Plants creeping or spreading; new shoots arising below middle of old shoots; roots not very fleshy; floral bracts prominent 6

3(2). Lip kidney-shaped, not lobed, with 2 basal calli and 2 fleshy ridges along midline; C: S, P green, lip flushed with maroon; D: CR, Pma (Mt); M: pl to 60 cm, S 7–12 mm .... *Epidendrum bisulcatum*
3. Lip distinctly lobed 4

4(3). Lip 2-lobed; C: greenish brown; D: CR (Mt); M: pl to 60 cm, S 7–12 mm .... *Epidendrum brenesii*
4. Lip 3-lobed 5

5(4). Lip shallowly lobed, without basal calli, midlobe rounded; C: waxy yellowish green; D: CR, Pma (Mt, Pc); M: pl to 70 cm, S 11–15 mm .... *Epidendrum incomptum*
5. Lip deeply lobed, with basal calli, midlobe notched; D: No, CR?; M: pl 50 cm, S 13 mm .... *Epidendrum nubium*

6(2). Flowers largely hidden by dense, conelike cluster of bracts; C: greenish cream; D: CR, Pma (Mt); M: pl to 40 cm, S 10–11 mm .... *Epidendrum cryptanthum*
6. Flowers exposed 7

7(6). Flowers with distinct external spur, tip free from ovary; C: apricot yellow; D: CR; M: pl 35 cm, S 6–8 mm .... *Epidendrum puteum*
7. Flowers with swollen nectary beneath flower, but this completely united with ovary or floral tube 8

8(7). Lip oblong, much longer than wide; D: Pma (Mt); M: pl 20 cm, S 1.4–1.7 cm .... *Epidendrum lutheri*
8. Lip ovate to subcircular; C: pale green, waxy; D: CR, Pma (Mt, Pk); M: pl 30 cm, S 2.3–2.7 cm .... *Epidendrum polychlamys*
    The plants that have been called *E. polychlamys* are rather variable, and there may be 2 or 3 species involved, including *E. barbae*. Typical *E. polychlamys*, with its waxy, fragrant flowers, is very attractive.

Key 9

Stems unbranched; bracts of peduncle laterally flattened or keeled, similar to floral bracts.

1. Leaves laterally flattened.................................. 2
1. Leaves dorsoventrally flattened 3

2(1). Leaves and bracts more or less triangular and obtuse; C: green; D: CR, Pma, (Cb, Mt, Pc); M: pl 25 cm, S 6–8 mm .... *Epidendrum lockhartioides*

2. Leaves slender and acute; C: green; D: rPma, SA (Mt); M: pl 20 cm, S 8 mm............................... *Epidendrum aporum*

3(1). Lip wider than long 4

3. Lip longer than wide or subcircular 8

4(3). Leaves very fleshy, U-shaped in cross section; C: plant purplish, fls dark purplish green; D: CR, Pma (Mt); M: pl 30 cm, S 14 mm .... *Epidendrum fuscopurpureum*

4. Leaves flat 5

5(4). Petals oblanceolate to spatulate; lip deeply notched; C: pale apple green; D: CR (Mt, Pc); M: pl 40 cm, S 12–15 mm............... ....*Epidendrum bilobatum*

5. Petals linear; lip shallowly notched........................... 6

6(5). Leaves elliptic or elliptic-obovate, 1 or rarely 2 per shoot; C: green or brownish green; D: Pma (Mt); M: pl 30 cm, S 12–16 mm ....*Epidendrum pleurothalloides*

6. Leaves straplike, 2 or 3 on each shoot........................ 7

7(6). Peduncle with bracts about 5 mm wide, usually longer than rachis; column slender; C: green or pale green; D: Pma (Cb, Mt, Wt); M: pl 30 cm, S 1.1–1.7 cm...................*Epidendrum luerorum*

7. Peduncle with bracts 10–15 mm wide, usually shorter than rachis; column thick and fleshy; C: green or bronzy green; D: No, CR, Pma, SA (Mt); M: pl to 40 cm, S 1.5–2.5 cm .... *Epidendrum coriifolium*

This name is currently used for a complex of superficially similar plants, including *E. palmense* and *E. summerhayesii*. Further study may show that several species are involved.

8(3). Leaves about half as wide as long; lip concave; C: pale apple green; D: CR, Pma (Mt); M: pl 30 cm, S 11–13 mm.................. ....*Epidendrum concavilabium*

8. Leaves less than half as wide as long 9

9(8). Plant large; lip > 2 cm wide, widest above middle; sepals and petals **curled back**; C: pale green; D: CR, Pma (Mt); M: pl to 60 cm, S 1.4–2.2 cm .... *Epidendrum circinatum*

9. Plants smaller; lip < 1 cm wide, widest basally or in middle; sepals and petals spreading.. 10

10(9). Lateral sepals apically keeled externally; lip slightly constricted in middle on each side, widest basally; **C:** green; **D:** No, CR?; **M:** pl 30–40 cm, S 11–12 mm .... *Epidendrum nitens*
10. Lateral sepals not keeled; lip not constricted in middle, widest near middle; **C:** greenish yellow-brown; **D:** No, CR, Pma, SA (Cb); **M:** pl to 25 cm, S 7 mm .... *Epidendrum rigidum*

Key 10

Plants with branched stems, branches irregular or unequal; plants usually either bushy or pendent.

Note: *Epidendrum insulanum,* of the *ramosum* complex, is known only from Cocos Island. I have seen no specimens or drawings of *E. insulanum*, so I do not attempt to key it out. There is also at least 1 unnamed species on Cocos.

1. Leaves subcylindric; leaf sheath rough and warty 2
1. Leaves dorsoventrally flattened; sheath smooth or warty 3

2(1). Sepals 5–6 mm long; lip widest near base, apex narrow; **D:** CR (Mt); **M:** pl to 60 cm, S 6–7.5 mm ... *Epidendrum ramosissimum*
2. Sepals 8–10 mm long; lip widest near middle; **C:** dull purple or dark red-brown to nearly black; **D:** CR, Pma (Mt); **M:** pl to 40 cm, S 8–10 mm .... *Epidendrum lancilabium*
Plate 5(4)

3(1). Leaves very narrowly lanceolate (15–20 times as long as wide), acute or acuminate 4
3. Leaves wider or blunt 5

4(3). Flowers usually solitary; lip lanceolate; **C:** yellow-green; **D:** No, CR, Pma (Cb); **M:** pl to 1.5 m, S 11–15 mm .... *Epidendrum isomerum*
4. Flowers usually 2–3; lip oblong-ovate; **C:** greenish or brownish yellow; **D:** CR, Pma (Pc); **M:** pl to 2 m, S 11–15 mm .... *Epidendrum guanacastense*
(Syn. *E. cocleense*)

5(3). Flowers 1–2 per inflorescence, if 2, flowers borne close together (lateral sepals together) 6
5. Flowers 2 to many per inflorescence, if 2, flowers well separated on rachis 12

6(5). Flowers usually 2 in each inflorescence 7
6. Flowers usually 1 in each inflorescence 9

7. Plants < 10 cm tall; **C:** green; **D:** CR, Pma (Cb); **M:** pl to 10 cm, S 5–5.5 mm .... *Epidendrum exiguum*
7. Plants much more than 20 cm long, pendent 8

8(7). Lip about 6 mm wide, weakly 3-lobed, lateral lobes curled upward and nearly tubular; **C:** green or bronzy green; **D:** No, CR, Pma, SA (Cb, Pc); **M:** pl to 50 cm, S 10–13 mm..... *Epidendrum sculptum*
8. Lip > 10 mm wide, base not markedly tubular; **C:** rose-pink; **D:** CR (Mt); **M:** pl to 2 m, S 3 cm............ *Epidendrum mirabile*
Plate 4(4)

9(6). Plant erect, sepals and petals with narrow, acuminate tips; **C:** pale green or yellow-green; **D:** CR, Pma (Mt); **M:** pl to 30 cm, S 1.6–1.8 cm .... *Epidendrum oxyglossum*
(Syn. *E. exile*)
9. Plants pendent; sepals and petals obtuse or acute but not acuminate 10

10(9). Leaf blades not jointed at base (do not fall off when dead); branching irregular; **C:** translucent green; **D:** CR, Pma (Cb, Mt); **M:** pl to 50 cm, S 7–9 mm..................... *Epidendrum confertum*
10. Leaf blades jointed at base; main stem pendent, with indefinite growth (if not injured); flowers borne on short lateral shoots, usually with smaller leaves, these may form chains of several short stems.. 11

11(10). Leaves on main stem 10–15 cm long; **C:** bronzy green; **D:** No, CR, Pma (Cb); **M:** pl to 2 m, S 1.8–2 cm .... *Epidendrum acunae*
Usually grows near or over water.
11. Leaves on main stem < 3 cm long; **C:** light green; **D:** No, CR, Pma, SA (Cb, Mt, Pk); **M:** pl to 35 cm, S 7.5–8 mm ....
.... *Epidendrum repens*

12(5). Lateral sepals basally united, falcate, curving outward from center; **C:** pale green; **D:** Pma (Cb, Mt); **M:** pl to 30 cm, S 6–7 mm
.... *Epidendrum curvisepalum*
12. Lateral sepals straight or curving inward..................... 13

13(12). Leaf sheaths rough and warty............................. 14
13. Leaf sheaths smooth....................................... 18

14(13). Lip rounded or abruptly acute 15
14. Lip tapering or narrowly acute 16

15(14). Margins of lip thick and fleshy, lip about as wide as long, apex rounded; **C:** yellow-green or bronze-green; **D:** CR (Mt); **M:** pl to 40 cm, S 4.1–5.5 mm .... *Epidendrum sanchoi* var. *exasperatum*
The shape of the lip is quite distinctive; this supposed variety may be a distinct species.

15. Edge of lip not thick and fleshy, lip wider than long; C: greenish white; D: No, CR, Pma (Mt); M: pl 20 cm, S 2.5 mm............
.... *Epidendrum trachythece*

16(14). Basal angles of lip acute, reaching above column, lip wider than long; D: CR; M: pl 10 cm, S 8–10 mm... *Epidendrum cordiforme*
16. Basal angles of lip rounded, not surpassing top of column...... 17

17(16). Margins of lip thick and fleshy, lip much wider than column; C: pale green; D: CR, Pma (Mt); M: pl to 50 cm, S 5–8 mm............
.... *Epidendrum sanchoi*
17. Margins of lip not fleshy, lip only slightly wider than column; C: greenish yellow; D: CR, Pma (Cb, Mt); M: pl 20 cm, S 5–6 mm
.... *Epidendrum rugosum*

18(13). Plants large, leaves mostly > 1.5 cm wide.................... 19
18. Plants smaller, leaves mostly < 1.5 cm wide................. 23

19(18). Plant pendent; floral bracts conspicuous, > 2 cm long......... 20
19. Plant usually erect; floral bracts inconspicuous, < 1.5 cm long
....22

20(19). Floral bracts tubular, not overlapping; C: green-yellow; D: No, CR (Mt); M: pl to 1.5 m, S 1–1.8 cm...... *Epidendrum santaclarense*
20. Floral bracts inflated, overlapping.......................... 21

21(20). Leaves straplike; C: cream-white; D: CR, Pma (Mt); M: pl to 1.5 m, S 10–13 mm .... *Epidendrum mora-retanae*
21. Leaves elliptic or ovate-elliptic; C: lemon yellow; D: CR, Pma (Mt); M: pl > 40 cm, S 1.9–2 cm........... *Epidendrum rafael-lucasii*

22(19). Flowers green, not very fragrant (at least by day), rounded; C: green; D: CR, Pma (Mt); M: pl to 2 m, S 9–11 mm.............
.... *Epidendrum platystigma*
22. Flowers white, fragrant by day; C: white; D: CR, Pma (Mt); M: pl to 1.5 m, S 7–8 mm................. *Epidendrum ramonianum*

23(18). Sepals and petals long-acuminate; flowers short-stalked; C: green; D: CR (Mt); M: pl to 40 cm, S 5–7 mm.....................
.... *Epidendrum anoglossoides*
23. Sepals and petals acute or obtuse but not long-acuminate, flowers distinctly stalked... 24

24(23). Plants generally erect, irregularly branched; inflorescences mostly from lateral branches; main stem may have larger leaves than branches 25

24. Plants sprawling, new stems arising near tips of older stems; flowering at tips of all stems 30
25(24). Lip shallowly 3-lobed; C: white to pale green; D: CR, Pma (Mt); M: pl to 50 cm, S 6 mm.................. *Epidendrum nutantirachis*
25. Lip ovate or triangular-ovate, not at all lobed 26
26(25). Floral bracts wide and conspicuous relative to flower size, overlapping; C: greenish yellow; D: No, CR, Pma, SA (Cb, Pc); M: pl to 20 cm, S 3–4.5 mm .... *Epidendrum strobiliferum*
26. Floral bracts inconspicuous, not overlapping 27
27(26). Plants up to 10 cm tall, few-branched; lip triangular; C: green; D: CR, Pma (Cb); M: pl to 10 cm, S 5–5.8 mm .... *Epidendrum exiguum*
27. Plants much larger or many-branched, lip usually ovate........ 28
28(27). Lip with 3 longitudinal keels; D: CR (Mt); M: pl 15 cm, S 6 mm ....*Epidendrum microdendron*
28. Lip with basal callus 29
29(28). Callus 3- or 5-toothed; C: pale green; D: No, CR, Pma (Mt); M: pl to 40 cm, S 5 mm................. *Epidendrum pseudoramosum*
29. Callus V-shaped; C: green; D: No, CR, Pma, SA (Cb); M: pl to 60 cm, S 5–6 mm......................... *Epidendrum ramosum*
30(25). Ovary > 16 mm long; sepals > 15 mm long; C: green; D: Pma (Cb, Wt); M: pl to 60 cm, S 16 mm .... *Epidendrum dosbocasense*
30. Ovary and sepals both < 13 mm long....................... 31
31(30). Leaves relatively short and wide, mostly < 5 cm long, < 1 cm wide 32
31. Leaves relatively long and narrow, mostly > 5 cm long, < 8 mm wide.. 33
32(31). Lip constricted below middle, about 6.4 by 4.5 mm; nectary narrow and deep, not swollen; C: green; D: CR (Cb, Mt); M: pl 18–40 cm, S 11 mm.......................... *Epidendrum modestiflorum*
32. Lip constricted above middle, about 8 by 6 mm; nectary swollen; C: white; D: Pma (Cb, Mt); M: pl to 30 cm, S 11–12 mm........... .... *Epidendrum veraguasense*
33(31). Sepals 7–9 mm long, petals linear, obtuse; apex of lip acute or obtuse; C: green; D: CR (Cb, Mt, Pc); M: pl to 50 cm, S 7–9 mm ....*Epidendrum stevensii*
33. Sepals 9–12 mm long, petals narrowed basally, wider apically; apex of lip rounded; C: green becoming orange-brown with age; D: CR (Pc, Mt); M: pl to 30 cm, S 9–12 mm..... *Epidendrum flexicaule*

Key 11

Small epiphytes; stems slender, new stems may arise well above bases of old stems, plants thus sprawling or climbing; leaves very narrow, sometimes fleshy; inflorescence terminal, somewhat zigzag; flowers tiny, with only 2 pollinia; *Epidendrum* (*Epidanthus*) group.

This group is easily recognized but clearly a subgroup of *Epidendrum*, so they are treated as such. *Epidendrum miserrimum* and *E. selaginella*, dwarf species with distinctly flattened leaves, are not of this group and are treated in Key 4.

1. Lip simple, rachis strongly zigzag; leaf blade 1.4–2.1 cm long; **C**: white to brownish yellow, sometimes speckled with dark red-brown; **D**: CR (Mt, Wt); **M**: pl to 20 cm, S 3–3.5 mm............ .... *Epidendrum goniorhachis*
1. Lip 3-lobed; rachis slightly zigzag........................... 2

2(1). Sepals 5–6 mm long; leaf apices folded and boatlike; **C**: white to purplish; **D**: CR (Pk); **M**: pl to 30 cm, S 4.5–7 mm.............. .... *Epidendrum talamancanum*
2. Sepals 2–5 mm long; leaf apices flat or cylindrical............. 3

3(2). Leaf blades thick and fleshy, widest near middle; **C**: dull orange; **D**: Pma (Cb, Mt); **M**: pl 4–6 cm, S 2.8–3 mm..................... .... *Epidendrum insolatum*
(Syn. *Epidanthus crassus*)
3. Leaf blades thin or leathery, sides parallel.................... 4

4(3). Lip 1–2 mm long; sepals 2–2.5 mm long; leaf apices rounded to obtuse; **C**: white to cream, lip yellow sometimes marked with red-purple; **D**: CR, Pma (Mt, Wt); **M**: pl to 30 cm, S 2–2.5 mm .... *Epidendrum sancti-ramoni*
Costa Rican and Panamanian plants have been treated as *E. paranthicum*, which is found farther north.
4. Lip 2–3.5 mm long; sepals 3–5 mm long; leaf apex acute or subacute; petals with basal lobules; **C**: S cream, P and lip cream to purple; **D**: CR, Pma (Mt, Pk, Wt); **M**: pl 15–20 cm, S 4–4.5 mm .... *Epidendrum muscicola*

Key 12

A distinctive group with flattened peduncles, hard caudicles, and the opening between lip and column narrowing above to a slit. New stems often arise well above bases of old stems, so plants are quite irregular; *Epidendrum* (*Neowilliamsia*) group. All occur in wet habitats.

1. Lip more or less triangular, not divided into distinct lobes 2
1. Lip with distinct lateral lobes................................ 3
2(1). Flowers tiny, sepals < 4 mm long; **C:** S, P green, lip white with 2 orange spots; **D:** CR (Mt); **M:** pl to 45 cm, S 3.8–4 mm..........
....*Epidendrum nervosiflorum*
2. Flowers larger, sepals > 5 mm long; **C:** light yellow to green; **D:** CR, Pma (Mt, Pk); **M:** pl to 30 cm, S 10–11 mm...............
.... *Epidendrum anoglossum*
3(1). Lateral lobes of lip not divided 4
3. Lateral lobes of lip deeply divided........................... 5
4(3). Lateral lobes of lip rounded, midlobe linear; sepals about 12 mm long, petals rugose within; **C:** S, P red-brown below, yellow above, lip white; **D:** CR (Cb, Wt); **M:** pl to 30 cm, S 12 mm ....
.... *Epidendrum epidendroides*
4. All lobes of lip triangular; sepals 6–7 mm long, petals smooth; **C:** S, P brown, lip white; **D:** Pma (Mt); **M:** pl 7–9 cm, S 6–7 mm
.... *Neowilliamsia* (= *Epidendrum*) *tenuisulcata*
5(3). Lateral lobes of lip divided into 6–9 slender linear divisions; **C:** S, P brown or greenish brown, lip white; **D:** CR, Pma (Mt, Pk); **M:** pl to 30 cm, S 6–8 mm........................*Epidendrum alfaroi*
5. Lateral lobes of lip divided into thick, finger-like divisions 6
6(5). Midlobe of lip projecting beyond deeply divided lateral lobes; **C:** white or yellow spotted and barred with brown or red-brown; **D:** CR, Pma (Mt, Pk); **M:** pl to 50 cm, S 4.5 mm...................
.... *Epidendrum wercklei*
Plate 4(3)
6. Midlobe of lip scarcely longer than lateral lobes, lateral lobes shallowly divided; **C:** cream with purple spots near ends of segments; **D:** Pma (Pk); **M:** pl to 20 cm, S 5–5.5 mm ....
.... *Neowilliamsia* (= *Epidendrum*) *cuneata*

Helleriella. Epiphytic or terrestrial; stems clustered, sometimes branched, 1–3 m long; leaves 2-ranked, leathery; inflorescences both terminal and lateral, with few flowers; flower with short but prominent chin; sepals elliptic or lanceolate; lip elliptic-obovate, hinged to column foot; pollinia 4; 1 species in our area; **C:** pale green with purplish veins;

D: No, Pma (Mt); M: pl to 3 m, S 2–2.3 cm ....
.... *Helleriella nicaraguensis*
   This genus is closely allied to *Platyglottis*, and the genera should perhaps be merged.

Hexisea. Epiphytic; with new pseudobulbs arising from tips of old ones, ellipsoid, usually with 2 apical leaves; flowers bright orange-red, densely clustered at stem tips; sepals and petals similar; lip basally united with column and then abruptly bent up against column and down again. Hexiseas are essentially hummingbird-pollinated species of *Scaphyglottis*. Several species that were once treated as *Hexisea* are now classified in *Scaphyglottis*, where these, too, should perhaps be placed.

Key to *Hexisea*

1. Pseudobulbs round in section, with narrow grooves, surface dull green; callus of lip maroon; C: orange-red; D: No, CR, Pma, SA (Mt); M: pl to 40 cm, S 10–14 mm .... *Hexisea bidentata*
Plate 8(5)
1. Pseudobulbs somewhat flattened, with wide, shallow grooves, surface shiny; callus of lip yellow; C: orange-red; D: No, CR, Pma, SA (Cb, Mt); M: pl to 35 cm, S 10–14 mm .... *Hexisea imbricata*

Homalopetalum. Small epiphyte; pseudobulbs loosely clustered on a creeping or pendent rhizome, ovoid, each with 1 fleshy leaf; inflorescence terminal, slender, each with a single proportionately very large flower; sepals and petals narrowly lanceolate, long-acuminate; lip elliptic, with 2 high keels near base; column arched; 4 large and 4 minute pollinia; 1 species in our area; flower membranous; C: translucent green or suffused with pink; D: No, CR, Pma, SA (Mt); M: pl to 15 cm, S 2.5 cm ....
.... *Homalopetalum pumilio*
Plate 3(4)

Isochilus. Epiphytic; stems slender but roots fleshy; leaves thin, narrow, tips retuse; inflorescence terminal, usually curved, with flowers on upper side of inflorescence (secund); flowers relatively thin, usually pink or rose-purple, rather bell-shaped.

There are clearly a number of *Isochilus* species in Mexico and Central America, but herbarium workers have lumped most of them under *I. linearis*. Recent workers recognize more species, but a careful study of the genus is still needed.

Key to *Isochilus*

1. Bracts wide and conspicuous 2
1. Bracts small and inconspicuous 3
2(1). Bracts and upper leaves reddish or wine-colored at flowering; C: light pink with 2 dark red spots on lip; D: No, CR, Pma (Mt); M: pl to 50 cm, S 11–12 mm .... *Isochilus major*
   Plate 7(4)
2. Bracts and upper leaves green; C: red-violet; D: No, CR (Mt); M: pl to 30 cm, S 9 mm .... *Isochilus latibracteatus*
3(1). Leaf sheaths warty; flowers loosely clustered; C: light carmine; D: No, CR (Cb, Mt); M: pl to 30 cm, S 8 mm ....
   .... *Isochilus carnosiflorus*
3. Leaf sheaths smooth 4
4(3). Flowers about 6 mm long, loosely clustered; C: flowers and bracts red-carmine; D: No, CR, Pma, SA (Mt); M: pl to 30 cm, S 10 mm
   .... *Isochilus linearis*
4. Flowers about 10 mm long, densely clustered, bracts green; C: purple, lip white with 2 maroon spots; D: No, CR (Mt); M: pl to 40 cm, S 10 mm .... *Isochilus amparoanus*

Jacquiniella. Epiphytic; reed-stem, leaves several, fleshy, **cylindrical** or **laterally flattened**, 2-ranked; flowers fascicled, usually developing 1 or a few at a time; midlobe of lip fleshy; flowers green, brownish green, or bronzy.

Key to *Jacquiniella*

1. Leaves laterally flattened 2
1. Leaves cylindrical or subcylindrical 3

2(1). Leaves 15–25 cm long, resembling peduncle; lip united to basal two-thirds of column, with a strongly developed callus at base of concavity; C: brownish ocher or green; D: No, CR, Pma (Mt); M: pl to 25 cm, S 12–14 mm............. *Jacquiniella equitantifolia*
2. Leaves 4–8 cm long, not closely resembling much longer peduncle; lip united to basal third of column, without strongly developed callus; C: brownish ocher; D: CR, Pma (Mt); M: pl to 10 cm, S 10–11 mm .... *Jacquiniella aporophylla*
Plate 7(5)
3(1). Leaves 1–2 cm long; flowers 2–3 mm long; C: green or bronzy green; D: No, CR, Pma, SA (Mt); M: pl to 10 cm, S 3 mm .... *Jacquiniella globosa*
3. Leaves mostly 3–5 cm long; flowers about 1 cm long 4
4(3). Lip united to basal third of column or less 5
4. Lip united to basal two-thirds of column 7
5(4). Uppermost internode (peduncle) much longer than others; pedicels exposed; C: greenish yellow to bronzy green; D: Pma, SA (Cb); M: pl to 15 cm, S 10–12 mm............. *Jacquiniella pedunculata*
5. Uppermost internode not much longer than others; pedicels and ovary concealed by conspicuous sheathing bracts................ 6
6(5). Midlobe of lip oblong, tongue-like, obtuse, lateral lobes entire, enfolding column nearly to base; C: brownish ocher; D: No, CR?; M: pl to 20 cm, S 12–15 mm............. *Jacquiniella cobanensis*
6. Midlobe of lip triangular, acute, lateral lobes toothed, enfolding column only at apex; C: ocher; D: CR, Pma (Mt); M: pl to 30 cm, S 10–13 mm............................. *Jacquiniella standleyi*
7(4). Diameter of leaves < 2 mm; sepals with narrow, subulate apices; often self-pollinating; C: pale green; D: No, CR, Pma, SA (Mt); M: pl to 30 cm, S 9–11 mm................ *Jacquiniella teretifolia*
7. Diameter of leaves about 2 mm; fleshy apices of sepals short and blunt; C: green; D: No, CR (Mt); M: pl to 25 cm, S 8–10 mm .... *Jacquiniella teres*
Usually self-pollinating; may be a high-altitude form of *J. teretifolia*.

Laelia. Epiphytic; pseudobulbs clustered, subcircular, strongly flattened, overlapping, each with 1 leathery leaf; inflorescence terminal, long and slender, flowers clustered near tip; flowers similar to *Cattleya* but with 8 pollinia; 1 species in our area; C: pale pink with darker throat, rarely

white; **D:** No, CR, wPma (Pc); **M:** pl to 20 cm, S 3.8 cm ....
.... *Laelia rubescens*
Plate 1(4)

Myrmecophila. Epiphytic; pseudobulbs clustered, thick basally and tapering upward, strongly **ridged, hollow**, normally occupied by ants, each with 2–3 leathery leaves; inflorescence terminal, very long, flowers clustered at apex; sepals and petals wavy, spreading, lip 3-lobed; at least 1 species in our area; **C:** white heavily flushed and stained with red-purple, midlobe of lip dark purple; **D:** No, CR (Pc); **M:** pl to 50 cm, S to 4.5 cm .... *Myrmecophila tibicinis*
Plate 1(5)

Myrmecophila is often treated as a subgroup of *Schomburgkia*, yet seems more distinct from *Schomburgkia* than *Schomburgkia* is from *Laelia*. A smaller-flowered *Myrmecophila* with yellow flowers has been reported from the Caribbean coast of Central America, but its distribution is not well documented. The names *M. brysiana* and *M. thomsoniana* have been used for such plants both in the Cayman Islands and in Central America. It is not clear whether these act as a distinct species on the mainland or intergrade with *M. tibicinis*. A zoologist colleague reports *Myrmecophila* in northwestern Panama (Bocas del Toro), but no specimens have been collected.

Nidema. Creeping plants; pseudobulbs ellipsoid, stalked; leaves solitary, nearly straplike; inflorescence terminal and shorter than leaf; flowers white, lip hinged to a short column foot.

Key to *Nidema*

1. Flowers large, about 15 mm long, opening widely; ovary scurfy; **C:** white; **D:** No, Pma, SA (Cb, Mt, Pc); **M:** pl to 20 cm, S 1.6 cm .... *Nidema boothii*
1. Flowers smaller, < 10 mm long, often not opening widely; ovary smooth; **C:** white; **D:** No, CR, Pma, SA (Cb, Pc); **M:** pl to 20 cm, S 8 mm .... *Nidema ottonis*
Plate 8(4)

Oerstedella. Reed-stem plants resembling *Epidendrum*, but leaf sheaths with prominent warts (very small in *O. exasperata*); leaves 2-ranked, many; often with lateral as well as terminal inflorescences; lip not united with column to apex; apex of column forming prominent, usually 4-lobed hood; anther beaked; pollinia without viscidium. Plate 6(3–6).

Key to *Oerstedella*

1. Lip entire, crenate but not distinctly 3- or 4-lobed; lobes at apex of column fleshy and hornlike; **C:** S, P yellowish brown or bronze-green, lip white or yellow with red-purple markings; **D:** CR, Pma (Mt); **M:** pl to 1.5 m, S 12–16 mm .... ***Oerstedella tetraceros***
1. Lip distinctly 3- or 4-lobed.................................. 2

2(1). Midlobe of lip oblong to lanceolate, not expanded or bilobed from a narrower stalk or bilobed apically........................... 3
2. Midlobe of lip expanded apically, usually distinctly bilobed 4

3(2). Lobes of lip not more than twice as long as wide; **C:** S, P pale green, lip white; **D:** CR, Pma (Mt); **M:** pl to 70 cm, S 6– 8 mm.........
....*Oerstedella intermixta*
3. Lobes of lip 3 or more times as long as wide; **C:** S, P greenish cream, lip white; **D:** Pma (Mt); **M:** pl to 35 cm, S 11–15 mm.....
.... ***Oerstedella fuscina***
Possibly only a large form of *O. intermixta*.

4(2). Sepals with prominent warts or teeth externally............,... 5
4. Sepals smooth or with few small warts externally............. 6

5(4). Leaf sheaths distinctly warty; each flower with 3 prominent appendages external to and alternating with sepals; **C:** S, P pink, lip white with maroon spots; **D:** Pma (Pk); **M:** pl to 1.5 m, S 1.6–1.9 cm
.... ***Oerstedella ornata***
5. Leaf sheaths smooth (warts minute); flower without appendages alternating with sepals; lateral lobes of lip deeply toothed to nearly fringed; **C:** S brown, P brownish green or white, lip white with purple markings; **D:** CR, Pma (Mt, Pk); **M:** pl to 3 m, S 10–13 mm
.... ***Oerstedella exasperata***
This species is common at higher elevations and becomes very large.

6(4). Lateral lobes of lip each deeply divided into 3 or more narrow lobes
.... 7

        6. Lateral lobes of lip entire, crenate, or shallowly bilobed 8

   7(6). Midlobe of lip about 2 mm wide, crenate; lateral lobes each divided
        into 3 fleshy, finger-like lobes; C: S, P greenish or pinkish brown, lip
        cream with yellow callus; D: CR (Mt); M: pl to 80 cm, S 3.5–5 mm
        .... *Oerstedella pentadactyla*
        7. Midlobe of lip about 6 mm wide, fringed, lateral lobes deeply
        fringed; C: pale pink; D: Pma, SA (Pc); M: pl to 1 m, S 5–7 mm
        .... *Oerstedella caligaria*

   8(6). Lateral lobes of lip each distinctly larger than either division of bilo-
        bed midlobe; C: S, P yellowish brown, lip bluish lavender; M: CR,
        Pma (Pc); M: pl to 2 m, S 1.6–2 cm .... *Oerstedella pinnifera*
        Plate 6(5)
        8. Lateral lobes of lip subequal to divisions of midlobe or smaller ....
       9

   9(8). Hood of column (over anther) equal to column or longer; flowers
        rose-pink 10
        9. Hood of column much shorter than column.................. 13

  10(9). Lip without a distinct parallel-sided isthmus between lateral lobes
        and midlobe, sinuses narrow or broadly V-shaped............. 11
        10. Lip with distinct, parallel-sided isthmus, sinuses broad or squarish
       12

 11(10). Midlobe deeply notched, sinuses broad; C: pink with white center;
        D: No, CR, Pma (Mt); M: pl to 60 cm, S 8–11 mm.............
        .... *Oerstedella centradenia*
        Plate 6(6)
        11. Midlobe not deeply notched, prominent apicule present; sinuses nar-
        row; C: pink with white center; D: No, CR, Pma (Mt); M: pl to 50
        cm, S 10–12 mm .... *Oerstedella pansamalae*

 12(10). Both lateral lobes and divisions of midlobe recurved; C: pink; D: CR
        (Mt); M: pl to 40 cm, S 10–12 mm ..... *Oerstedella crescentiloba*
        12. Segments of lip spreading, not recurved; C: S, P pink, lip white; D:
        CR, Pma (Mt, Pk); M: pl to 60 cm, S 6–9 mm ....
        .... *Oerstedella centropetala*

  13(9). Lateral lobes of lip smaller than each division of midlobe 14
        13. Lateral lobes of lip each subequal to divisions of midlobe in length
        and usually in width.. 17

14(13). Sepals spotted; sinus between lateral lobes and midlobe subequal to lateral lobes in width; large plants.......................... 15

14. Sepals not spotted; sinus between lateral lobes and midlobe wider than lateral lobes; dwarf plants............................ 16

15(14). Sepals and petals white or pale blue, with red-brown or blue spots, lip blue; D: CR (Cb, Mt); M: pl to 1.5 m, S 11–14 mm ....*Oerstedella schummaniana*

15. Sepals and petals ocher within and blue externally, spotted with red-brown within, lip blue-violet; D: Pma (Cb, Mt); M: pl to 1.5 m, S 8–12 mm.................... *Oerstedella pseudoschummaniana*
    This species is best known from the hills above El Valle de Antón but is by no means limited to that area. It is very similar to the Costa Rican *O. schummaniana* except in color and may prove to be a geographic race of that species.

16(14). Callus bidentate; C: white with **blue** on lip; D: CR, wPma (Mt, Pk); M: pl to 35 cm, S 10–12 mm................*Oerstedella endresii*
    Plate 6(4)

16. Callus tridentate; C: S, P pale yellow, lip yellowish cream; D: CR, Pma (Mt); M: pl to 25 cm, S 7–9 mm........ *Oerstedella pumila*
    Similar to *O. endresii* but smaller and differently colored.

17(13). Lip minutely warty; lobes of column hood fleshy; very large flowers; C: S, P orange-yellow with purple spots, lip white or yellow with purple specks and streaks; D: CR, Pma, SA (Cb, Mt); M: pl to 1 m, S 1.7–2.2 cm .... *Oerstedella wallisii*
    Plate 6(3)
    The name *O. pseudowallisii* has been used in Central America, but the plants range from Costa Rica to Colombia and are nowhere divided into 2 distinct sorts.

17. Lip smooth or only slightly warty; lobes of column hood petaloid... 18

18(17). Sinuses of lip reaching callus; C: S, P green, lip white; D: Pma (Mt); M: pl to 65 cm, S 11–12 mm.............. *Oerstedella pajitensis*
    Two forms appear in the market of El Valle de Antón at different seasons and may prove to be distinct species.

18. Sinuses of lip reaching about halfway to callus; C: S, P greenish cream, lip white; D: Pma (Mt, Wt); M: pl to 45 cm, S 14–16 mm ....*Oerstedella lactea*
    Known only from Veraguas; this may be the same species as the larger "form" of *O. pajitensis*.

Platyglottis. Epiphytic; stems clustered, sometimes branched; leaves 2-ranked, leathery; inflorescences both terminal and lateral, with few flowers; flower with short but prominent chin, sepals 11–13 mm long, lip hinged at base; pollinia 6; 1 species known; C: pale green; D: Pma (Mt); M: pl to 50 cm, S 10 mm .... ***Platyglottis coriacea***

Ponera. Epiphytic or lithophytic; with slender stems and fleshy roots; stems long, with many leaves; inflorescence lateral or terminal, flowers in short, dense clusters; flower with definite chin, sepals 6–7 mm long, lip longer, column short, lip hinged at tip of column foot; 1 species in our area; C: pale green with reddish veins; D: No, CR, Pma, SA (Mt); M: pl to 1 m, S 6 mm .... ***Ponera striata***
Plate 7(6)

Reichenbachanthus. Epiphytic; pendent, with slender pseudobulbs or none, stems clustered, new stems arising from apices of older ones, so that chains of stems are formed; each stem with 1 very fleshy or nearly cylindrical terminal leaf; inflorescence terminal, flowers densely clustered, produced 1 or a few at a time from a cluster of bracts; sepals and petals recurved; with definite, deep nectary between lip and column, lip hinged, fleshy, with apex recurved and notched. Though easily distinguished, this small group is closely allied to *Scaphyglottis*.

Key to ***Reichenbachanthus***

    1. Plants with definite club-shaped pseudobulbs; plants erect, mostly < 15 cm tall; leaves mostly < 10 cm long; lip rhombic when flattened; C: S, P green, lip white; D: CR, Pma (Mt); M: pl 20 cm, S 4–5 mm .... ***Reichenbachanthus cuniculatus***
    1. Plants pendent, with slender stems, mostly > 20 cm long; leaves mostly > 10 cm long; lip 3-lobed............................ 2
       2(1). Older inflorescences with bell-shaped clusters of bracts; lip deeply 3-lobed; C: S, P green, lip white; D: CR (Mt); M: pl 25 cm, S 6–7 mm .... ***Reichenbachanthus lankesteri***
Plate 8(6)
       2. Bracts few, closely surrounding ovary, not forming a prominent bell-shaped cluster; lip shallowly 3-lobed; C: S, P pale green, lip

white; **D:** CR?, Pma, SA (Cb, Mt); **M:** pl 30 cm, S 5–6 mm . . . . . .
. . . . *Reichenbachanthus reflexus*

Scaphyglottis. Epiphytic; stems usually forming pseudobulbs, these commonly formed in chains (new ones on top of older ones); leaves conduplicate, usually terminal on stems; inflorescence terminal, racemose or densely clustered; lip usually hinged to a definite column foot; pollinia 4 or 6. Species with 6 pollinia have been treated as *Hexadesmia*, but no other feature is correlated with number of pollinia to indicate that *Hexadesmia* is a valid genus. Plate 8(1–3).

Key to *Scaphyglottis*

1. Flowers racemose; racemes may be short, but either peduncle or rachis is visible 2
1. Flowers fascicled; peduncle or rachis, if present, hidden by bracts
. 7
2(1). Pseudobulbs with distinct narrow basal stalks 3
2. Stems or pseudobulbs without narrow basal stalks 6
3(2). Leaves elliptic or oblong, usually > 10 mm wide. 4
3. Leaves linear, usually much less than 10 mm wide 5
4(3). Sepals about 8 mm long; lip weakly 3-lobed; **C:** cream or pale green; **D:** No, CR, Pma, SA (Cb, Mt, Pc); **M:** pl to 25 cm, S 8–11 mm . . . . *Scaphyglottis lindeniana*
Plate 8(3)
4. Sepals 12–15 mm long; **C:** cream to greenish yellow, may have purple streaks; **D:** CR; **M:** pl to 30 cm, S 12 mm . . . .
. . . . *Scaphyglottis bifida*
5(3). Sepals 3–4 mm long; column < 2 mm long; **C:** white to pink; **D:** CR, Pma (Mt); **M:** pl 20 cm, S 3–4 mm . . . . *Scaphyglottis acostaei*
5. Sepals 5–7 mm long; column about 5 mm long; **C:** whitish or pinkish; **D:** No, CR, Pma, SA; **M:** pl to 30 cm, S 5–7 mm . . . .
. . . . *Scaphyglottis crurigera*
6(2). Sepals < 3 mm long; flowers tiny; pseudobulbs fusiform; **C:** whitish or pale green with purplish nerves; **D:** No, CR, Pma (Cb, Mt, Pc); **M:** pl to 10 cm, S 2 mm . . . . *Scaphyglottis micrantha*
    This curious little plant is not closely related to any other *Scaphyglottis*.

6. Sepals > 1 cm long; stems long and slender; C: greenish, lip white with yellow center; D: CR (Mt); M: pl to 80 cm, S 11–13 mm .... *Scaphyglottis corallorhiza*

7(1). Column very short and wide, less than twice as long as wide; pseudobulbs slender; always shrublike, with new stems on top of older ones.. 8

7. Column more than twice as long as wide; plants various....... 11

8(7). Stems shiny (as though varnished); lip without basal calli; petals widest above middle; column wings not surpassing anther; C: S, P greenish white, lip white with red-brown center; D: No, rCR; M: pl to 45 cm, S 4–6 mm .... *Scaphyglottis minuta*
(Syn. *S. confusa*)

8. Stems not markedly shiny; lip more or less thickened basally; petals straplike or widest below middle; column wings surpassing anther 9

9(8). Stem sheaths warty and truncate; sepals about 5 mm long; C: pale yellow; D: CR (Mt, Wt); M: pl to 35 cm, S 5 mm .... *Scaphyglottis jimenezii*

9. Stem sheaths smooth, acute; sepals 8–10 mm long............ 10

10(9). Column about three-fourths length of lip; C: greenish tinged with purple; D: Pma (Mt); M: pl to 45 cm, S 9–9.5 mm .... *Scaphyglottis chlorantha*

10. Column about half length of lip; C: S, P light green to maroon, lip pale yellow fading to pale purple; D: CR, Pma (Mt); M: pl to 25 cm, S 8 mm .... *Scaphyglottis densa*

11(7). Column with narrow toothlike or acute wings 12

11. Column wings lacking or rounded, not acute 16

12(11). Leaves linear, scarcely wider than thick; D: CR (Cb); M: pl 15 cm, S 3.5 mm .... *Scaphyglottis subulata*

12. Leaves much wider than thick 13

13(12). Column wings near middle of column; C: white; D: CR, Pma (Mt, Pc); M: pl to 50 cm, S 5.5–8 mm .... *Scaphyglottis mesocopis*

13. Column wings near apex of column 14

14(13). Leaves broad, usually > 10 mm wide; pseudobulbs thick and sausage-like; lip obtuse or notched; C: S, P greenish white or pale pink, lip darker; D: CR, Pma, SA (Pc); M: pl to 20 cm, S 11–13 mm .... *Scaphyglottis stellata*
(Syn. *S. amethystina*)

14. Leaves narrow, < 10 mm wide; lip acute 15

15(14). Lip with definite callus; C: white to greenish white; D: No, CR, Pma, SA (Cb, Mt, Pc); M: pl to 40 cm, S 5–6 mm .... *Scaphyglottis boliviensis*
(Syn. *S. huebneri*)
15. Lip without obvious callus; C: white with purple streaks on lip; D: Pma (Pc); M: pl to 20 cm, S 7–8 mm .... *Scaphyglottis laevilabia*
16(11). Lip sharply curved upward or thickened near base, forming a partially closed nectary 17
16. With nectary at base of column and lip, but lip not sharply curved upward or thickened in front of nectary 23
17(16). Flowers red or orange 18
17. Flowers white, greenish, brownish, or marked with rose-purple, never red or orange 20
18(17). Pseudobulbs markedly thickened, usually grooved ..... See *Hexisea*
18. Pseudobulbs long and slender, only slightly thickened.......... 19
19(18). Flowers with vertical slit in front of nectary; C: orange-red, upper column yellow; D: Pma (Mt, Pk); M: pl to 40 cm, S 11–13 mm .... *Scaphyglottis arctata*
19. Flowers without vertical slit in front of nectary; C: orange or reddish orange; D: CR, Pma (Mt, Pk); M: pl to 30 cm, S 14–15 mm .... *Scaphyglottis sigmoidea*
(Syn. *Hexisea sigmoidea*)
20(17). Sepals > 2 cm long; C: S, P green flushed with brown, lip white or with purplish streak along midline; D: CR, wPma (Mt, Pk); M: pl to 1 m, S 2.5–3 cm .... *Scaphyglottis gigantea*
20. Sepals < 1.5 cm long... 21
21(20). Leaves broad, < 10 times as long as wide; with very short nectary at base of lip; C: white or greenish; D: CR, Pma (Cb); M: pl to 40 cm, S 7–9.5 mm .... *Scaphyglottis gracilis*
21. Leaves narrow, > 10 times as long as wide; with prominent nectary at base of lip 22
22(21). Column slender basally; basal portion of lip (before sharp bend) about one-third length of lip; C: greenish, lip white with yellow center; D: CR (Mt); M: pl to 80 cm, S 11–13 mm .... *Scaphyglottis corallorhiza*
22. Column thick throughout; basal portion of lip about one-fourth length of lip; C: S, P brownish green, lip white with yellow streak; D: CR, Pma (Mt); M: pl to 80 cm, S 12–14 mm .... *Scaphyglottis amparoana*
Plate 8(2)
23(16). Leaves very wide, elliptic or at least one-fourth as wide as long;

plants with definite stalked pseudobulbs; **C:** pale green with purple streaks; **D:** CR, Pma (Mt); **M:** pl to 25 cm, S 10 mm............. .... *Scaphyglottis spathulata*
23. Leaves narrower, width less than one-fourth length............ 24
24(23). Pseudobulbs with definite basal stalks, stalks at least one-fourth length of pseudobulb 25
24. Pseudobulbs cylindrical, without definite stalks, or these very short 30
25(24). Base of lip firmly united to base of column; flowers fleshy, sepals and petals spreading in a single plane, petals and lateral sepals markedly falcate; **C:** S, P pale green, laterals rose-purple ventrally, lip orange-yellow; **D:** CR (Cb, Mt); **M:** pl to 25 cm, S 13–15 mm .... *Scaphyglottis geminata*
Plate 8(1)
25. Lip usually hinged at base; without the above combination of features 26
26(25). Each pseudobulb normally with a single leaf; lip widest near apex; pseudobulbs often not superposed; **C:** S, P pale green with pinkish nerves, lip white; **D:** CR, SA (Cb, Mt, Pc); **M:** pl to 20 cm, S 8–9 mm................................. *Scaphyglottis fusiformis*
26. Each pseudobulb normally with 2 or 3 leaves................. 27
27(26). Flowers tiny, < 7 mm long; **C:** white; **D:** No, CR, Pma, SA (Cb, Pc); **M:** pl to 30 cm, S 3–4.5 mm .... *Scaphyglottis behrii*
27. Flowers large, > 15 mm long.............................. 28
28(27). Lip definitely constricted above middle; petals obtuse or abruptly acute; **C:** white; **D:** No, Pma (Mt); **M:** pl to 50 cm, S 8–9 mm .... *Scaphyglottis tenella*
28. Lip not constricted; petals acute or acuminate 29
29(28). Pseudobulbs not markedly thickened above; sepals and petals acuminate; **C:** pale green or yellow, often tinged with dark red; **D:** CR (Mt); **M:** pl to 90 cm, S 1.7–2 cm........................... .... *Scaphyglottis sessiliflora*
29. Pseudobulbs distinctly thickened above; sepals and petals acute but not acuminate; **C:** cream or pale green, may have purple tinge or lines on lip; **D:** CR, Pma (Cb, Mt); **M:** pl to 90 cm, S 1.8–2 cm .... *Scaphyglottis pulchella*
30(24). Base of lip very narrow, with sides parallel or nearly so for about half length of lip 31
30. Base of lip expanding gradually or narrow for only about one-fourth length of lip... 33

31(30). Lip distinctly 3-lobed, lateral lobes subequal to midlobe or longer; C: white, apex of lip maroon; D: No, CR, Pma, SA (Cb, Mt, Pc); M: pl to 30 cm, S 6–7 mm .... *Scaphyglottis longicaulis*
31. Lip not 3-lobed, or midlobe projecting beyond lateral lobes 32
32(31). Column much wider near apex, with definite wings; surface of midlobe papillose; C: white or greenish; D: CR, Pma (Cb); M: pl to 40 cm, S 7–9.5 mm .... *Scaphyglottis gracilis*
32. Column slightly wider near apex, not winged; surface of midlobe smooth; C: usually white, sometimes violet or flushed with violet; D: No, CR, Pma, SA (Cb, Mt, Pc); M: pl to 20 cm, S 4–6 mm .... *Scaphyglottis prolifera*
(Syn. *S. cuneata*)
33(30). Pseudobulbs very narrow, with leaves along stem (when young); column with prominent wings; C: green marked with maroon; D: Pma, SA (Mt); M: pl to 50 cm, S 10 mm .... *Scaphyglottis punctulata*
33. Pseudobulbs with leaves only at apex, usually distinctly thickened 34
34(33). Pseudobulbs thick and sausage-like throughout, diameter 5–10 mm; column with squarish wings; C: S, P pale green with pinkish nerves, lip white; D: Pma (Cb, Mt, Pc); M: pl to 50 cm, S 7–9.5 mm .... *Scaphyglottis robusta*
34. Pseudobulbs slender; column wings lacking or rounded 35
35(34). Flowers relatively large, sepals 0.6–2 cm long 36
35. Flowers small or tiny, sepals 2–4 mm long 38
36(35). Blade of lip obovate, apex weakly 3-lobed, wider across lateral lobes than across midlobe; C: greenish white, purplish streaks on lip; D: CR, SA; M: pl to 35 cm, S 8–10 mm ..... *Scaphyglottis leucantha*
36. Blade of lip constricted above lateral lobes, midlobe wider than across lateral lobes 37
37(36). Pseudobulbs distinctly thickened; leaves relatively broad throughout; C: white, tip of column purple; D: No, CR (Cb, Mt, Pc); M: pl to 30 cm, S 5 mm .... *Scaphyglottis bilineata*
37. Stems thin and flexible; leaves widest near base, tapering above; C: S, P greenish or yellowish, may be tinged with purple, lip white; D: Pma (Mt); M: pl to 30 cm, S 6–10.5 mm .... *Scaphyglottis panamensis*
38(35). Leaves wide, to 15 cm long; inflorescence bracts numerous, forming large fibrous tufts with age; lip subquadrate, wide near base; C: whitish green with violet streak on lip; D: No, Pma (Cb, Mt); M: pl to 75 cm, S 2.5 mm .... *Scaphyglottis minutiflora*

38. Leaves narrower, to 10 cm long; inflorescence bracts few, not forming large, fibrous tufts; lip ovate-spatulate, not 3-lobed; C: white; D: CR (Cb); M: pl to 40 cm, S 2.5–3 mm ....
.... *Scaphyglottis limonensis*

Schomburgkia. Large, epiphytic or on rocks; pseudobulbs loosely clustered, stalked, rather club-shaped, each with 2 or 3 leaves; inflorescence terminal, very long, with flowers clustered at apex; sepals and petals with wavy or crisped margins; pollinia 8.

Schomburgkia is closely related to both *Cattleya* and *Laelia*. There appear to be 2 distinct species in Panama and Costa Rica, as indicated in the key.

Key to *Schomburgkia*

1. Lip with 5 keels; C: S, P dark brownish purple, lip rose-purple; D: CR, Pma, SA (Pc); M: pl to 60 cm, S 3.3–3.5 cm ....
.... *Schomburgkia undulata*
1. Lip with 3 keels; C: S, P orange-brown or bronze, lip pinkish cream to purple shaded with yellow and edged with brown; D: CR, Pma (Pc); M: pl to 60 cm, S 4–5 cm ..... *Schomburgkia lueddemannii*
Plate 1(6)

5 *Oncidium* and Its Relatives: Subtribe Oncidiinae

Terrestrial or usually epiphytic, with or without pseudobulbs; **pseudobulbs, when present, always of a single internode; leaves conduplicate or fleshy, often with sheathing leaves clasping the pseudobulb; inflorescence lateral,** usually basal (upper lateral or even terminal in *Lockhartia*), with few to many flowers; pollinia always 2.

Though the Oncidiinae are considered a subgroup of the Maxillarieae, they are nonetheless a major group and relatively easy to recognize. The inflorescence usually has several to many flowers, though the flowers may be produced 1 at a time. The 2 pollinia have a definite stipe and viscidium. Some species of *Maxillaria* may be taken for Oncidiinae when they have no flowers, but one can usually find the remains of small 1-flowered inflorescences; the Oncidiinae have fewer but stouter inflorescences with several or many flowers. Within these limitations, the Oncidiinae are quite variable. Pseudobulbs may be very small or lacking, the leaves may all be lateral (sheathing), and they may be very fleshy or laterally flattened. In some genera, such as *Fernandezia*, *Pachyphyllum*, and some species of *Macroclinium* and *Psygmorchis*, the habit may be monopodial, like that of the Old World Vandeae; that is, the stem may continue apical growth indefinitely.

Most genera of the Oncidiinae are rather clearly delimited, but the complex that includes *Miltonia*, *Odontoglossum*, and *Oncidium* poses a problem. These genera have been characterized by quite superficial features, and similar and closely related species have been placed in different genera. *Odontoglossum* has been broken up into smaller groups, with *Odontoglossum* in the strict sense being limited to the South American

Andes. *Miltonia* is restricted to a small South American group, but even so, it is not clearly distinguished from *Oncidium*. Some of the more distinctive groups now called *Oncidium* may be treated as distinct genera in the future. *Oncidium* is heterogeneous, and attempts to break it into more natural groups have been unsatisfactory.

Key to Genera of Oncidiinae

1. Plants without pseudobulbs; leaves scattered along slender stems *or* laterally flattened (or both) 2
1. Plants with large or small pseudobulbs; leaves fleshy or dorsoventrally flattened, not scattered along a slender stem. 7

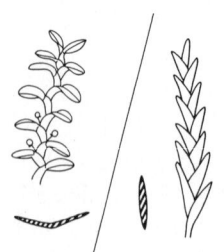

2(1). Leaves fleshy but dorsoventrally flattened 3
2. Leaves laterally flattened 4

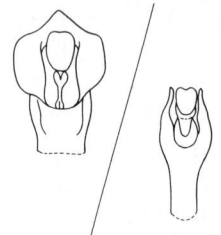

3(2). Flowers rose-purple or red; column hooded over anther .... *Fernandezia*
3. Flowers white or greenish; column not winged or hooded; flowers tiny .... ***Pachyphyllum***

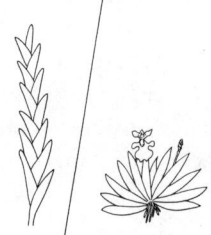

4(3). Stems elongate; leaves relatively short; plants not fanlike .... ***Lockhartia***
4. Stems short; plants fanlike 5

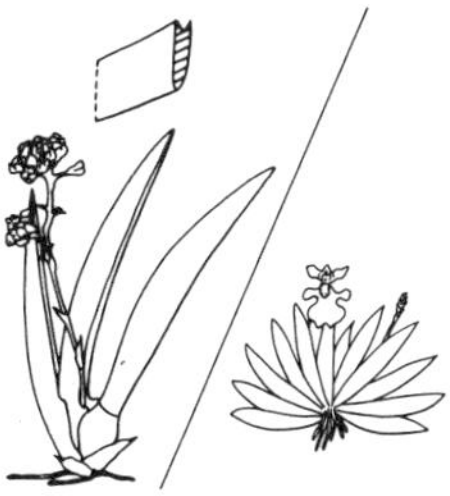

Key to Genera of Oncidiinae

5(4). Inflorescence branched, flowers tiny, bell-shaped, and densely clustered; leaves distinctly grooved to tips on upper side; with small but distinct pseudobulb .... *Trizeuxis falcata*

5. Inflorescence simple or branched, flowers larger and more open, not densely clustered; leaves not grooved to tips; with or without pseudobulbs 6

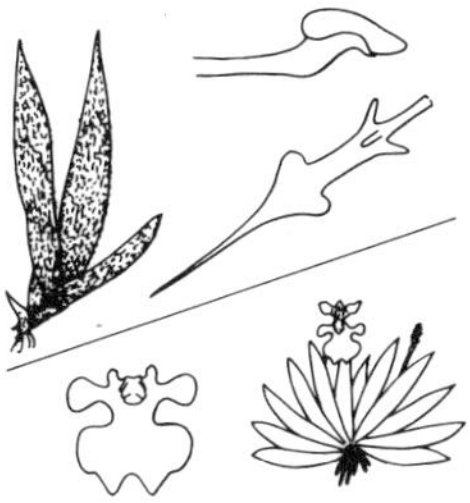

6(5). Leaves warty, reddish, apices acute; flowers racemose or umbellate; column slender and wingless; lip with narrow basal stalk .... *Macroclinium*

6. Leaves smooth and green, apices obtuse; flowers produced 1 at a time on a condensed inflorescence, yellow; column short and winged .... **Psygmorchis**

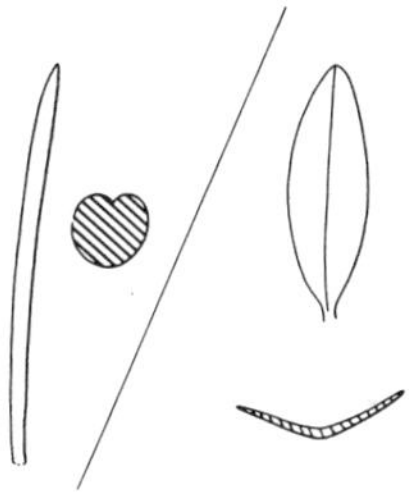

7(1). Leaves thick and fleshy 8

7. Leaves leathery but not markedly succulent or fleshy, < 3 mm thick 14

8(7). Leaves cylindrical, subcylindrical, or laterally flattened 9

8. Leaves dorsoventrally flattened 11

9(8). Larger plants, > 10 cm tall. .... **Oncidium**

9. Dwarf plants < 6 cm tall 10

10(9). Flowers white, lip much longer than sepals, few flowers open at a time .... **Ionopsis** (*satyrioides*)

10. Flowers yellow-green, lip slightly longer than sepals, dense cluster of tiny flowers................*Trizeuxis falcata*

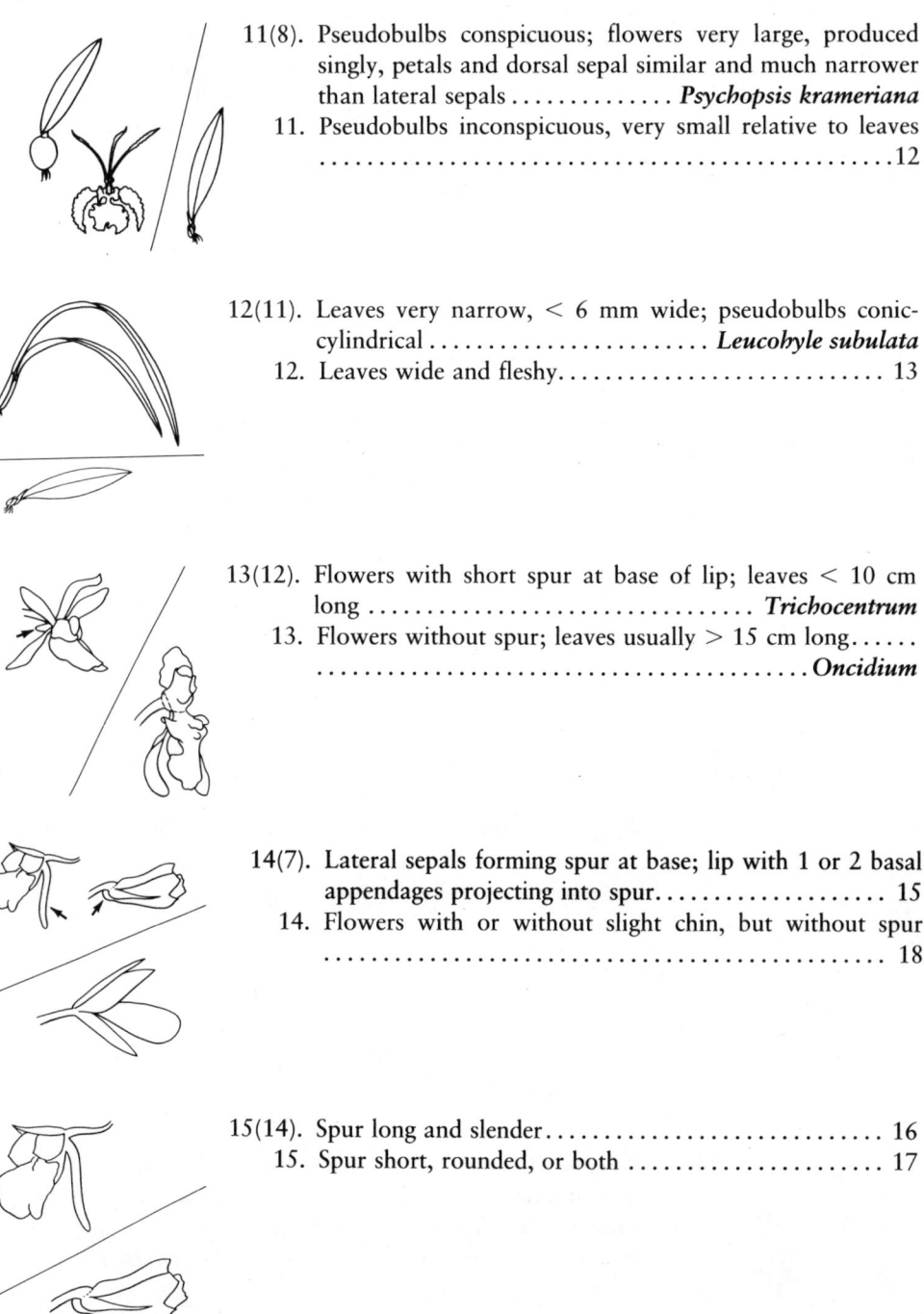

11(8). Pseudobulbs conspicuous; flowers very large, produced singly, petals and dorsal sepal similar and much narrower than lateral sepals .... ***Psychopsis krameriana***
11. Pseudobulbs inconspicuous, very small relative to leaves 12

12(11). Leaves very narrow, < 6 mm wide; pseudobulbs conic-cylindrical .... ***Leucohyle subulata***
12. Leaves wide and fleshy........................... 13

13(12). Flowers with short spur at base of lip; leaves < 10 cm long .... ***Trichocentrum***
13. Flowers without spur; leaves usually > 15 cm long......
....***Oncidium***

14(7). Lateral sepals forming spur at base; lip with 1 or 2 basal appendages projecting into spur.................... 15
14. Flowers with or without slight chin, but without spur 18

15(14). Spur long and slender............................. 16
15. Spur short, rounded, or both 17

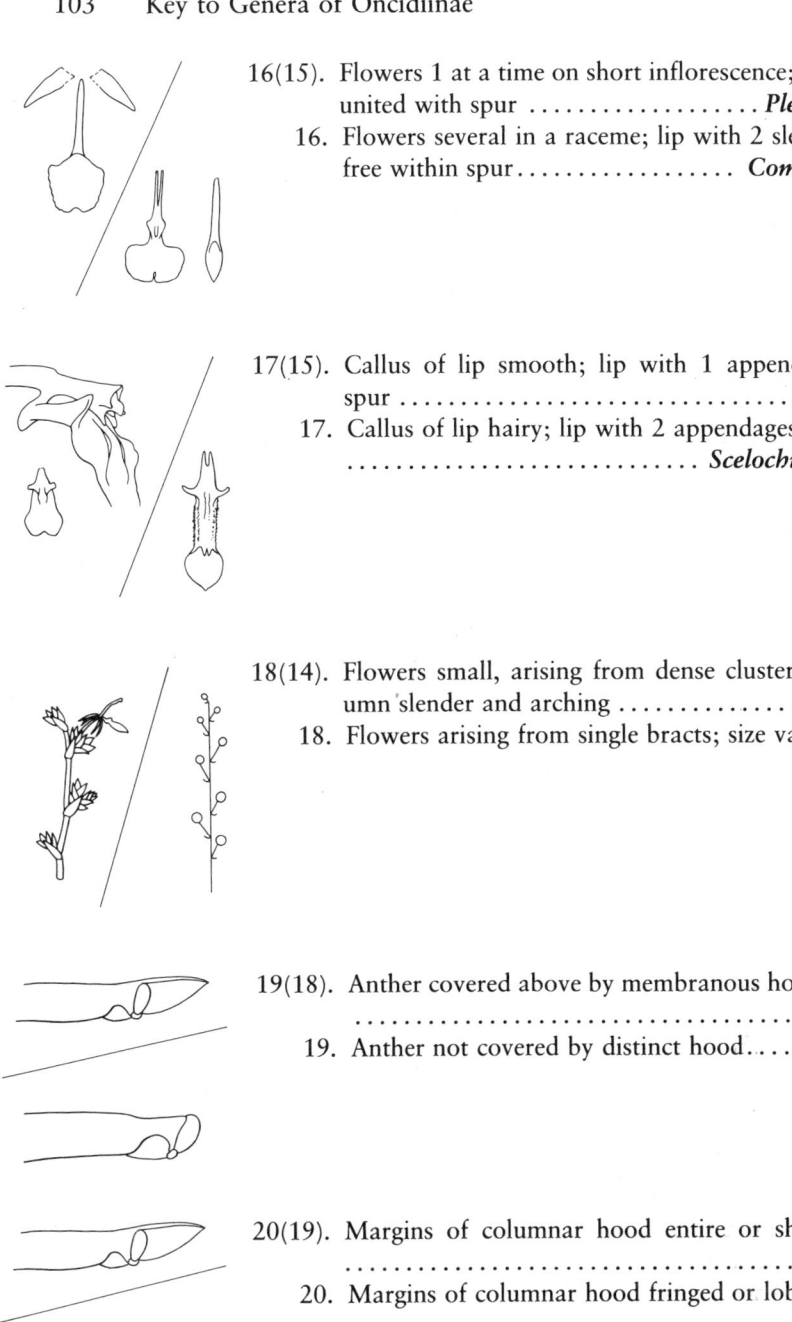

16(15). Flowers 1 at a time on short inflorescence; extension of lip united with spur .... ***Plectrophora alata***
16. Flowers several in a raceme; lip with 2 slender extensions free within spur................. ***Comparettia falcata***

17(15). Callus of lip smooth; lip with 1 appendage in sepaline spur .... ***Rodriguezia***
17. Callus of lip hairy; lip with 2 appendages in spur .... ***Scelochilus tuerckheimii***

18(14). Flowers small, arising from dense clusters of bracts; column slender and arching .... ***Sigmatostalix***
18. Flowers arising from single bracts; size various 19

19(18). Anther covered above by membranous hood (clinandrium) 20
19. Anther not covered by distinct hood................ 24

20(19). Margins of columnar hood entire or shallowly toothed 21
20. Margins of columnar hood fringed or lobed 22

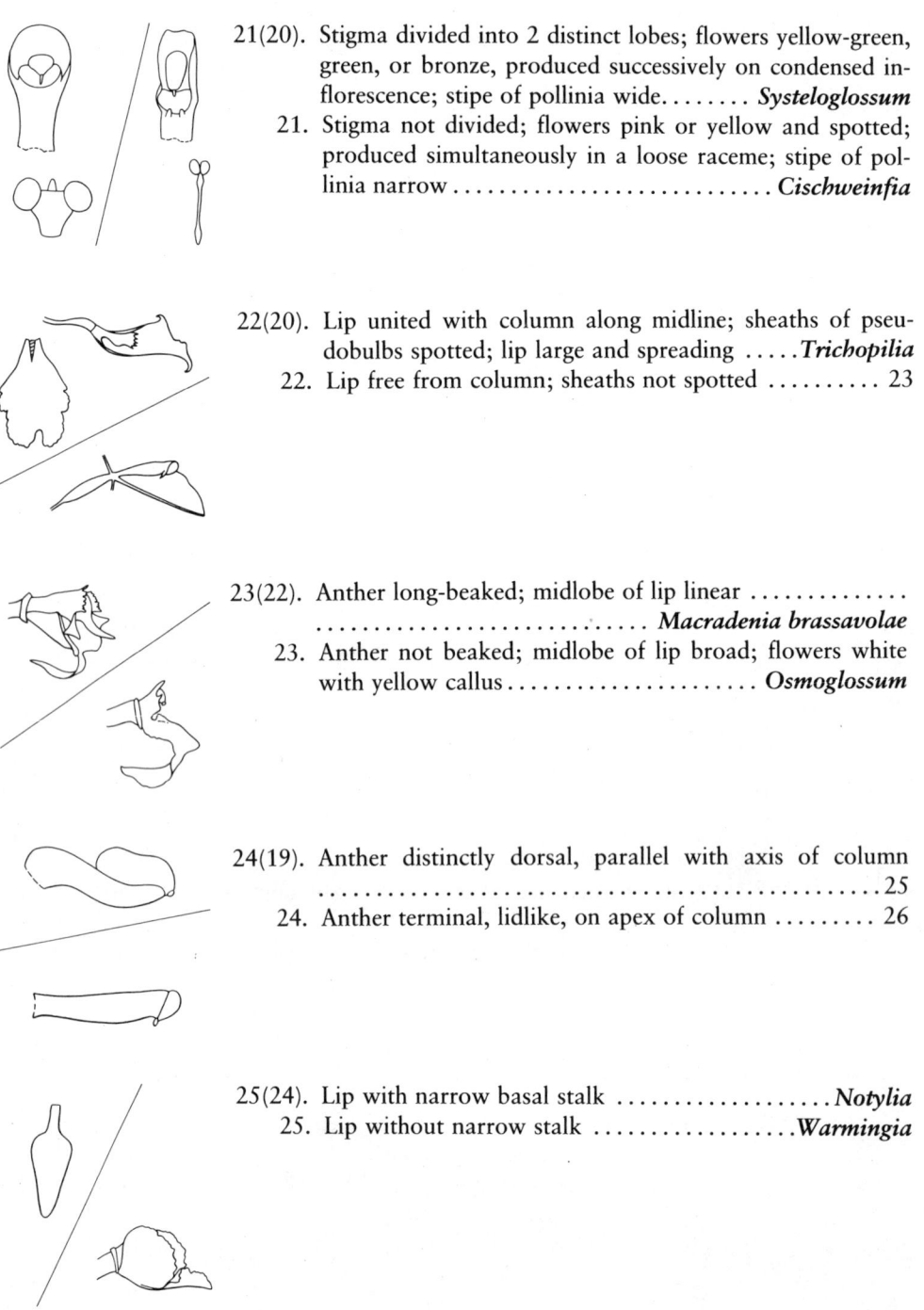

21(20). Stigma divided into 2 distinct lobes; flowers yellow-green, green, or bronze, produced successively on condensed inflorescence; stipe of pollinia wide........ ***Systeloglossum***
21. Stigma not divided; flowers pink or yellow and spotted; produced simultaneously in a loose raceme; stipe of pollinia narrow............................ ***Cischweinfia***

22(20). Lip united with column along midline; sheaths of pseudobulbs spotted; lip large and spreading ..... ***Trichopilia***
22. Lip free from column; sheaths not spotted 23

23(22). Anther long-beaked; midlobe of lip linear ....
.... ***Macradenia brassavolae***
23. Anther not beaked; midlobe of lip broad; flowers white with yellow callus..................... ***Osmoglossum***

24(19). Anther distinctly dorsal, parallel with axis of column
.... 25
24. Anther terminal, lidlike, on apex of column 26

25(24). Lip with narrow basal stalk .... ***Notylia***
25. Lip without narrow stalk .... ***Warmingia***

Key to Genera of Oncidiinae

26(24). Lip united with basal half of column by its edges (forming a nectary-like tube) and then diverging sharply; column without wings .... *Aspasia*
26. Lip free or united only at base 27

27(26). Column with broad wings from base, where wings are united to column foot, column thus markedly concave beneath; flowers small .... *Mesospinidium*
27. Wings, if present, only on upper part of column; without a prominent column foot; column not markedly concave beneath 28

28(27). Flower with small chin, *or* with distinct concave nectary near base of lip; small flowers (3–15 mm long), generally cupped basally 29
28. Flowers without chin or nectary on lip; parts usually spreading widely 32

29(28). Flowers with short chin; flowers white, pinkish, or purplish; lip notched, much longer than sepals and petals .... *Ionopsis*
29. Chin small or indistinct; flowers yellow or green with darker spots 30

30(29). Flowers tiny, 3–3.5 mm long; inflorescence branched .... *Hybochilus inconspicuus*
30. Flowers > 5 mm; inflorescence branched or not 31

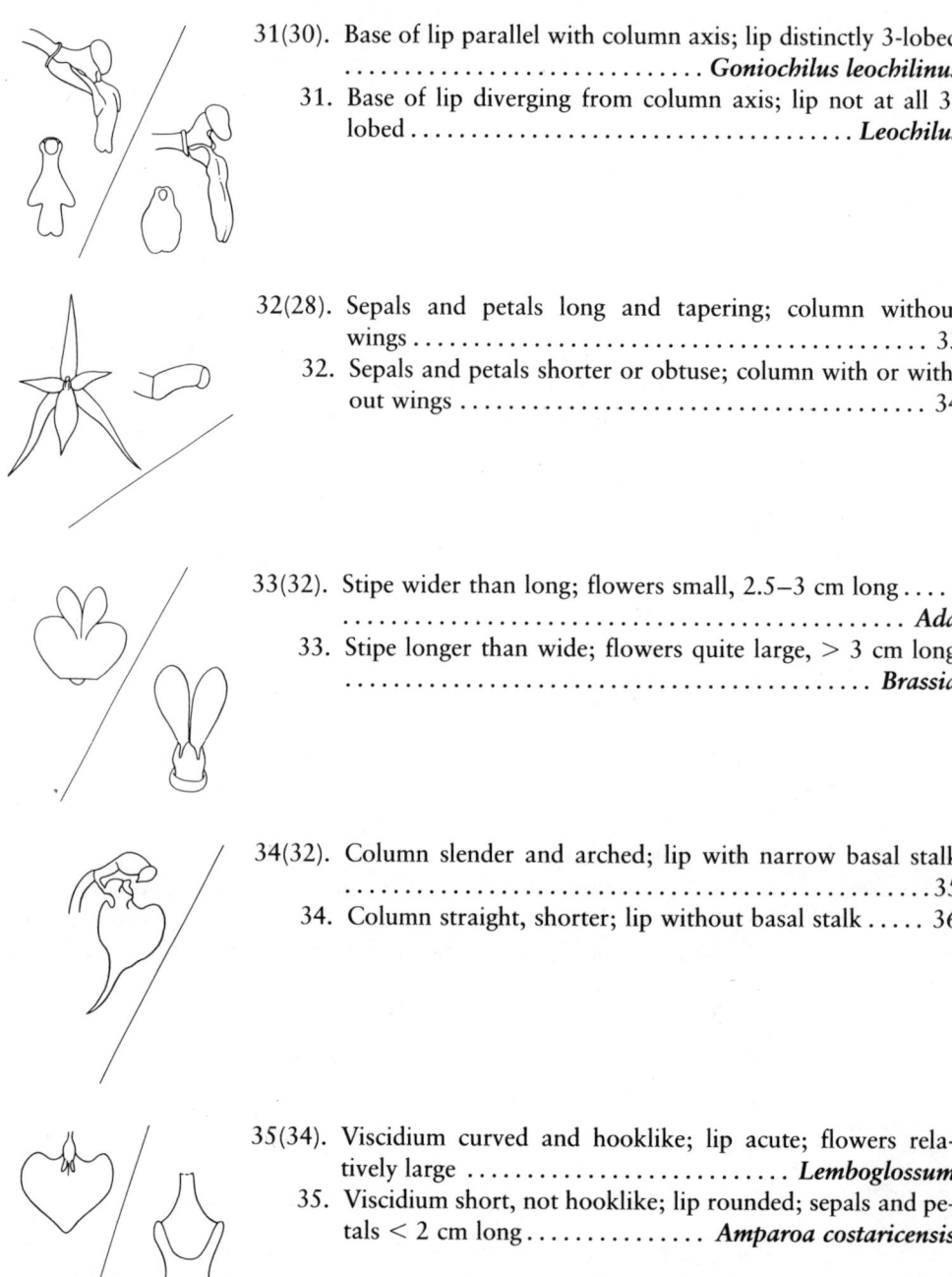

31(30). Base of lip parallel with column axis; lip distinctly 3-lobed .... ***Goniochilus leochilinus***
31. Base of lip diverging from column axis; lip not at all 3-lobed .... ***Leochilus***

32(28). Sepals and petals long and tapering; column without wings 33
32. Sepals and petals shorter or obtuse; column with or without wings 34

33(32). Stipe wider than long; flowers small, 2.5–3 cm long .... ***Ada***
33. Stipe longer than wide; flowers quite large, > 3 cm long .... ***Brassia***

34(32). Column slender and arched; lip with narrow basal stalk 35
34. Column straight, shorter; lip without basal stalk 36

35(34). Viscidium curved and hooklike; lip acute; flowers relatively large .... ***Lemboglossum***
35. Viscidium short, not hooklike; lip rounded; sepals and petals < 2 cm long .... ***Amparoa costaricensis***

36(34). Column without wings; flowers very flat, sepals, petals, and lip overlapping, callus of lip small and basal; flowers white or white and purple .... **Miltoniopsis**

36. Column with definite wings; flowers usually cupped or parts not overlapping............................. 37

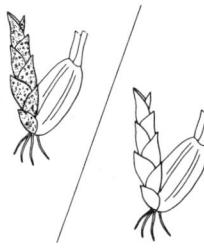

37(36). Pseudobulbs dark grayish green; sheaths spotted; large yellow and red-brown flowers 38

37. Pseudobulbs light or dark green; sheaths not spotted, or flowers not as above.............................. 40

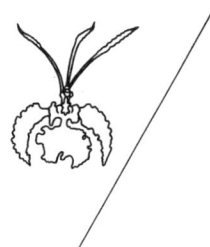

38(37). Petals and dorsal sepal much narrower than lateral sepals; flowers produced successively on a condensed inflorescence .... ***Psychopsis krameriana***

38. Petals and dorsal sepal similar to lateral sepals; flowers produced simultaneously in a raceme 39

39(38). Column wings narrow and recurved; pseudobulbs clustered.................. ***Rossioglossum schlieperianum***

39. Column wings wide and toothed; pseudobulbs widely separated on a creeping rhizome ... ***Otoglossum chiriquense***

40(37). Flowers white or pink with a yellow callus; lip with tuft of hairs at base of callus; flowers few, relatively flat........ .... ***Ticoglossum***

40. Not with above combination of features ..... **Oncidium**

Ada. Epiphytic; pseudobulbs clustered; leaves and leaf sheaths thin; inflorescence racemose; sepals and petals narrow and acute; column short, without wings. Sometimes treated as members of *Brassia*.

Key to *Ada*

1. Pseudobulbs hidden by sheathing leaves; plants fanlike; inflorescence shorter than leaves; **C:** yellow with brown blotches; **D:** Pma, SA (Pc, Mt); **M:** pl to 35 cm, S 3–4 cm .... ***Ada allenii***
Plate 11(5)
1. Pseudobulbs not hidden by leaf sheaths; plants not fanlike; inflorescence subequal to leaves; **C:** S, P greenish yellow with red-brown spots, lip yellow; **D:** No, CR, wPma (Mt); **M:** pl to 30 cm, S 2–2.3 cm .... ***Ada chlorops***

Amparoa. Epiphytic; pseudobulbs loosely clustered on creeping rhizome, flattened with sharp edges, each with 1 apical leaf and 2 or 3 sheathing leaves; inflorescence with few to several flowers; flowers small, column without wings, lip stalked, with a U-shaped callus at base of blade; 1 species in Central America; **C:** greenish yellow, base of lip orange-red; **D:** No, CR (Mt); **M:** pl to 35 cm, S 1.6–1.8 cm ... ***Amparoa costaricensis***

Aspasia. Epiphytic; up to 50 cm tall; pseudobulbs oblong-elliptic, flattened, somewhat stalked, well separated on rhizome, apical leaves 2, narrowly elliptic, 10–30 cm long; inflorescence lateral, few-flowered; sepals and petals similar, lip united with basal half of column, forming a nectary-like tube, quadrate-oblong, constricted in middle, notched, diverging abruptly from column; column without wings; anther fleshy.

Key to *Aspasia*

1. Lip large, spreading, nearly 2 cm across; **C:** S, P yellow-green with brown stripes, lip white, becoming yellow with age; **D:** No, CR, Pma, SA (Cb, Pc); **M:** pl to 30 cm, S 2.8 cm ....
.... ***Aspasia principissa***
Plate 11(3)

1. Lip smaller, midlobe rather cupped, about 1 cm across; C: S yellow-green with brown or red-brown spots, P pink, lip white with red spots; D: No, CR, Pma, SA (Cb, Pc); M: pl to 40 cm, S 2–2.4 cm .... *Aspasia epidendroides*

Brassia. Epiphytic; pseudobulbs more or less clustered, each with 1 or 2 terminal leaves; inflorescence basal, many-flowered; sepals and petals long and tapering, lip with simple callus; column without wings.

Key to *Brassia*

1. Each pseudobulb with a single leaf; C: yellow or greenish yellow with red-brown spots and blotches; D: CR, Pma, SA (Cb, Pc); M: pl to 50 cm, S 10–30 cm .... *Brassia arcuigera*
(Syn. *B. longissima*)
Plate 11(6)
1. Each pseudobulb with 2 leaves 2
2(1). Pseudobulbs markedly flattened, oblong, smooth when young; calli of lip ending in 2 slender teeth; C: yellow or greenish yellow, usually spotted with red-brown; D: No, CR,Pma, SA (Cb, Pc); M: pl to 30 cm, S 5–10 cm .... *Brassia caudata*
2. Pseudobulbs weakly flattened, ovoid, ridged.................. 3
3(2). Lip smooth; C: greenish yellow with brown blotches; D: CR, Pma (Mt); M: pl to 30 cm, S 4–6 cm .... *Brassia gireoudiana*
3. Lip with prominent green warts; C: S, P green with brown spots basally, lip whitish with green warts; D: No, rCR; M: pl to 25 cm, S 4–6 cm .... *Brassia verrucosa*

Cischweinfia. Small epiphytes; pseudobulbs ellipsoid, markedly flattened, loosely clustered; inflorescence basal, few-flowered; lip basally united to column along midline, column with definite hood over anther.

Key to *Cischweinfia*

1. Beak of anther shorter than anther cells; viscidium rounded or squarish at both ends; C: S brown with yellow tips, lip cream, D: CR, Pma (Mt); M: pl 20 cm, S 13–13.5 mm ....
 .... *Cischweinfia pusilla*

1. Beak of anther about as long as anther cells; viscidium elongate and pointed at both ends; C: S, P pale green, lip white with faint pink specks; D: CR, Pma (Mt); M: pl 20 cm, S 12–13.5 mm .... *Cischweinfia dasyandra*
Plate 13(1)

Comparettia. Small epiphyte, usually hanging on twigs; pseudobulbs small, clustered, largely hidden by sheaths, each with 1 terminal leaf; inflorescence arching, with few to many flowers; flowers cupped, with long sepaline spur, lip large and notched; 1 species in Central America; C: magenta-pink; D: No, CR, Pma, SA (Cb, Mt); M: pl 12 cm, S 10 mm .... *Comparettia falcata*
Plate 13(3)

Fernandezia. Epiphytic or growing in moss at high elevations; without pseudobulbs, stems slender, with continued apical growth; leaves conduplicate, 2-ranked; inflorescence of few flowers; flowers bell-shaped, red or rose-purple; 1 species in Costa Rica; C: bright rose-purple; D: CR (Pk); *Pachyphyllum costaricensis* was first described as *Centropetalum costaricensis*, and *Centropetalum* is a synonym of *Fernandezia*. A Costa Rican *Fernandezia* was illustrated by Rafael Lucas Rodríguez, but the species has not yet been named.

Goniochilus. Small epiphyte; pseudobulbs clustered, flattened, each with 1 terminal leaf; inflorescence pendent, with several to many flowers; flowers bell-shaped, with lip distinctly longer than sepals and petals; 1 species; C: S, P green-yellow with red-brown or brown spots, lip white with red or rose-red spots; D: No, CR, Pma (Mt, Pc); M: pl to 15 cm, S 6–9 mm .... *Goniochilus leochilinus*
Plate 12(2)

Hybochilus. Small epiphyte; pseudobulbs clustered, somewhat flattened, each with 1 terminal leaf and 1 or 2 small sheathing leaves; inflorescence pendent, branched, with many tiny flowers; flowers bell-shaped, lip longer than sepals and petals; 1 species; C: S green with red-purplish margins, P whitish with red-purplish margins, lip white with red-purple spots; D: CR, Pma (Mt); M: pl to 10 cm, S 2.5–3 mm .... *Hybochilus inconspicuus*
Plate 12(3)

Ionopsis. Small twig epiphytes; small pseudobulbs partially hidden by leaf sheaths, usually without terminal leaves; inflorescence slender, often branching; sepals and petals subparallel, much shorter than lip, lip notched or 2-lobed.

Key to *Ionopsis*

1. Leaves dorsoventrally flattened, to 16 cm long; inflorescence an open panicle of many flowers; **C:** white, pink, or pink-violet; **D:** No, CR, Pma, SA (Cb, Pc); **M:** pl 12 cm, S 3–6 mm .... *Ionopsis utricularioides*
Plate 13(4)
1. Leaves subcylindric or laterally flattened, to 10 cm long; inflorescence subequal to leaves and producing few flowers at a time; **C:** white or pinkish; **D:** No, CR, Pma, SA (Cb); **M:** pl 6 cm, S 4–6 mm .... *Ionopsis satyrioides*

Lemboglossum. Epiphytic or on rocks; pseudobulbs clustered, somewhat flattened, each with 1–3 terminal leaves and few sheathing leaves; inflorescence basal, with few to many flowers; column arched, with definite wings; lip stalked, with notched callus between stalk and blade of lip. This is the main group of *Odontoglossum*-like plants of Mexico and Central America.

Key to *Lemboglossum*

1. Pseudobulbs usually with 2 or 3 terminal leaves; generally terrestrial or lithophytic; inflorescence erect and much taller than leaves; **C:** S, P pale green with red-brown bars, lip pink; **D:** No, CR, Pma (Mt); **M:** pl to 50 cm, S 2–2.7 cm .... *Lemboglossum bictoniense*
Plate 11(4)
1. Pseudobulbs each with a single terminal leaf 2
2(1). Sepals and petals similar; ovary 3-angled; **C:** S, P orange-brown, lip white shading to pale lavender at base; **D:** No, CR, SA (Mt); **M:** pl to 18 cm, S 2.5–2.6 cm............... *Lemboglossum stellatum*
2. Sepals and petals dissimilar; ovary cylindrical 3
3(2). Sepals acute; lip generally broadly cordate, margins irregularly toothed; **C:** S brown or red-brown, P yellow with red spots near base, lip white with brown spots; **D:** No, CR (Mt); **M:** pl to 35 cm, S 2.5–4.5 cm..................... *Lemboglossum maculatum*
3. Sepals long-acuminate 4

4(3). Lip cordate, acuminate, margins entire; **C:** S, P pale yellow spotted or stained with red-brown, lip white with red-brown spots; **D:** No, CR (Mt); **M:** pl 15 cm, S 3.5–4 cm . . . . . **Lemboglossum cordatum**

4. Lip trullate or rhombic, acute, margins irregularly toothed; **C:** S yellow with brown blotches, P and lip white with brown spots; **D:** CR (Mt); **M:** pl 20 cm, S 3.5–5 cm . . . . **Lemboglossum hortensiae**

Leochilus. Small epiphytes; pseudobulbs clustered, 1-leaved, with few sheathing leaves; inflorescence basal, racemose or branched, of few to several flowers; sepals and petals similar, cupped; lip much larger, with deep nectary near base; column with 2 armlike wings. These twig epiphytes are often found on coffee, guava, or calabash.

Key to *Leochilus*

1. Mature pseudobulb reduced, largely hidden by sheaths, strongly flattened; **C:** green-yellow with red or red-brown spots; **D:** CR, Pma (Mt); **M:** pl to 10 cm, S 4–7 mm . . . . *Leochilus tricuspidatus*

1. Mature pseudobulbs prominent, somewhat compressed 2

2(1). Plants green; lip callus larger than nectary; nectary front wall much lower than side walls or nearly absent; **C:** yellow to yellow-green with red or rose spots; **D:** No, CR, Pma, SA (Cb, Pc); **M:** pl to 15 cm, S 7–12 mm . . . . **Leochilus scriptus**
Plate 12(4)

2. Plants red to red-purple; lip callus equal to or smaller than nectary; nectary front wall nearly as high as side walls; **C:** yellow with red spots; **D:** No, CR, Pma, SA (Cb, Pc, Mt); **M:** pl to 8 cm, S 4–8 mm . . . . **Leochilus labiatus**

Leucohyle. Epiphytic; pseudobulbs clustered, narrowly conic-cylindric, each with 1 terminal leaf; leaves narrow, fleshy, more or less 3-angled; inflorescence pendent, with few to several flowers; sepals and petals narrow and spreading, lip concave, not lobed; column with definite hood over anther; related to *Trichopilia* and sometimes included in that genus; 1 species in Central America; **C:** translucent white, lip with red or pink specks; **D:** No, CR, Pma, SA (Cb); **M:** pl 10–25 cm, S 2–2.3 cm . . . . **Leucohyle subulata**

Lockhartia. Epiphytic; without pseudobulbs, stems slender but with limited growth; leaves 2-ranked, short, laterally flattened; inflorescence upper lateral or terminal, simple or branched, often with conspicuous bracts; flowers 1 to many, sepals and petals usually ovate or oblong, lip simple or complex, column short, usually winged. These are sometimes called the braided orchids because of the characteristic aspect of the overlapping, laterally flattened leaves.

Key to *Lockhartia*

1. Lip with definite, narrow or acute lateral lobes 2
1. Lip entire or at most shallowly lobed 5
2(1). Lateral lobes borne in middle of lip; leaves acute, not spreading; inflorescence large and spreading; **C:** white with yellow on lip; **D:** Pma, SA (Cb, Pc); **M:** pl 15–50 cm, S 3–5 mm . . . . *Lockhartia acuta*
2. Lateral lobes basal; leaves obtuse or acute; inflorescence relatively small 3
3(2). Midlobe of lip shallowly notched; leaves obtuse; **C:** pale yellow with red spots on base of lip; **D:** No, CR, Pma, SA (Cb, Mt, Pc); **M:** pl 8–40 cm, S 3–4 mm . . . . *Lockhartia micrantha*
3. Midlobe deeply notched 4
4(3). Midlobe of lip subquadrate, as wide basally as apically; bracts conspicuous; **C:** yellow with red-brown markings on base of lip; **D:** No, CR, Pma (Mt); **M:** pl 12–40 cm, S 4–6 mm . . . .
. . . . *Lockhartia amoena*
Plate 13(5)
4. Midlobe narrow basally, widest near apex; bracts not especially conspicuous; **C:** yellow, base of lip spotted and barred with dark red; **D:** No, CR, Pma (Mt); **M:** pl 10–40 cm, S 5–8 mm . . . .
. . . . *Lockhartia oerstedii*
5(1). Lip subcircular, as wide as long 6
5. Lip oblong, longer than wide 7
6(5). Leaves obtuse; callus with single tooth; **C:** yellow with orange callus; **D:** Pma (Mt); **M:** pl 10–35 cm, S 9–10 mm . . . .
. . . . *Lockhartia obtusata*
6. Leaves acute; callus with several teeth in front; **C:** white with orange callus; **D:** No, CR, Pma (Cb, Mt, Pc); **M:** pl 10–30 cm, S 5–6 mm . . . . *Lockhartia hercodonta*
7(5). Lip not at all 4-lobed; **C:** yellow with orange-brown spot on callus; **D:** No, CR, SA (Pc); **M:** pl 10–20 cm, S 3.5–5 mm . . . .

.... *Lockhartia integra*
7. Lip shallowly 4-lobed 8
8(7). Lip narrow basally; leaves acute and spreading; C: pale yellow; D: No, CR, Pma (Cb); M: pl 12–20 cm, S 4–5 mm ....
.... *Lockhartia pittieri*
8. Lip wide basally; leaves obtuse and ascending; C: yellow; D: ePma, SA (Mt) .... *Lockhartia chocoensis*

Macradenia. Small epiphyte; pseudobulbs clustered, narrowly conic, each with 1 terminal leaf; inflorescence pendent, of many flowers; lip 3-lobed, column with fringed hood over anther, anther long-beaked, asymmetrical; 1 species in our area; C: S, P dark red with yellow borders, lip white with red spots; D: No, CR, Pma, SA (Cb, Pc); M: pl 20 cm, S 1.8–2 cm .... *Macradenia brassavolae*
Plate 13(6)

Macroclinium. Small epiphytes; pseudobulbs small or lacking; plant fan-shaped, leaves laterally flattened, usually somewhat warty; inflorescence arching or pendent, of few to many flowers, commonly flowering again from axillary buds; sepals and petals narrow, spreading, lip narrow, often arrowhead-shaped; column slender, anther parallel with column.

Key to *Macroclinium*

1. Lip sublinear, without distinct lateral lobes................... 2
1. Lip not sublinear, with distinct lateral lobes.................. 3
2(1). Sepals about 5 mm long; C: S greenish cream, P and lip paler, each P with lilac median streak; D: Pma (Mt); M: pl to 4 cm, S 4.8–5 mm .... *Macroclinium simplex*
2. Sepals about 10–12 mm long; C: rose-purple with darker spots on P; D: CR, Pma (Cb, Mt); M: pl 1.5–4 cm, S 12–13 mm ....
.... *Macroclinium lineare*
3(1). Inflorescence racemose, elongate; C: S greenish, P pink with dark spots, lip rose-purple; D: CR, Pma (Cb, Mt, Wt); M: pl 3–4 cm, S 10 mm ....*Macroclinium ramonense*
Plate 14(2)
The name *M. bicolor*, based on a Guatemalan species, has been misapplied to Costa Rican *M. ramonense*.

3. Inflorescence subumbellate, condensed....................... 4

4(3). Claw of lip at least partly united with column................ 5
4. Claw of lip completely free from column 6

5(4). Claw of lip basally united with column, occupying half the length of column; C: S whitish, P white or pale pink, lip pink or rose-pink; D: Pma, SA (Mt); M: pl to 1.8 cm, S 4–6 mm .... *Macroclinium junctum*
5. Claw of lip basally united with column but occupying much less than half length of column; D: CR (Pc)......................... .... *Macroclinium "cordesii"* of Pacific slope
The name *M. cordesii* has been misapplied to this unnamed species which is similar to *M. glicensteinii* (see below).

6(4). Apex of column abruptly bent upward; C: pale lavender with dark spots mostly on petals; D: CR (Mt); M: pl 2–4 cm, S 11–14 mm .... *Macroclinium glicensteinii*
6. Apex of column more or less curved, but not abruptly bent upward 7

7(6). Lip about 8 mm long, long-acuminate, sharply constricted above basal lobes; D: CR, Pma (Cb); M: pl 4–5.5 cm, S 10–13 mm .... *Macroclinium cordesii*
7. Lip about 5 mm long, shortly acuminate, gradually constricted above basal lobes; C: rose, P and lip with darker stripes; D: No, CR (Mt); M: pl 5 cm, S 6 mm .... *Macroclinium paniculatum*

Mesospinidium. Small epiphytes; pseudobulbs clustered, narrowly ovoid or ellipsoid, smooth, with several sheathing leaves and 1 or 2 terminal leaves; inflorescence erect or arching, many-flowered, several flowers open at a time; flowers small, sepals and petals cupped; column broadly winged, wings attached to distinct column foot.

Key to *Mesospinidium*

1. Lip rhombic in outline, apex shallowly notched, margins wavy; C: S, P greenish yellow with red-brown spots, lip white with greenish

yellow spots; D: No, CR, Pma (Mt); M: pl to 30 cm, S 7.5–8 mm
.... *Mesospinidium warscewizcii*
Plate 12(5)
1. Lip elliptic or obovate in outline, apex acute or acuminate, margins not wavy 2
2(1). Callus of lip widest in forward half; petals 3-veined; C: yellow, S, P with red-brown spots or P with red-brown margins, lip with red-brown specks; D: CR, Pma (Mt); M: pl to 35 cm, S 5 mm........
.... *Mesospinidium endresii*
(Syn. *M. horichii*)
2. Callus of lip widest at base; petals 1-nerved; C: S, P greenish yellow with brown spots, lip cream with orange-yellow spots; D: Pma (Mt); M: pl to 30 cm, S 7–8 mm..... *Mesospinidium panamense*

Miltoniopsis. Epiphytic; pseudobulbs clustered, somewhat flattened, each with 1 terminal leaf and several sheathing leaves; leaves thin; inflorescence of few to several flowers, flowers flat and open; lip large with basal callus.

Key to *Miltoniopsis*

1. Lip fiddle-shaped, constricted above base; C: white with yellow callus, may have rose blotches by callus; D: CR, wPma (Mt); M: pl to 35 cm, S 2.5–3 cm .... *Miltoniopsis warscewiczii*
(Syn. *Miltonia endresii*)
1. Lip narrow at base, not constricted above; C: white, callus yellow, often with red lines, P often basally red-purple; D: Pma, SA (Mt); M: pl 35 cm, S 5 cm .... *Miltoniopsis roezlii*
Plate 10(5)
This handsome species is well known in the region of El Valle de Antón, but it ranges much farther to the west. It requires a moist habitat and occurs near sea level in the wet Chocó of Colombia.

Notylia. Small epiphytes; pseudobulbs clustered, smooth or grooved, each with 1 terminal leaf; inflorescence pendent, with many flowers;

sepals and petals narrow, spreading; lip more or less arrowhead-shaped, column straight, wingless, anther oblong, parallel with column. There are undoubtedly more than 4 species of *Notylia* in Costa Rica and Panama, but most of them are treated as "more or less *barkeri*" in herbaria and floras. All are very similar and only careful fieldwork will clarify the species of this group.

Key to *Notylia*

    1. Dorsal sepal rounded, about 5 mm wide, wider than lip; **C**: white; **D**: CR, Pma, SA (Cb); **M**: pl to 18 cm, S 7–8 mm . . . . *Notylia albida*
    (Syn. *N. panamensis*)
    Plate 14(3)
    Most species of *Notylia* are rather plain (if not ugly), but this one is quite pretty.
    1. Dorsal sepal narrow, 2–3 mm wide; green or greenish 2

    2(1). Column papillose; leaves large, to 4 cm wide and 20 cm long; **C**: greenish white with orange spot on each P; **D**: Pma (Cb, Pc); **M**: pl to 25 cm, S 8–10 mm . . . . *Notylia pentachne*
    This species is often found in hedges.
    2. Column smooth 3

    3(2). Lip about 5 mm wide; leaves 4–5 cm wide and 20 cm long; **C**: S orange?, P and lip white; **D**: Pma (Cb); **M**: pl to 25 cm, S 9–10 mm . . . . *Notylia latilabia*
    3. Lip < 3 mm wide; leaves to 3 cm wide, 15 cm long; **C**: greenish white, S with orange spots; **D**: No, CR, Pma, SA (Cb, Mt, Pc); **M**: pl to 20 cm, S 3–6 mm . . . . *Notylia barkeri*

Oncidium. Small to large, usually epiphytic; with or without pseudobulbs; leaves thin to very fleshy; inflorescence often branched; flowers usually yellow marked with brown; column usually winged; lip usually with fleshy callus near base. This is a very diverse genus, but attempts to divide it into more natural groups have not been successful. Plate 9(1–6).

Key to *Oncidium*

1. Leaves thick and fleshy; pseudobulbs small and inconspicuous... 2
1. Leaves thin or leathery; pseudobulbs well developed, conspicuous.... 8

2(1). Leaves fleshy and flattened ("mule ears") 3
2. Leaves subcylindric ("rat tails")............................. 4

3(2). Lip > 1 cm wide; midlobe wider than across lateral lobes; C: S, P yellow with brown spots, lip reddish brown; D: No, CR; M: pl to 60 cm, S 13 mm ....*Oncidium luridum*
(Syns. *O. altissimum* [misapplied], *O. guttatum*)
3. Lip < 1 cm wide, widest across lateral lobes; C: white or pale pink with pink or purple spots; D: No, CR, Pma, SA (Pc); M: pl to 40 cm, S 8–10 mm .... *Oncidium carthagenense*
Plate 9(1)

4(2). Lateral lobes of lip rounded, wider above base; lip often with pink spots on reverse; C: S, P yellowish green with red-brown spots, lip yellow with red spots near callus; D: No, CR, Pma?, SA (Pc); M: pl to 40 cm, S 10 mm........................ *Oncidium cebolleta*
4. Lateral lobes of lip oblong or finger-like, not markedly expanded above bases 5

5(5). Callus between lateral lobes, with very narrow isthmus between callus and midlobe (not more than one-fourth width of midlobe) 6
5. Callus on isthmus, isthmus at least a third as wide as midlobe... 7

6(5). Callus simple, not divided; C: S, P yellow with brown spots, lip yellow with chestnut brown spots near callus; D: No, CR, Pma (Cb, Pc); M: pl to 70 cm, S 5–7 mm ....*Oncidium stipitatum*
6. Callus with 3 parallel ridges; C: S, P greenish brown or brownish yellow with maroon markings; lip yellow, base pale brown with maroon-pink markings; D: SA, ePma; M: pl to 50 cm, S 4–8 mm .... *Oncidium nudum*

7(5). Column wings curved; C: S, P yellowish green with red-brown spots, lip yellow with red-brown spots near callus; D: No, CR (Cb, Pc); M: pl 15 cm, S 7 mm .... *Oncidium ascendens*
7. Column wings straight; C: S, P yellow with red-brown spots, lip yellow; D: No, CR, Pma (Pc); M: pl to 60 cm, S 5–6 mm ....*Oncidium teres*

8(1). Leaves leathery; pseudobulbs bumpy even when young; column with petal-like lobe over anther (between column wings); C: yellow, S, P, and callus with brown or reddish brown spots; D: No, CR,

Pma, SA (Cb, Pc); **M:** pl to 30 cm, S 6–10 mm.................
.... *Oncidium ampliatum*
Sometimes called *la tortuga* or "turtle orchid" because of the flattened, bumpy pseudobulbs; flower size is variable.
8. Leaves thin; pseudobulbs smooth or grooved; column without a lobe over anther 9
9(8). Most flowers abortive, with straplike sepals and petals, without a column; only a few flowers normally developed; C:yellow, S, P, and lateral lobes of lip marked with red-brown, callus whitish; **D:** CR, Pma, SA (Mt); **M:** pl to 25 cm, S 6–10 mm....................
.... *Oncidium heteranthum*
Plate 9(2)
This species may flower when the plant is very small.
9. Inflorescence normally without abortive flowers 10
10(9). Anther with long beak, beak usually longer than body of anther
....11
10. Anther without beak or with short beak 16
11(10). Pseudobulbs hidden by leaf sheaths; **C:** light yellow; **D:** CR, Pma (Mt); **M:** pl 12 cm, S 8.5–10 mm .... *Oncidium luteum*
11. Pseudobulbs not concealed by leaf sheaths 12
12(11). Flowers rose-pink, callus orange, column wings white; **D:** No, CR, Pma?; **M:** pl 35 cm, S 8–11.5 .... *Oncidium ornithorrhynchum*
12. Flowers yellow... 13
13(12). Column with prominent wings............................. 14
13. Column without wings, or wings rudimentary 15
14(13). Isthmus about as wide as callus; sinuses narrow; **C:** bright yellow; **D:** No, CR, Pma (Mt, Pk, Pc); **M:** pl 15 cm, S 5 mm ....
.... *Oncidium cheirophorum*
14. Isthmus narrower than callus; sinuses broad and rounded; **C:** yellow; **D:** CR (Mt); **M:** pl 20 cm, S 1.7 cm .... *Oncidium pittieri*
15(13). Column with recurved hook at base (in front of stigma), column wings rudimentary; **C:** bright yellow; **D:** No, CR (Pc); **M:** pl 20 cm, S 4.5–5.5 mm .... *Oncidium exauriculatum*
15. Column with low, rounded knob near base, slender, without wings; **C:** S, P yellowish green, lip greenish yellow, callus white; **D:** Pma (Mt); **M:** pl 20 cm, S 6–9 mm............... *Oncidium exalatum*
16(10). Lip not clearly 3- or 4-lobed, without distinct lateral lobes or only shallowly lobed; column and lip may be subparallel or diverge at an acute angle 17
16. Lip with distinct lateral lobes, deeply 3- or 4-lobed; column and lip diverging, not forming an acute angle 23

17(16). Lip subcircular; column only about twice as long as wide; C: S, P brownish or purplish, often with white margins, lip brown or rose-purple, or with white margins; D: Pma, SA (Cb, Mt); M: pl 30 cm, S 1.8–2.2 cm............................*Oncidium fuscatum*
(Syn. *Miltonia warscewiczii*)
Plate 9(5)

This is an unusual *Oncidium*, not at all related to *Miltonia* or *Miltoniopsis*. Because of the quirks of botanical nomenclature, *Miltonia warscewiczii* became *O. fuscatum*, and *Miltonia endresii* became *Miltoniopsis warscewiczii*.

17. Lip not subcircular.. 18

18(17). Lip oblong, shallowly constricted laterally, margins curled back in middle; C: S, P yellowish green heavily blotched with red-brown, lip pink, rose-purple, or pink with white margins; D: CR, wPma (Mt); M: pl 40 cm, S 5–6 cm...............*Oncidium schroederianum*
(Syn. *Miltonia schroederiana*)
Plate 9(3)

This species is similar to *Miltonia clowesii* of Brazil; if *Miltonia* is maintained as a distinct genus, this should probably be a *Miltonia*.

18. Lip narrow or tapering basally............................. 19

19(18). Column slender with very small wings; lip white.............. 20

19. Column not more than twice as long as wide, usually with prominent wings; lip yellow or white 21

20(19). Lip oblanceolate, apex only slightly expanded, base subparallel with column, almost without callus; C: S, P greenish spotted with red-brown, lip white with purplish or violet markings on lower half; D: No, CR (Mt); M: pl 25 cm, S 2.1 cm..... *Oncidium stenoglossum*

20. Lip anchor-shaped (sides of midlobe recurved), with prominent callus on claw, base not parallel with column; C: S, P green stained with purple-brown, lip white, basally rose-purple; D: CR (Mt); M: pl 50 cm, S 2.5 cm..................... *Oncidium cariniferum*

21(19). Lip about 1 cm long; callus ridges diverging apically; C: S, P greenish with purple spots, lip yellow; D: No, CR (Mt); M: pl 20 cm, S 9 mm....................................... *Oncidium endocharis*

21. Lip 2–3 cm long, callus ridges not diverging apically 22

22(21). Column with rounded wings in lower half; callus ending in finger-like extension; C: S, P yellow blotched with brown, lip yellow; D: No, CR; M: pl 25 cm, S 11–12 mm..... *Oncidium graminifolium*

22. Column without prominent rounded wings; callus of low ridges ending in acute angles; **C:** S, P greenish yellow with brown blotches, lip white, often with brown spots or lines near base, midlobe yellow; **D:** No, rCR; **M:** pl 50 cm, S 1.8–3 cm ..... *Oncidium maculatum*

23(16). Lateral sepals more or less united 24
23. Lateral sepals free to bases 25

24(23). Lateral sepals united to apices; **C:** clear yellow or with small red streak on each side of lip; **D:** CR, Pma (Mt); **M:** pl 25 cm, S 8 mm .... *Oncidium warscewiczii*
24. Lateral sepals united for less than half length; **C:** S, P yellowish green, base of lip brownish yellow, callus orange-yellow, midlobe yellow-cream; **D:** CR (Mt); **M:** pl 30 cm, S 8–11 mm .... *Oncidium storkii*

25(23). Lateral sepals distinctly longer than lip, sepals brown with yellow margins 26
25. Lateral sepals subequal to lip or shorter, sepals yellow with brown spots 28

26(25). Lip wider across midlobe than across lateral lobes, with a narrow isthmus; **C:** S, P shiny chocolate brown with yellow margins, lip yellow with brown spots basally; **D:** Pma (Cb); **M:** pl 45 cm, S 1.5–2.5 cm .... *Oncidium powellii*
Similar to the Colombian *O. anthocrene*.
26. Lip wider across lateral lobes, or lateral lobes and midlobe subequal 27

27(26). Callus of lip with 9 crests; **C:** S, P brown with yellow margins, lip yellow, isthmus red-brown; **D:** No, CR, Pma, SA (Mt); **M:** pl 50 cm, S 1.5–2 cm .... *Oncidium stenotis*
Plate 9(4)
27. Callus of lip with 7 crests; **C:** S, P green-brown-olive with yellow tips, lip yellow with red-brown isthmus; **D:** No, CR (Mt); **M:** pl 50 cm, S 1.6–1.9 cm .... *Oncidium paleatum*
Similar to *O. stenotis* and probably a form of that species.

28(25). Pseudobulbs without apical leaves; dwarf plant (< 5 cm tall), with 1 or 2 flowers at a time on each inflorescence, inflorescence subequal to leaves; **C:** yellow with orange spots; **D:** No, CR, Pma, SA (Mt); **M:** pl 5 cm, S 5–6 mm .... *Oncidium crista-galli*
28. Pseudobulbs with 1–3 terminal leaves; plants usually > 10 cm tall 29

29(28). Pseudobulbs small, solitary or in small clusters on long, wiry stems (inflorescences); flowers proportionately very large 30
29. Pseudobulbs relatively large, clustered on a thick rhizome, never scattered on a long, wiry stem 31

30(29). Midlobe of lip from distinct narrow isthmus, lateral sepals largely exposed; C: bright yellow, S, P barred with red-brown; D: No, CR, Pma, SA (Mt); M: pl to 6 cm (individual unit), S 10–12 mm, lip to 2.5 cm wide .... *Oncidium globuliferum*
This species forms large, tangled mats in the treetops.
30. Midlobe of lip from short, wide isthmus, lateral sepals largely concealed by lip; C: S, P pale brownish yellow with red-brown markings, lip yellow with red-brown markings basally; D: rCR, SA; M: pl to 8 cm, S 1.5–1.7 cm, lip to 4 cm wide .... *Oncidium scansor*

31(29). Petals nearly twice as wide as sepals 32
31. Petals subequal to sepals 33

32(31). Sepals and petals obtuse, petals widest near apices; C:yellow with brown spots toward center; D: CR, Pma, SA (Mt); M: pl 35 cm, S 6–14 mm .... *Oncidium obryzatum*
Plate 9(6)
This species is quite variable in flower size and season; 2 or more species may be confused under this name.
32. Sepals and petals acute, petals widest near middle; midlobe of lip somewhat anchor-shaped; large plant with prominent leaf sheaths and smooth pseudobulbs; inflorescence twining; C: S, P greenish yellow marked with red-brown, base of lip orange, midlobe white; D: No, CR, Pma (Cb); M: pl 70 cm, S 1.5–2 cm ....
.... *Oncidium ochmatochilum*

33(31). Sheathing leaves prominent, not jointed basally; pseudobulbs thick and smooth when young; usually terrestrial; C: yellow, S, P with bronze or olive-brown spots; D: No, CR, Pma (Pc); M: pl 80 cm, S 10–15 mm .... *Oncidium ensatum*
33. Sheathing leaves few, jointed basally; pseudobulbs thinner and ridged 34

34(33). Flowers white with pale purple to rose spots; D: No, rCR; M: pl 40 cm, S 10–13 mm......................... *Oncidium incurvum*
Doubtful in Costa Rica.
34. Flowers yellow and brown 35

35(34). Column without wings, or wings reduced to a ridge or small triangle
.... 36

35. Column with prominent wings, these spreading and fanlike 40

36(35). Bracts of inflorescence very prominent, lower floral bracts longer than ovary with pedicel; C: yellow, S, P greenish yellow with brown or maroon spots, lip yellow; D: CR, Pma? (Mt); M: pl 50 cm, S 1–1.7 cm.................................*Oncidium bracteatum*
36. Bracts smaller, floral bracts shorter than ovary and pedicel..... 37

37(36). Sepals and petals tapering, widest near bases; callus finely fuzzy, 3-toothed; C: yellow with brown to purplish spots; D: No, CR; M: pl 25 cm, S 14–15 mm .... *Oncidium guttulatum*
37. Sepals widest near or above middle......................... 38

38(37). Dorsal sepal expanding abruptly from a narrow claw; column without wings; C: yellow with red-brown spots; D: No, Pma, SA (Mt); M: pl 35 cm, S 1.3–1.7 cm................. *Oncidium planilabre*
38. Dorsal sepal expanding gradually from base.................. 39

39(38). Callus with several toothlike lobes; column wings small but definite; sheathing leaves few; C: S, P yellow with red-brown spots, lip yellow; D: CR, Pma, SA; M: pl 35 cm, S 13–15 mm................
.... *Oncidium klotzschianum*
39. Callus with a few blunt lobes; column without wings; sheathing leaves several; C: yellowish brown with yellow tips; D: Pma (Mt); M: pl 50 cm, S 12–13 mm.............. *Oncidium panduriforme*

40(35). Pseudobulbs strongly flattened, with sharp edges, more than two-thirds as wide as long, smooth when young; C: S, P brown with yellow tips, lip yellow; D: No, CR, Pma (Mt); M: pl 30 cm, S 1.2–1.7 cm.................................. *Oncidium ansiferum*
40. Pseudobulbs thick with rounded edges, grooved, usually twice as long as wide... 41

41(40). Flowers < 1.6 cm across; callus 4-lobed, fuzzy; pseudobulbs marked with dark spots; C: S, P yellow marked with brown, lip yellow; D: Pma (Mt); M: pl 40 cm, S 5–6 mm .... *Oncidium parviflorum*
41. Flowers mostly > 1.8 cm in diameter; callus not fuzzy......... 42

42(41). Lip about as wide across lateral lobes as across midlobe; callus complex, with 9 distinct lobes; C: yellow with dark maroon spots; D: No, CR, Pma, SA (Pc); M: pl 80 cm, S 1–1.7 cm ....
.... *Oncidium baueri*
> There is some doubt that the Central American plant is truly *O. baueri*, a species of the West Indies and northern South America.

42. Lip wider across midlobe than across lateral lobes 43
43(42). Petals more than half as wide as long; moderate-sized plant with pseudobulbs heavily spotted with dark brown; **C:** S, P yellow densely spotted with brown, lip yellow with broad band of chestnut brown near callus, *or* S, P white spotted with pinkish purple, lip white with band of pinkish purple; **D:** CR, Pma (Mt, Pc); **M:** pl 30 cm, S 8–12 mm........................... *Oncidium cabagrae*
43. Petals less than half as wide as long; plants becoming very large ...
.... 44
44(43). Lip with narrow isthmus about one-fourth width across lateral lobes; **C:** S, P yellow heavily spotted with brown, lip yellow, heavily spotted on isthmus; **D:** CR (Pc); **M:** pl 50 cm, S 9–13 mm........
.... *Oncidium isthmii*
44. Lip with wide isthmus, lateral lobes broad; **C:** S, P yellow heavily spotted with brown, lip yellow basally spotted with brown; **D:** Pma (Cb, Pc); **M:** pl 80 cm, S 10–12 mm .... *Oncidium panamense*
This species resembles O. *sphacelatum*, which is not native in our area but is common to the north and may be in cultivation here. O. *sphacelatum* forms very large plants, and the base of the lip is shorter (much less than half the length of the lip) and wider.

Osmoglossum. Epiphytic; pseudobulbs clustered, ovate or narrowly ovate, somewhat flattened, each with 1 or (usually) 2 terminal leaves, leaves narrow, thin; inflorescence basal, erect, of several flowers; column short, with 3 fringed lobes around anther. Closely related to *Palumbina*, of northern Central America; the 2 genera should probably be united (as *Palumbina*).

Key to *Osmoglossum*

1. Lateral sepals united for about half their length; lip slightly narrower in lower half; rachis of inflorescence zigzag; **C:** white with red spots on yellow callus; **D:** No, CR, Pma (Mt); **M:** pl 30 cm, S 7–10 mm............................. *Osmoglossum egertonii*
Plate 10(2)

O. *anceps*, described from Costa Rica, is very similar, if not the same.

1. Lateral sepals united for about 5 mm basally; lip markedly narrower in lower half; rachis of inflorescence nearly straight 2

2(1). Lip markedly bent, 10–18 mm long; wings of column 3-lobed, toothed; C: white with red spots on yellow callus; D: No, CR? (Mt); M: pl 25 cm, S 1.8 cm .... *Osmoglossum pulchellum*
2. Lip flat, not bent, about 10 mm long; wings of column slightly 3-lobed, margins subentire; C: white with red spots on yellow callus; D: No, CR, Pma (Mt); M: pl 45 cm, S 9–12 mm..........
.... *Osmoglossum convallarioides*

Otoglossum. Epiphytic; thick rhizome long, pseudobulbs widely separated, flattened, nearly hidden by sheathing leaves, each with 1 terminal leaf, sheaths spotted; inflorescence erect, with several flowers; sepals and petals wide, spreading, lip with thick, fleshy callus at base; column with short-toothed hood over anther, more or less continuous with toothed column wings; 1 species in Central America; C: S, P red-brown with yellow edges and some spots, lip yellow with red-brown spots on base; D: CR, Pma, SA (Mt, Pk); M: pl 30 cm, S 2–3 cm...................
.... *Otoglossum chiriquense*
Plate 10(3)

Pachyphyllum. Small epiphytes; without pseudobulbs, stems slender, with indefinite apical growth, sometimes branched; leaves 2-ranked, conduplicate or fleshy; inflorescence lateral, short, of few to several tiny white or pale green flowers; pollinia 2.

Key to *Pachyphyllum*

1. Plants usually < 10 cm tall; leaves acute, < 2 mm wide; sepals and petals united into a tube; C: cream or pale green; D: CR, Pma, SA (Pk); M: pl 5–8 cm, S 2.5 mm .... *Pachyphyllum hispidulum*
Plate 14(6)
1. Plants 10–30 cm tall; leaves obtuse or notched, 4–8 mm wide; sepals and petals free....................................... 2
2(1). Leaves 4–5 mm wide; margins of sepals not distinctly toothed; C: white; D: CR (Pk); M: pl to 12 cm, S 3.5–4 mm ....
.... *Pachyphyllum costaricense*
2. Leaves to about 8 mm wide; margins of sepals distinctly toothed; C: pale green; D: CR, SA (Pk); M: pl to 30 cm, S 1.5–2 mm ....
.... *Pachyphyllum crystallinum*
(Syn. *P. pastii*, misapplied)

Plectrophora. Small epiphyte; pseudobulbs small, loosely clustered, each with 1 terminal leaf, leaves fleshy; inflorescence short, with 1 or few flowers; trumpet-shaped flower has long slender spur at base, large lip; **C:** cream with yellow in throat; **D:** CR, Pma, SA (Mt); **M:** pl 7–10 cm, S 1.6–1.7 cm.....................................*Plectrophora alata*

Psychopsis. Epiphytic; pseudobulbs flattened, bumpy, nearly horizontal and overlapping, each with 1 fleshy terminal leaf, leaves and sheaths with dark spots; inflorescence erect, slender, with thick joints; flowers produced 1 at a time at apex; petals and dorsal sepal narrow and antenna-like, lateral sepals wide, arching and ruffled; lip 3-lobed with fleshy callus between lateral lobes; 1 species in Central America; **C:** P and dS dark red-brown, lS and lip yellow marked with red-brown; **D:** CR, Pma, SA (Cb, Pc); **M:** pl 25 cm, dS 5.5–8 cm, lS 3–5 cm ....
....*Psychopsis krameriana*
Plate 10(6)

Psygmorchis. Small fan-shaped epiphytes; without pseudobulbs; leaves laterally flattened; inflorescence subequal to leaves or longer, producing 1 flower at a time; flowers similar to *Oncidium*, anther triangular.

Key to *Psygmorchis*

    1. Margins of callus smooth; **C:** yellow with red-brown spots on petals and base of lip; **D:** No, CR, Pma, SA (Cb, Pc); **M:** pl 4–8 cm, lip 14 mm ....*Psygmorchis pusilla*
    1. Margins of callus deeply notched or fringed................... 2
  2(1). Lateral lobes of lip quadrangular, lip longer than wide; **C:** pure yellow without spots; **D:** No, CR, Pma (Mt); **M:** pl 2.3 cm, lip 10 mm
....*Psygmorchis pumilio*
Plate 14(4)
    2. Lateral lobes of lip rounded; lip about as wide as long; **C:** yellow with brown spots on bases of petals and lip; **D:** No, CR, Pma; **M:** pl 5 cm, lip 14 mm ....*Psygmorchis glossomystax*

Rodriguezia. Epiphytic; pseudobulbs loosely clustered, 1-leaved, with few leaf sheaths surrounding pseudobulb; inflorescence of few to many flowers; sepals and petals similar, lateral sepals united, saccate or spurred at base, lip with solid basal extension in spur; anther fleshy, dorsal.

Key to *Rodriguezia*

1. Inflorescence shorter than leaves, few-flowered; **C**: cream or pale yellow; **D**: No, CR, Pma (Cb); **M**: pl 7 cm, S 2.5–2.8 cm
.... *Rodriguezia compacta*
1. Inflorescence longer than leaves, many-flowered; **C**: rose or magenta; **D**: Pma, SA (Cb); **M**: pl 20 cm, S 14 mm..................
.... *Rodriguezia lanceolata*
(Syn. *R. secunda*)
Plate 13(2)

Rossioglossum. Large epiphyte; pseudobulbs clustered, somewhat flattened, each with 2 terminal leaves; inflorescence erect, with several large flowers; sepals and petals spreading, lip 3-lobed with fleshy callus between lateral lobes; column with 2 spreading, curved wings; 1 species in Costa Rica and Panama; **C**: yellow blotched and barred with red-brown; **D**: CR, Pma (Mt); **M**: pl 35 cm, S 3–4.5 cm ....
.... *Rossioglossum schlieperianum*
Plate 10(4)

Scelochilus. Small epiphyte, often pendent; pseudobulbs clustered but small and nearly hidden by sheaths, each with a single leaf; inflorescence with few flowers, flowers somewhat bell-shaped, with short chin or spur at base; **C**: yellow or orange-yellow; **D**: No, CR, Pma (Mt); **M**: pl 10 cm, S 10 mm .... *Scelochilus tuerckheimii*
Plate 12(6)

Sigmatostalix. Small epiphytes; pseudobulbs flattened, clustered, with terminal leaves and several sheathing leaves; inflorescence erect or arching, with many flowers, each flower produced from a cluster of bracts; sepals and petals similar, lip usually stalked, with open oil gland near base; column slender, usually arched.

Key to *Sigmatostalix*

1. Lip strongly stalked, stalk at least a third as long as blade; lateral lobes of lip narrow.. 2
1. Lip not stalked, or stalk less than one-fourth as long as blade; lateral lobes of lip wide and flat 4

2(1). Lateral lobes of lip short and pointed toward base of lip, stalk of lip shorter than blade, blade subquadrate-ovate; C: yellow to greenish yellow with distinct red or red-purple spots; D: No, CR, Pma, SA (Mt, Wt); M: pl 15 cm, S 5 mm ..... *Sigmatostalix guatemalensis*
    The name *S. picta* has also been used, but that appears to be a distinct species of South America.
2. Lateral lobes of lip narrow, rather like cow horns, as long as blade of lip, stalk of lip much longer than blade 3

3(2). Lateral lobes of lip tapering to narrow points; C: yellow to yellow-green, lip yellow; D: CR (Pc); M: pl 10 cm, S 5 mm .... *Sigmatostalix unguiculata*
3. Lateral lobes of lip narrow but obtuse, lip white................. .... New species, or form of *Sigmatostalix unguiculata*?

4(1). Inflorescence paniculate with distinct lateral branches; C: yellow or yellow-green, often tinged with red; D: CR, Pma, SA (Cb, Mt, Wt); M: pl 15 cm, S 3–5 mm .... *Sigmatostalix hymenantha*
4. Inflorescence appearing racemose, lateral branches reduced to clusters of bracts 5

5(4). Lip with 2 teeth projecting from near base of front of callus; C: S, P yellow or pale green with rose-purple bands, lip rose-purple; D: CR, wPma (Pc); M: pl 12 cm, S 3 mm.......... *Sigmatostalix brownii*
    Plate 14(1)
5. Lip without teeth projecting from callus 6

6(5). Lip blade with short claw; C: S yellow to pale green or marked with maroon spots, P and lip white or pale green; D: Pma, SA (Cb); M: pl 10 cm, S 3.2 mm .... *Sigmatostalix abortiva*
6. Lip without claw 7

7(6). Lip blade constricted near midsection; C: lip greenish cream marked with orange-brown; D: CR, Pma, SA (Cb, Mt, Wt); M: pl 15 cm, S 3 mm......................... *Sigmatostalix picturatissima*
    (Syn. *S. racemifera*)
7. Lip blade not constricted near middle; C: S, P greenish yellow, lip yellow; D: CR, wPma (Mt, Wt); M: pl 15 cm, S 2 mm.......... .... *Sigmatostalix macrobulbon*

Systeloglossum. Epiphytic; pseudobulbs clustered, oblong, strongly flattened, with single terminal leaf and leaf-bearing sheaths on each side, leaves relatively thin; inflorescence subequal to leaves or longer, producing 1 flower at a time over a long period, rachis condensed; flowers green or bronzy, with distinct column foot and chin, lip strongly united with column and column foot, lateral sepals united; column with prominent hood; stigma with 2 distinct lobes; stipe very broad.

Key to *Systeloglossum*

1. Lip deeply bilobed; C: green or bronzy green; D: CR, Pma (Mt); M: pl 20–25 cm, dS 11–15 mm, synS 1.5–1.8 cm .... *Systeloglossum panamense*
Plate 12(1)
1. Lip shallowly lobed or apiculate at apex 2
2(1). Column foot much shorter than column; sepals and petals obtuse or broadly acute; C: green or brownish green; D: CR, (Mt); M: pl 20–25 cm, S 9 mm .... *Systeloglossum costaricense*
2. Column foot about as long as column or longer; sepals and petals acute to acuminate; C: green or bronzy green; D: CR (Mt); M: pl 20–30 cm, dS 8.6–10 mm, synS 11–12.5 mm.................. .... ***Systeloglossum acuminatum***

Ticoglossum. Epiphytic; pseudobulbs clustered, each with 1 leaf; inflorescence basal, few-flowered; sepals and petals similar, elliptic to obovate, lip ovate to fan-shaped, notched; column short with small wings.

Key to *Ticoglossum*

1. Pseudobulbs strongly flattened, edges sharp; leaves narrowed basally; inflorescence few-flowered; C: white or pink, with yellow callus; D: No, CR, Pma (Mt); M: pl 25 cm, S 1.8 cm .... ***Ticoglossum krameri***
1. Pseudobulbs thick, edges rounded; leaves stalked; inflorescence with 1 or 2 flowers; C: white with yellow callus; D: CR, Pma (Mt); M: pl 6 cm, S 16 mm...................... ***Ticoglossum oerstedii***
Plate 10(1)

Trichocentrum. Epiphytic; pseudobulbs small; leaves elliptic, very fleshy (like miniature mule-ear oncidiums); flowers with short, blunt spur at base of lip, sepals and petals similar, column short, winged.

Key to *Trichocentrum*

1. Midlobe of lip distinctly ruffled, flowers relatively large, about 2 cm wide; with single unlobed, conic or cylindric spur 2
1. Lip not distinctly ruffled; flowers usually smaller; spur 2- or 4-lobed 3

2(1). Midlobe of lip much wider than length of claw (tip of column to base of midlobe); sepals and petals obtuse; column wings acute above; **C:** S, P cream with brown blotches, lip white with rose-purple blotch; **D:** CR, Pma (Mt); **M:** S 2 cm .... *Trichocentrum pfavii*
Plate 11(2)
2. Claw of lip much longer than width of midlobe; sepals and petals acute; column wings obtuse or squarish; **C:** S, P red-brown with yellow tips, lip white with 2 rose-purple spots near base; **D:** CR (Mt); **M:** S 11–12 mm .... *Trichocentrum* species

3(1). Anther cap smooth; spur always 4-lobed; **C:** S, P pale green or brownish green, lip cream spotted with rose; **D:** CR (Pacific slope SE of Cerro de la Muerte), Pma (Pc, Mt); **M:** S 11–12 mm .... *Trichocentrum caloceras*
3. Anther cap papillose or hairy 4

4(3). Lip constricted near middle; flowers small (5–6 mm); lip no longer than sepals; **D:** CR (Wt) .... *Trichocentrum brenesii*
This distinctive species has not been found in recent years.
4. Lip not constricted near middle; flowers > 10 mm; lip surpassing sepals 5

5(4). Anther covered with thick hairs; column wings projecting only slightly beyond anther; **C:** S, P pale green, lip white; **D:** CR, Pma, SA (Cb, Mt); **M:** S 11–13 mm .... *Trichocentrum capistratum*
(Syn. *T. panamense*)
5. Anther papillose; column wings projecting beyond tip of anther (for about length of anther) 6

6(5). Tips of column wings slightly fringed or toothed; lip rhombic when flattened, 1.5–2 cm wide; **C:** S, P pale green, lip white; **D:** No, CR (Pc, Mt, Cb); **M:** S 2–2.5 cm .... *Trichocentrum candidum*

6. Tips of column wings rounded; lip oblong when flattened, about 1 cm wide; C: S, P pale green or darker green, lip cream spotted with rose; D: CR (Mt); M: S 11–12 mm........*Trichocentrum* species
Very similar to *T. caloceras*, but as yet unnamed.

Trichopilia. Epiphytic; pseudobulbs clustered, somewhat to strongly flattened, from wider than long to long and slender, each with 1 terminal leaf, leaves wide and short-stalked, sheaths on young pseudobulbs spotted; inflorescence lateral, of few flowers; lip trumpet-shaped, enfolding column, column with fringed hood over anther.

Key to *Trichopilia*

1. Pseudobulbs squarish, normally about as wide as tall.......... 2
1. Pseudobulbs elongate, normally at least twice as long as wide ... 4
2(1). Lip longer than wide when spread out, margins irregular but not crisped-wavy; C: S, P pale yellow or greenish yellow, lip white; D: No, CR, Pma (Cb, Pc); M: pl 12 cm, S 3–4 cm.................
....*Trichopilia maculata*
2. Lip about as wide as long when spread out, margins markedly crisped and wavy 3
3(2). Lip 4.5–6.5 cm long, usually marked with rose-pink; C: S, P white or creamy, sometimes spotted with rose-pink or red, lip white or creamy, usually heavily spotted with rose-pink; D: CR, wPma (Mt, Wt); M: pl 35 cm, S 3–5.5 cm ....*Trichopilia suavis*
Plate 11(1)
3. Lip 3–3.5 cm long, white; C: white with pale yellow blotch in throat; D: Pma (Mt); M: pl 25 cm, S 2.5–3.5 cm.................
...................................*Trichopilia leucoxantha*
Known only from near El Valle de Antón, where it is now scarce.
4(1). Pseudobulbs elliptic or narrowly ovate; sepals and petals strongly twisted; C: S, P brownish red with white or pale green margins, lip cream, throat yellow dotted with red-brown; D: No, CR; M: pl 15 cm, S 5.5–6 cm......................*Trichopilia tortilis*
4. Pseudobulbs long and narrow with parallel sides............... 5

5(4). Lateral sepals free or nearly so; C: S, P greenish yellow, lip pale yellow or white with orange-yellow center, sometimes with red spots; D: No, CR, wPma (Mt, Wt); M: pl 20 cm, S 5 cm .... *Trichopilia marginata*
Quite variable; hybrids with *T. suavis* have been found near San Ramón, Costa Rica.

5. Lateral sepals united for more than 1 cm; C: S, P pale yellow, lip white with yellow throat; D: No, CR, Pma (Mt); M: pl 25 cm, S 4 cm .... *Trichopilia turialbae*

Trizeuxis. Small fan-shaped epiphyte; small pseudobulb nearly hidden by sheaths, with 1 terminal leaf and several sheathing leaves, all fleshy and laterally flattened; inflorescence branched, with many tiny flowers, flowers somewhat bell-shaped; 1 species; C: greenish or yellowish cream; D: No, CR, Pma, SA (Pc); M: pl to 10 cm, S 2 mm .... *Trizeuxis falcata*
Plate 14(5)

Warmingia. Small epiphyte; pseudobulbs clustered, conic to ovoid, 7–8 mm long, each with a single leaf; leaves leathery, lanceolate, 3.5–6 cm long (including petiole 6–10 mm long), 1–1.7 cm wide; inflorescence pendent, to 3 cm long, with 2–4 flowers; sepals lanceolate, slightly concave, keeled, acute, 7–8 mm long; petals rhombic, wider than sepals; lip about 8 mm long, 9 mm wide (when spread), margins crisped and irregularly toothed, 3-lobed, lateral lobes erect, about 3 mm long, 5 mm wide, midlobe ovate, about 3 mm long and wide; 1 species in Central America; C: ivory white; D: CR (Cb); M: pl 6 cm, S 7–8 mm .... **Warmingia margaritacea**

6 The Bizarre Subtribes: Catasetinae and Stanhopeinae

The 2 groups treated in this chapter are not close relatives, but each is too small to justify a chapter by itself, and each includes bizarre and interesting flower structures. In fact, these 2 groups share an important ecological factor: their members are all pollinated by euglossine bees (see Chapter 2). The subtribes Catasetinae and Stanhopeinae both have pleated leaves and 2 pollinia, but in other features they are quite different: they are easily distinguished, even without flowers (see Table 3, Figs. 4, 5).

Catasetinae

In other orchids the flowers are virtually always "perfect," or bisexual, with a functional anther and stigma in each flower. Most members of the Catasetinae, however, have 2 distinct types of flowers: "male" flowers with functional anthers, and "female" flowers with functional stigmas. Seemingly bisexual flowers occur occasionally in these same genera, but they cannot function either as males or females. Even in the few bisexual Catasetinae, the pollinaria are released when the right part of the column is touched. In *Clowesia* the sticky pad attached to the pollinia merely swings out a bit when the "trigger" is touched. In *Dressleria* the whole pollinarium swings out so that the sticky pad is attached to whatever is close to the column. In the genera with unisexual flowers, the stipe (the strap between pollinia and the sticky pad) is elastic and the whole structure is thrown out with considerable force. Orchid growers often invite unsuspecting visitors to put a finger into a male *Catasetum* flower.

The Bizarre Subtribes

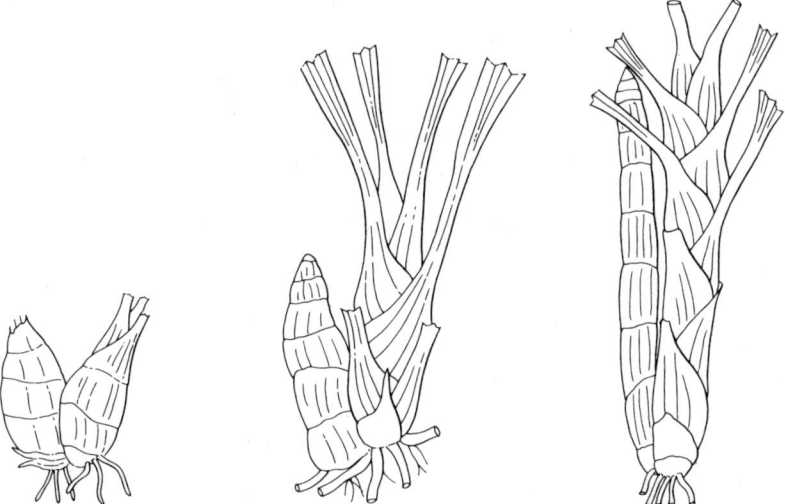

Figure 4. The pseudobulbs and leaf bases of Catasetinae.

Table 3. A comparison of the Catasetinae and Stanhopeinae

	Catasetinae	Stanhopeinae
Pseudobulbs	Several subequal internodes	1 internode or 1 major internode
Spines on pseudobulbs	Often	Never
Leaves	Several, lateral on upper pseudobulb	1–4, always terminal
Sheathing leaf bases	Always, whitish	Never
Deciduous in dry season	Often	Never
Narrow petioles	Never	Often
Flowers unisexual	Often	Never
Throwing pollinia	Usually	Never

Key to Genera of Catasetinae

1. Column and lip each twisted to 1 side; apex of male column held against lip until pollinia discharge............ .... ***Mormodes***
1. Column and lip not twisted; apex of column held away from lip 2

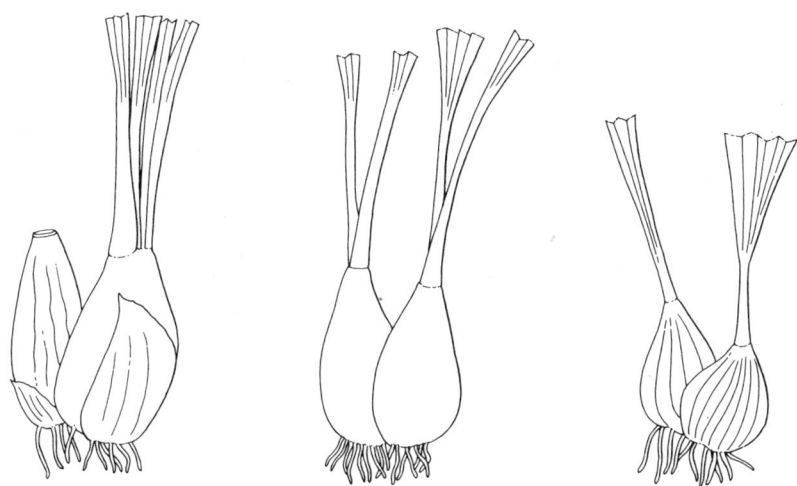

Figure 5. The pseudobulbs and leaf bases of Stanhopeinae.

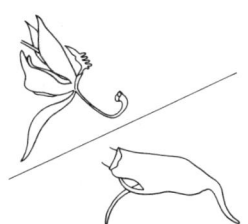

2(1). Flowers bisexual, functional pollinia and stigma in each flower; pollinia moved or thrown a short distance when triggered 3
2. Flowers normally unisexual, each with either functional pollinia or functional stigma; pollinia thrown farther than its own length; male and female flowers distinct in form 4

3(2). Flowers membranous; margins of lip fringed .... *Clowesia warscewiczii*
3. Flowers fleshy .... **Dressleria**

4(2). Male column obtuse; pseudobulbs narrow, not spiny; inflorescence from upper pseudobulb .... **Cycnoches**
4. Male column acute; pseudobulbs thicker, spiny (when leafless); inflorescences from lower part of pseudobulb .... **Catasetum**

Catasetum. Pseudobulbs thick, cigar-shaped, clustered, with several wide, pleated leaves in upper part; leaves deciduous, pseudobulbs **spiny** after leaves have dropped; inflorescence basal; flowers fleshy; column of male flowers pointed, with 2 slender antennae below anther. The female flowers are similar in all species of *Catasetum*, with the lip uppermost and hoodlike; they are much smaller in *C. bicolor* than in the other 2 species in our area.

Key to *Catasetum*

1. Lip of male flowers small, deeply 3–5 lobed; **C:** S, P greenish brown with maroon suffusion, lip pale brown, sometimes spotted with red; **D:** Pma, SA (Cb); **M:** ps 4–10 cm, S 2.5–4 cm .... ***Catasetum bicolor***
1. Lip of male flowers saclike, not lobed 2
2(1). Margins of male lip fringed; **C:** green more or less spotted with red-brown; **D:** No, CR, Pma, SA (Pc); **M:** ps 10–30 cm, S 3.5–4 cm ....***Catasetum maculatum***
(Syn. *C. oerstedii*)
Plate 15(1)
2. Margins of male lip not fringed............................ 3
3(2). Opening of male lip constricted near column, lip rounded beneath; **C:** green; **D:** No, rCR; **M:** ps 10–25 cm, S 3.2–5.4 cm........... .... ***Catasetum integerrimum***
3. Opening of male lip not constricted near column, lip with conic chin beneath near base; **C:** green or yellowish green; **D:** Pma (Cb, Pc); **M:** ps 10–30 cm, S 3.5–5 cm .... ***Catasetum viridiflavum***
The most common *Catasetum* in central Panama.

Clowesia. Epiphytic; pseudobulbs clustered, 3–9 cm long, more or less ovoid or conic, with several leaves, leaves pleated, deciduous, 12–40 cm long; inflorescence from base of bulb, pendent; flowers bisexual, lip lowermost, sepals and petals broadly ovate or obovate, 10–12 mm wide; lip deeply concave, 3-lobed, lateral lobes toothed or fringed, midlobe fringed; column somewhat hooded over anther, anther with short beak; pollinia not thrown when disturbed; 1 species in our area; **C:** cream or pale green with green veining; **D:** No, CR, Pma, SA (Cb); **M:** ps 3–9 cm, S 10–12 mm .... ***Clowesia warscewiczii***
Plate 15(2)

C. russelliana has been reported from our area and may occur here; it is much larger, with a deeply saccate lip that is not deeply 3-lobed or clearly fringed. The Mexican *C. rosea* has been reported from Panama, but its presence here is improbable.

Some people have problems spelling Warscewicz, and in this case the original spelling was *warczewitzii*. There are other versions in the literature, but I believe both they and the original may be considered misspellings.

Cycnoches. Epiphytic, usually on dead wood; pseudobulbs few, clustered, cigar-shaped to cylindrical, tapering above, with several pleated leaves along upper bulb, leaves deciduous; inflorescence from upper part of bulb, male inflorescence pendent; flowers with lip uppermost, female flowers always with unlobed, ovate or heart-shaped lip, with short column; male flowers variable, in many species with a lobed, more or less movable lip, always with a long, arched column.

Cycnoches aureum, *C. dianae*, and *C. warscewiczii* are all locally common, but other species appear to be very sporadic, and they are poorly represented in botanical collections. A distinctive species in the region of El Valle de Antón, Panama, fits none of the named species keyed out below. It has become rather scarce in that area.

Key to *Cycnoches*

1. Lip of male flowers not lobed or fringed, green; flowers very large, sepals and petals 5–7 cm long.................................. 2
1. Lip of male flowers distinctly lobed or fringed; flowers smaller, sepals and petals 2–4 cm long.................................. 3

2(1). Male lip with short, cylindrical stalk; **C:** S, P green, lip white with dark green callus; **D:** CR, wPma (Mt); **M:** ps 30–100 cm, S 4–7 cm .... **Cycnoches warscewiczii**
Plate 15(4)

The name *C. warscewiczii* may be based on a female flower of another species rather than the species for which the name is being used.

2. Male lip without distinct stalk; **C:** S, P green, lip white with dark green callus; **D:** ePma, SA (Mt); **M:** ps 30–80 cm, S 7–11 cm .... **Cycnoches chlorochilon**

3(1). Lip of male flowers fringed or lobed, but indentations shallow, less than one-fourth width of lip................................. 4
3. Lip of male flowers divided into finger-like appendages nearly one-third as long as width of lip................................. 5

4(3). Sepals and petals cream or yellowish green with pink specks, lip white; D: CR, Pma (Mt); M: ps 15–30 cm, S 3.5–4 cm .... *Cycnoches aureum*
4. Flowers rosy pink; D: wPma (Mt); M: ps 15–30 cm, S 2–2.6 cm .... *Cycnoches dianae*
Plate 15(3)
To be expected in adjacent Costa Rica.

5(3). Male inflorescence long, flowers well separated from their neighbors; C: green usually spotted with red-brown or dark red; No, CR (Cb); M: ps 15–25 cm, S 2.5–3 cm ..... *Cycnoches egertonianum*
5. Male inflorescence denser, flowers overlapping with their neighbors 6

6(5). Finger-like divisions of lip not or only slightly expanded at their tips; C: greenish, center of lip white; D: No, Pma (Cb, Mt); M: ps 10–15 cm, S 2–2.3 cm .... *Cycnoches stenodactylon*
6. Finger-like divisions of lip markedly broader at their tips 7

7(6). Finger-like divisions of lip with slender, narrow stalks much longer than their broad tips; C: green more or less spotted with dark red, lip white; D: No, Pma (Cb); M: ps 15–25 cm, S 2–2.3 cm........ .... *Cycnoches guttulatum*
7. Finger-like divisions of lip shorter, broad portion longer than slender base; C: green with white lip; D: No, Pma (Cb); M: ps 20–30 cm, S 2.5 cm .... *Cycnoches pachydactylon*

Dressleria. Pseudobulbs thick, cigar-shaped to narrowly ovoid, 8–12 cm long, with several wide, pleated leaves in upper part, leaves evergreen or deciduous, pseudobulbs without spines; inflorescence basal; flowers fleshy, bisexual, sepals and petals elliptic to ovate, lip ovate to obovate or subcircular, thick and fleshy, with deep cavity in front of anther; column short and thick, partly united with lip. All dresslerias are superficially very similar, and several species may yet be named.

Key to *Dressleria*

1. Flowers densely crowded, cluster nearly as wide as long; lip subcircular, cavity transversely oblong, much wider than long; **C:** S, P greenish cream, lip and column ivory, lip cavity orange with red dots within; **D:** No, CR (Mt); **M:** ps 10 cm, S 1.7 cm ....
.... *Dressleria dilecta*
Plate 15(5)
1. Flowers not densely crowded, cluster much longer than wide; lip cavity about as long as wide 2

2(1). Lip cavity subcircular; flowers very loosely scattered on rachis, internodes of rachis longer than pedicels; **C:** S, P greenish cream, lip and column ivory, cavity with red spots within; **D:** Pma (Mt)
.... *Dressleria* species
This plant, found near El Valle de Antón, fits none of the named species.
2. Lip cavity bilobed; pedicels longer than internodes of rachis 3

3(2). Lip united to full length of column; lip cavity about half width of lip, with thin flap of tissue along each side; **C:** flowers ivory, lip cavity orange with red spots within; **D:** No, rPma (Mt); **M:** ps 8–12 cm, S 2 cm................................. *Dressleria helleri*
This species has been reported from Panama, but the plants of Cerro Jefe are probably another unnamed species.
3. Lip united to basal half of column; cavity nearly as wide as lip, sides of cavity thickened; **C:** S, P greenish cream, lip and column ivory, cavity orange with red spots within, few red spots on blade of lip; **D:** No, CR, Pma (Mt); **M:** ps 10 cm, S 2.2 cm .. *Dressleria suavis*

Mormodes. Epiphytic, usually on rotten wood; pseudobulbs oblong, tapering above, never spiny, with several pleated leaves; inflorescences from base or sides of pseudobulb, erect or ascending; flowers fleshy; lip lobed or entire, concave or sides turned under; column twisted, pointed, with finger-like appendage at tip, this borne against surface of lip until pollinia are discharged. *Mormodes* flowers are quite variable, and most species have both smaller flowers that seem to function as male flowers ("staminoid") and larger flowers that seem to function as females ("pistilloid"). The lips of the female flowers are never hairy, though the male flowers are in some species. There are 3 easily distinguished species of *Mormodes* in central Panama. Several other species occur in western Panama and Costa Rica, where there appears to be some natural hybridization, and the variation is bewildering. *Mormodes* is not easy to classify.

Key to *Mormodes*

1. Lip of staminoid (male) flowers hairy........................ 2
1. Lip smooth, without long hairs of any sort.................... 3

2(1). Blade of lip abruptly widened from narrow stalk, margins turned down, widely separated near base and overlapping near apex; C: S, P dull yellow striped with purplish red, lip pale yellow with red spots; D: CR (Cb) .... ***Mormodes skinneri***
Plate 15(6)
2. Blade of lip gradually widening from narrow stalk, margins turned down, parallel or closer near base than near apex; C: purplish maroon; D: wPma (Mt) .... ***Mormodes hookeri***

3(1). Lip distinctly 3-lobed; C: heavily spotted with red-brown; D: CR, Pma (Pc, Mt)............................. ***Mormodes lobulata***
This has been called *M. atropurpurea*, a South American species.
3. Lip not or only faintly 3-lobed 4

4(3). Blade of lip widest in middle or near base.................... 5
4. Blade of lip widest above middle............................ 8

5(4). Blade of lip squarish, narrowing abruptly to apex; margins curled under and irregular (not straight); C: S, P pale yellow more or less dotted with red-brown, lip reddish or brownish yellow with dark red spots; D: Pma (Mt).................... ***Mormodes punctata***
The name *M. cartonii* has been used for this species but belongs to a South American species.
5. Lip tapering gradually to apex............................... 6

6(5). Lip triangular-ovate, base of blade folded down but not curled, so that lip appears triangular in side view; C: S, P, and lip greenish

yellow to brown with darker veins; D: CR, wPma (Cb, Mt) . . . .
. . . . *Mormodes colossa*
6. Lip ovate, sides curled downward, so that lip appears very narrow
. 7

7(6). Lip 4–6 cm long, base narrow; C: S, P greenish yellow to brown
with darker veins, lip brownish yellow to red-brown; D: Pma (Cb)
. . . . *Mormodes powellii*
7. Lip about 3 cm long, base rounded; C: clear yellow or lip ivory: D:
Pma . . . . *Mormodes lancilabris*
Known only from the market in El Valle de Antón; either
very rare or brought in from far away.

8(4). Apex of lip not curled upward, sides curled downward, lip fleshy,
margins evenly parallel beneath, lip either rounded or truncate; C:
S, P brown or yellow, lip red-brown to dark red; D: Pma (Mt)
. . . .*Mormodes fractiflexa*
The name *M. ignea* has been used for this species but belongs to a South American species.
8. Apex of lip curled upward and back, thus concave 9

9(8). Lip about half as wide as long; C: yellow; D: CR . . . .
. . . . *Mormodes flavida*
9. Lip more than half as wide as long; C: dark blood red; D: CR (Cb)
. . . . *Mormodes horichii*

Stanhopeinae

The Stanhopeinae, like the Catasetinae, are pollinated by euglossine bees. The pollination of *Acineta* is relatively simple. The bee crawls into the flower; when it crawls out again, it may receive pollinia from the flower, or, if it is already carrying pollinia, one of the pollinia may be left in the stigma. In most Stanhopeinae, however, the flower structure is unusual, if not bizarre, and pollination involves falling through the flower or into the lip, sliding down the column, or other unusual mechanisms. The adaptation to euglossine pollination is evident in the strong perfume, the bizarre flower structure, and the relatively short life of the individual flower.

The Bizarre Subtribes

Key to Genera of Stanhopeinae

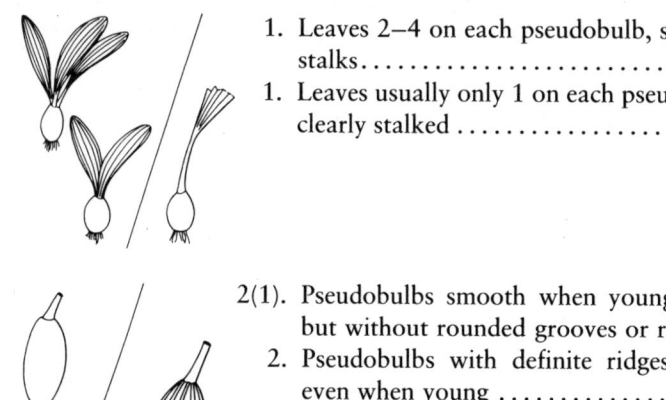

1. Leaves 2–4 on each pseudobulb, short-stalked or without stalks.. 2
1. Leaves usually only 1 on each pseudobulb (rarely 2), often clearly stalked 9

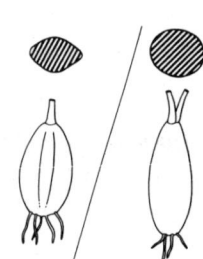

2(1). Pseudobulbs smooth when young, shriveled when older but without rounded grooves or ridges............... 3
2. Pseudobulbs with definite ridges and rounded grooves even when young 6

3(2). Pseudobulbs somewhat flattened, 4-angled; small epiphytic plants, 15–30 cm tall....................... 4
3. Pseudobulbs rounded, never 4-angled; large, often terrestrial plants, usually > 50 cm tall.................... 5

4(3). Flowers small, < 3 cm long; inflorescence with prominent black, bristly hairs; leaves purplish .... **Kegeliella**
4. Flowers much larger, > 5 cm long; hairs on inflorescence inconspicuous; leaves green .... **Paphinia**

5(3). Midlobe of lip hinged and movable; flowers globose..... .... **Peristeria**
5. Midlobe of lip not hinged or movable; flowers rather trumpet-shaped .... **Coeliopsis hyacinthosma**

Key to Genera of Stanhopeinae

6(2). Column bent upward near apex, with 2 glands at base; lip very complex, forming a fleshy cup next to column apex; pollinia thick, broadly ovoid, bent toward viscidium .... *Coryanthes*
6. Column straight, without basal glands; lip 3-lobed, not forming a cup; pollinia flat, more or less obovoid, straight 7

7(6). Flowers massive and fleshy, resupinate, lip lowermost .... *Acineta*
7. Flowers hanging downward, or lip uppermost 8

8(7). Flowers hanging downward; lip clearly 3-lobed, midlobe stalked. .... *Lacaena*
8. Flowers turned back toward rachis or upward; lip complicated, usually not clearly 3-lobed .... *Gongora*

9(1). Pseudobulbs with definite ridges and grooves 10
9. Pseudobulbs smooth, rounded or weakly 4-angled ... 11

10(9). Flower hanging downward; lip fleshy, base concave, midportion often with large horns or finger-like appendages .... *Stanhopea*
10. Flowers horizontal or nearly so, with lip uppermost; lip not fleshy or deeply lobed, never with horns .... *Sievekingia*

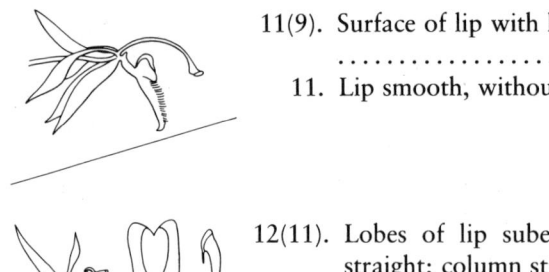

11(9). Surface of lip with hairs; column slender, arched ....
.... *Polycycnis*
11. Lip smooth, without hairs 12

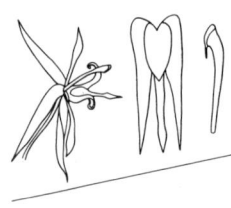

12(11). Lobes of lip subequal and similar; inflorescence erect, straight; column straight .... **Horichia dressleri**
12. Lobes of lip dissimilar, without above combination of features. 13

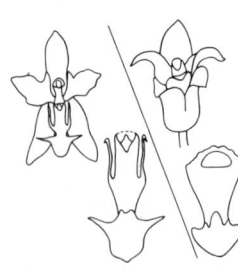

13(12). Lateral lobes of lip slender, arched, acuminate; flowers very large .... **Houlletia**
13. Lateral lobes short and broad; column short, without wings; flowers white or greenish 14

14(13). Lateral sepals free or only basally united ....
.... **Trevoria**
14. Lateral sepals completely united, forming a narrow, saclike or helmet-like structure .... **Schlimia jasminodora**

Acineta. Epiphytic; pseudobulbs large, clustered, ovoid, grooved even when young, 8–10 cm long, each with several pleated leaves without definite stalks; inflorescence lateral, pendent, of several to many fleshy flowers, flowers somewhat cupped; column thick, straight; lip stalked, deeply 3-lobed. The large, fleshy flowers of *Acineta* are rarely preserved for study, and this difficult genus is not well understood.

Key to *Acineta*

1. Lower profile of lip not indented, callus with suberect, finger-like appendages; **C**: orange-yellow with red spots; **D**: Pma, SA (Mt)
.... **Acineta superba**
1. Lower profile of lip markedly indented beneath lateral lobes, appendages of callus, if present, subparallel with column 2

2(1). Callus of lip simple, without prominent finger-like appendages; C: S, P yellow or greenish yellow speckled or spotted with red, lip orange marked with red; D: No, CR, wPma (Mt); M: pl 50 cm, S 4–4.5 cm .... *Acineta chrysantha*
Plate 16(1)
2. Callus of lip with short, finger-like appendages; D: rCR, rPma, SA .... *Acineta sella-turcica*

Coeliopsis. Epiphytic; pseudobulbs clustered, conic-ovoid, smooth when young, 6.5–8.5 cm long, each with 2 or 3 pleated leaves without definite stalks; inflorescence lateral, more or less pendent, short, with several or many crowded flowers, flowers somewhat bell-shaped, with parts spreading distally; 1 species recognized; C: white, lip marked with orange, yellow and violet within; D: CR, Pma, SA (Cb, Mt); M: pl 50 cm, S 1.8–2.2 cm .... *Coeliopsis hyacinthosma*
Plate 16(2)

A much larger plant found near the Caribbean coast of Panama may be the plant originally described as *C. hyacinthosma*; if so, the well-known plants of cloud forests in Costa Rica and Panama may be a distinct but unnamed species.

Coryanthes. Epiphytic; pseudobulbs clustered, conic or narrowly conic, with angled ridges, leaves 2 or 3, terminal on pseudobulb; inflorescence pendent, flowers hanging downward; column thick and fleshy with nipple-like glands near base; lip stalked, with hood at base of blade, blade cupped, with a small opening between column and lip.

The plants of *Coryanthes* normally grow on ant nests, where they have an unusually acidic medium and their roots and foliage are protected by the ants. These plants are hard to find and harder to grow. There seem to be at least 2 distinct species in central Panama, but whether this key separates distinct species and the names in use are correct is very hard to say.

Key to *Coryanthes*

1. Hood at base of lip less than one-third length of lip (top of hood to bottom of "bucket"); C: S, P buff, lip yellow or flushed with rose; D: No, CR, Pma, SA .... *Coryanthes speciosa*
1. Hood at base of lip much more than one-third length of lip 2

146   The Bizarre Subtribes

> 2(1). Back of hood markedly concave in outline; C: buff strongly spotted and stained with red-purple; D: Pma, SA (Cb, Pc) .... *Coryanthes maculata*
> Plate 17(3)
> 2. Back of hood flat or convex, not concave 3
> 3(2). Back of hood rounded, not flat; C: greenish cream heavily marked with red-brown; D: CR (Cb) .... *Coryanthes horichiana*
> 3. Back of hood more or less flat; C: cream or pale yellow, may have some red spots; D: CR, Pma (Cb) .... *Coryanthes hunteriana*
> C. *powellii* may be the same species as C. *hunteriana*. Both names were published at the same time and based on material from central Panama.

Gongora. Epiphytic; pseudobulbs clustered, ovoid, with definite grooves and rounded or rather sharp ridges, leaves 2 or 3, terminal on pseudobulb; inflorescence pendent, flowers borne with lip uppermost, lip either 3-lobed or rather conic in shape, usually with various horns, bristles, and lobules. All gongoras have medium-sized flowers, with the sepals 1.2–1.8 cm long, and the lip 0.8–2 cm.

Key to *Gongora*

> 1. Lip deeply 3-lobed, all lobes more or less flat; C: S, P green heavily spotted with red, lip orange-yellow; D: CR (Cb) .... *Gongora amparoana*
> 1. Lip more or less conical, not clearly 3-lobed 2
> 2(1). Lip with erect (toward column) bristles near middle (apex of hypochile), often with prominent lateral horns or knobs near base, apex of lip acute but not long-tailed 3
> 2. Lip without bristles near middle, without lateral horns or knobs near base, apex long-tailed 11
> 3(2). Terminal part of lip fleshy, not cupped 4
> 3. Terminal part of lip more or less cupped 10
> 4(3). Lateral horns at base of lip broad and rounded or lacking 5
> 4. Lateral horns at base of lip slender or flattened and turned upward 6
> 5(4). With rounded lateral knobs at base of lip; odor resinous; C: yellow usually more or less heavily spotted and blotched with red-brown,

rarely nearly solid red-brown; D: Pma (Cb) ......*Gongora tricolor*
Plate 17(2)

One author has suggested that *G. fulva* is the correct name for this species, but the original illustration (of a plant of unknown origin) is not convincing. In a group as difficult as *Gongora*, names based on plants of unknown origin should remain in limbo.

5. Base of lip without lateral horns or knobs, odor of clove oil; C: red-purple; D: No, CR (Pc, Mt)................*Gongora claviodora*

6(4). Ventral surface of hypochile convex; C: pale flesh more or less heavily spotted with red-brown; D: Pma (Cb, Mt) .... *Gongora gibba*

6. Ventral surface of hypochile flat or concave.................. 7

7(6). Petals attached to column to about middle, free portion short; horns very slender; C: red-purple, or pale yellow-green heavily spotted with red-purple; D: ePma, SA (Mt)........ *Gongora atropurpurea*

7. Petals attached well above middle of column, free portion longer; horns thick and fleshy...................................... 8

8(7). Sepals more or less heavily spotted with red-brown; C: greenish, yellow, or tan, more or less heavily spotted with red-brown, lip may be partly white or yellow; D: No, CR, Pma, SA (Cb, Mt, Pc) ....*Gongora quinquenervis*

This catchall name is used for any *Gongora* that doesn't fit clearly in another species. The species was originally named from a Peruvian plant (a very poor specimen). Careful work is still needed to sort out the species in this complex.

8. Flowers without spots 9

9(8). Flowers flesh-colored (pinkish), with odor of fresh-ground corn for tortillas (*nixtamal*); D: No, CR, Pma (Cb) ......*Gongora unicolor*

9. Flowers red-purple, odor of clove oil; D: No, CR (Pc, Mt)........ ....*Gongora claviodora*

10(3). Lip > 15 mm long; C: cream or yellowish, S and column spotted with red-brown; D: No, rCR ....*Gongora truncata*

10. Lip < 12 mm long; C: S, P yellow, lip yellow and cream, all with red-brown spots; D: ePma, SA (Mt) .... *Gongora charontis*

11(2). Petals reaching near apex of column; C: S green spotted and barred with red, P purple, lip yellow; D: CR, Pma (Cb, Mt) ....*Gongora horichiana*
(Syn. *G. armeniaca* var. *bicornuta*)
Plate 17(1)

11. Petals reaching about middle of column; C: yellow, orange, or salmon; D: No, CR, wPma (Cb, Mt) ....*Gongora armeniaca*
Plants from the Pacific slope are mostly the yellow or brownish yellow subsp. *armeniaca*, while plants from the Caribbean slope of Costa Rica are subsp. *cornuta*, in which the lip is more sharply angled above at the base, the sepals and petals are orange-yellow faintly spotted with red, and the lip is orange-yellow.

Horichia. Epiphytic; pseudobulbs clustered, ovoid, smooth when young, each with a single leaf; inflorescence erect, straight; flowers with lip lowermost, sepals lanceolate, petals narrowly oblanceolate, lip deeply 3-lobed into narrow, subequal parts, lateral lobes recurved, column straight; 1 species known; C: S, P red-brown, lip yellow; D: Pma (Cb, Mt); M: pl 30–35 cm, S 2.5–2.7 cm............... *Horichia dressleri*
Plate 18(2)

Houlletia. Terrestrial or epiphytic, large; pseudobulb smooth, conic-ovoid, each with a single, stalked leaf, leaves elliptic, pleated; flowers large, hanging downward, lip deeply 3-lobed, lateral lobes slender, curved toward apex of lip, midlobe widest near base.

Key to *Houlletia*

1. Plants terrestrial; inflorescence erect, several-flowered; petals not lobed; C: S, P mahogany barred with greenish cream within near base, lip white, lateral lobes wine red without and apically; D: Pma, SA (Mt); M: pl 40–50 cm, S 3.5 cm ...... *Houlletia odoratissima*
1. Plants epiphytic, inflorescence short, pendent, with 2 or 3 flowers; C: S, P yellow-green densely spotted with red, lip white with violet-purple spots and bars; D: No, CR, Pma, SA (Mt); M: pl 40 cm, S 5 cm... *Houlletia tigrina*
(Syn. *H. landsbergii*)
Plate 16(3)
This species is very rare in accessible areas but still frequent in parts of eastern Panama.

Kegeliella. Epiphytic; plants small or medium, flushed with purple; pseudobulbs smooth, slightly flattened or 4-angled, each with 2 or 3 broad, pleated leaves; inflorescence pendent with dark, bristly hairs; flowers with lip lowermost, lip 3-lobed, with prominent callus, column broadly winged.

Key to *Kegeliella*

1. Base of lip spreading abruptly, without narrow basal stalk; lip and column parallel, midlobe of lip much shorter than lateral lobes; column wings abruptly expanded in upper three-fifths of column, widest in middle of hood; **C**: column green, S, P green-tan more or less spotted and barred with red-brown, lip yellow; **D**: CR, Pma (Cb, Mt); **M**: pl 10–20 cm, S 1.8 cm.......... *Kegeliella kupperi*
Plate 18(5)
1. Base of lip expanding gradually, with narrow basal stalk; lip diverging from column, midlobe subequal to lateral lobes; column wings gradually expanded from near base, hood widest distally; **C**: column green, S, P tan-yellow with or without red-brown spots and bars, lip greenish yellow; **D**: No, CR, Pma (Cb, Mt); **M**: pl 10–20 cm, S 1.7 cm .... *Kegeliella atropilosa*
(Syn. *K. houtteana*, misapplied)

Lacaena. Epiphytic, large; pseudobulbs ovoid, ridged, each with 2–4 large, pleated leaves; inflorescence pendent, flowers hanging downward, lip deeply 3-lobed.

Key to *Lacaena*

1. Stalk of midlobe about as wide as long; **C**: pale green with dark purple on lip; **D**: No, CR; **M**: pl 70 cm, S 2–3 cm............... .... *Lacaena bicolor*
1. Stalk of midlobe long and slender; **C**: cream spotted with red, lip densely spotted; **D**: No, CR, wPma (Cb, Mt); **M**: pl 40 cm, S 2.3–3 cm .... *Lacaena spectabilis*

Paphinia. Plants small or medium; pseudobulbs somewhat flattened, smooth, each with 2 or 3 wide, pleated leaves; inflorescence pendent; flowers with lip uppermost, sepals and petals long and narrow, acute or acuminate, lip 3-lobed, with fleshy, hairlike appendages; column broadly winged above.

Key to *Paphinia*

1. Sepals 5–7 cm long; flower opening widely; C: S, P greenish white with maroon veins and bars, lip maroon-red; D: Pma, SA (Cb, Mt) .... ***Paphinia cristata*?**
Plate 16(4)
    The Panamanian plant grows on tree trunks in very wet (and remote) areas and may be an undescribed species.
1. Sepals 2.5–3.5 cm long; flower opening only slightly; C: white or greenish white; D: CR (Cb) .... ***Paphinia* "clausula"**
    The name *P. clausula* is based on *P. cristata* var. *modiglianiana*, which proves to be a different, South American plant, so the Costa Rican plant does not have a valid name.

Peristeria. Plants large; pseudobulbs ovoid, smooth, each with 2–4 large, pleated leaves; flowers with lip lowermost, sepals and petals cupped, midlobe of lip hinged and movable; column with or without wings.

Key to *Peristeria*

1. Plant epiphytic; inflorescence pendent, all flowers open at once; C: whitish or pale tan densely spotted with red-brown; D: Pma (Mt); M: pl 60–90 cm, S 2.2–2.5 cm................ ***Peristeria* species**
Plate 16(6)
    This species is locally common in the cloud forests above El Valle de Antón and ranges westward to Chiriquí. It may be the same as *P. cochlearis*, described from the Chocó of Colombia. If not, it is probably an unnamed species.
1. Plant terrestrial; inflorescence erect, flowering over a long period; C: white or with some red speckling; D: CR, Pma, SA (Cb, Pc, Mt); M: pl 1 m, S 2.5–3 cm........................ ***Peristeria elata***
Plate 16(5)
    This, the famous *flor de Espíritu Santo*, dove orchid, or Holy Ghost orchid, is the national flower of Panama. The

plants are much sought by gardeners, who usually do not know how to care for them. These plants grow in open, rocky areas and on steep, grassy slopes. Such habitats are now more common than they once were, and the dove orchids would be rather common if people would just leave them there.

Polycycnis. Epiphytic; pseudobulbs ovoid, somewhat flattened and 4-angled, smooth, each with a single stalked leaf (in our species); inflorescence pendent or arching, many-flowered; lip simple or 3-lobed, hairy; column slender, arched, winged only near apex.

Key to *Polycycnis*

1. Midlobe of lip very narrow, 4–6 times as long as wide 2
1. Midlobe ovate or 3-lobed, up to twice as long as wide 3

2(1). Midlobe elliptic-lanceolate, not widened at base; lateral lobes oblong, rounded; C: S pale green, P and lip pink; D: ePma, SA (Mt); M: lip 11 mm .... *Polycycnis ornata*
   Plate 18(3)
2. Midlobe widened near base, lateral lobes triangular; C: S, P yellowish green spotted with chestnut, or pure yellow; D: Pma (Mt); M: lip 15–16 mm .... *Polycycnis tortuosa*

3(1). Midlobe clearly 3-lobed, wider than long, lateral lobes clearly longer than stalk of lip; C: S, P pale green, lip cream, all spotted with red-brown; D: ePma, SA (Mt) .... *Polycycnis lehmannii*
3. Midlobe subentire or slightly 3-lobed, longer than wide; lateral lobes subequal to stalk of lip 4

4(3). Stalk of lip slender, about 3 times as long as wide; lip about 2.5 cm long; C: pink with dark spots; D: CR, wPma (Mt) ....
   .... *Polycycnis barbata*
   Plate 18(4)
4. Stalk of lip subquadrate, at least half as wide as long; lip up to about 2 cm long 5

5(4). Lip about 2 cm long, lateral lobes oblong and rounded, base of lip somewhat hairy; C: pale yellow with red-brown spots; D: CR, Pma (Mt) .... *Polycycnis gratiosa*
5. Lip about 1.5 cm long, lateral lobes triangular; base of lip without long hairs; C: yellow with pale brown spots; D: CR, Pma, SA (Mt)
   .... *Polycycnis muscifera*

Schlimia. Epiphytic; pseudobulbs smooth, narrowly conic, not flattened, each with 1 broad, stalked leaf; inflorescence pendent; lateral sepals completely united, concealing the 3-lobed lip; column small and straight, wingless, or wings very small; **C:** greenish white; **D:** rCR, SA; **M:** pl to 35 cm, S 2 cm .... *Schlimia jasminodora*

A single plant was reported from Costa Rica in 1992, otherwise known only from Colombia.

Sievekingia. Small epiphytes; pseudobulbs clustered, ovoid, strongly ridged, each with a single more or less stalked leaf; inflorescences short, pendent, flowers borne with lip uppermost; lip concave, simple or 3-lobed; column broadly winged. The plants of *Sievekingia* look much like miniature stanhopeas. There is at least 1 unnamed species of *Sievekingia* on the Caribbean coast of Panama; its pollinia have been found on bees captured in the area, but the rest of the plant is unknown to science.

Key to *Sievekingia*

1. Petals and lip toothed or short-fringed; **C:** pale yellow, lip orange in center and marked with red spots; **D:** CR, Pma (Mt) .... *Sievekingia fimbriata*
Plate 18(1)
1. Petals and lip not toothed or fringed 2
2(1). Lip with many keels or teeth basally; flowers 2.5–3 cm long; **C:** cream, lip basally orange and with violet spots up to callus; **D:** Pma (Mt); **M:** pl to 25 cm, S 2.5–3 cm .... *Sievekingia butcheri*
   A rare species known only from the area of El Valle de Antón.
2. Lip with 1–3 basal keels, 8–11 mm long; **C:** yellow with red spots in lip; **D:** CR, Pma (Cb); **M:** pl 12–15 cm, S 12–14 mm .... *Sievekingia suavis*

Stanhopea. Epiphytic; pseudobulbs subglobose or conic-ovoid, ridged, each with a single stalked leaf; inflorescence pendent, flowers hanging downward; column broadly winged; sepals ovate or elliptic, spreading, petals narrower, usually recurved, lip fleshy, from faintly lobed to strongly 3-lobed and complex.

The name *torito* (little bull) is generally used for the stanhopeas because the prominent horns on the lip are reminiscent of bull horns. There is at least 1 undescribed *Stanhopea* growing in the Rio Indio area north of El Valle de Antón. It has broad, flat horns like the South American *S. platyceras* but is not that species. Plate 17(4–6).

Key to *Stanhopea*

1. Lip clearly of 3 parts: hypochile, mesochile, and epichile, with long, curved lateral horns; usually 3–7 flowers per inflorescence 2
1. Epichile of lip not jointed at attachment to hypochile, lip obscurely divided into 2 parts, without horns or these short and blunt; inflorescences 2-flowered... 7
2(1). Hypochile long and slender in lateral view, more than twice as long as broad 3
2. Hypochile quadrate or globose in lateral view, less than twice as long as broad 4
3(2). Hypochile strongly bent, distinctly L-shaped in profile, ventral plates flared, extending beyond edges of lateral plates in dorsal view; **C:** S, P cream with red spots, lip basally orange-yellow with purple "eyes," rest of lip cream flushed and spotted with red-purple; **D:** No, CR, Pma, SA (Mt); **M:** lip 4.5–6.5 cm .... *Stanhopea oculata*
3. Hypochile only slightly bent in profile; ventral plates of hypochile not flared; **C:** creamy white, bases of S, P, and lip yellow, S, P sparsely spotted with red-purple; **D:** Pma (Mt); **M:** lip 5.5–7.5 cm .... *Stanhopea panamensis*
4(2). Hypochile globose in profile; **C:** greenish cream, horns and epichile white, base of lip may have dark purple eyes, column and epichile may be speckled with red; **D:** CR (Cb) .... *Stanhopea warscewicziana*
4. Hypochile squarish in profile 5
5(4). Underside of hypochile flattened, not at all bilobed; **C:** S, P yellow or greenish yellow speckled with red dots, hypochile yellow with 2 dark purple eyes, horns cream, epichile cream with red dots; **D:** No, CR, Pma, SA; **M:** lip 5–6 cm................. *Stanhopea wardii*
Plate 17(6)
5. Hypochile conspicuously bilobed on underside................ 6
6(5). Underside of hypochile deeply bilobed; **C:** S, P yellowish cream or yellow with purple spots, P marked with **rings**, lip cream flushed,

spotted, and blotched with purple; **D:** CR, Pma (Cb, Mt); **M:** lip 5 cm .... *Stanhopea costaricensis*
Plate 17(5)
6. Underside of hypochile shallowly bilobed; **C:** cream or yellowish cream marked with **solid** red-purple spots, lip with or without purple eyes; **D:** No, CR (Cb, Mt); **M:** lip 5 cm .... *Stanhopea gibbosa*

7(1). Hypochile with fleshy lateral projection (horn) on each side; **C:** yellow with purple markings on hypochile; **D:** CR, wPma (Pc); **M:** lip 3.6 cm...................................*Stanhopea cirrhata*
Plate 17(4)
7. Hypochile without fleshy lateral projections 8

8(7). Hypochile with thickened ring ventrally and laterally; **C:** cream, base of lip yellow; **D:** ePma, SA (Mt); **M:** lip 4.5 cm ....
.... *Stanhopea avicula*
8. Hypochile without thickened ring........................... 9

9(8). Fleshy plate in midportion of lip separated from fleshy epichile by a transverse groove; **C:** S, P cream, sepals lightly spotted with red, base of lip orange-yellow spotted with red; **D:** No, CR, wPma (Cb); **M:** lip 4 cm .... *Stanhopea ecornuta*
This is sometimes called the *torito sin cacho* (hornless bull) because it is obviously one of the toritos, even though it lacks the characteristic horns.
9. Fleshy plate in midportion of lip joined to fleshy epichile; **C:** cream or pale yellow, with purple markings in lip; **D:** CR, Pma, SA (Cb); **M:** lip 2.5–3 cm................................ *Stanhopea pulla*

Trevoria. Epiphytic or growing in moss; pseudobulbs smooth, ovoid to narrowly ovoid, not flattened, each with 1 broad, stalked leaf; inflorescence pendent; lateral sepals more or less united at base, lip 3-lobed, narrow, concave; column short and broad. In theory, there are 2 distinct species in Central America, but A. H. Heller's sketches of Nicaraguan material show bewildering variation rather than 2 clear species. Typical *T. zahlbruckneriana* has wider sepals and its aspect is very similar to *Schlimia*. Some plants have narrower sepals and less clearly cupped flowers and may be *T. glumacea*. Careful study may show a single variable species.

Key to *Trevoria*

1. Flowers loosely clustered; lateral sepals 6–7 mm wide, midlobe acuminate; column with ring about middle; **C:** greenish white or greenish yellow; **D:** No, CR, Pma (Mt, Wt); **M:** pl to 40 cm, S 1.5 cm
.... *Trevoria glumacea*
Plate 18(6)
1. Flowers densely clustered; lateral sepals 12–13 mm wide; midlobe obtuse or subacute; column with ring near base; **C:** white or greenish white; **D:** CR, SA (Mt); **M:** pl to 25 cm, S 1.8 cm ....
.... *Trevoria zahlbruckneriana*

7 Mostly Miniatures: Subtribe Pleurothallidinae

The Pleurothallidinae is the largest subtribe in the orchid family, with at least 4000 species. They are mostly small, and some are tiny plants with minuscule flowers. As a group they are easily recognized. They never have pseudobulbs and there is only 1 leaf on each stem (in Central America). Note, though, that some species of *Myoxanthus* and *Trichosalpinx* may have chains of several stems, one on top of the other, just as *Scaphyglottis* does. One of these chains looks superficially like a single stem with several leaves, but there are several sheaths between each pair of adjacent leaves. The inflorescence is usually produced near the tip of the stem—that is, near the base of the leaf. An unusual feature of the Pleurothallidinae is that the ovary separates from the pedicel, which remains on the inflorescence. In *Lepanthes*, for example, the persistent pedicels give older inflorescences a characteristic fish-bone pattern. The number of pollinia is considered important in determining the genera, but this is hard to see in the field, even in the largest pleurothallids. Except for *Pleurothallis*, the genera are fairly easy to recognize. If you have trouble with the key, review the photographs of Pleurothallidinae to see if one has the same aspect as the plant to be identified.

Key to Genera of Pleurothallidinae

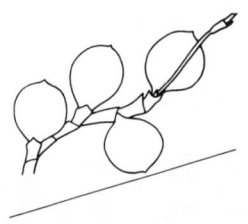

1. Leaves coinlike, nearly as wide as long and flat on bark 2
1. Leaves not coinlike, not flat on bark 3

Key to Genera of Pleurothallidinae

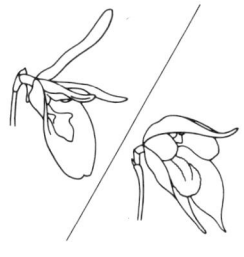

2(1). Dorsal sepal and petals narrow and finger-like, thickened apically . . . . **Barbosella orbicularis**
2. Sepals and petals all broad and rounded or tapering, not finger-like . . . . **Pleurothallis**

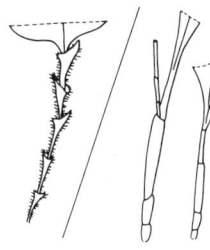

3(1). Sheaths of stem lepanthiform (see *Lepanthes*), ribbed and somewhat funnel-shaped, with ribs and margins bristly 4
3. Sheaths of stem not as above 6

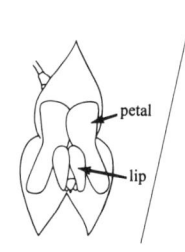

4(3). Column cylindrical, without foot; petals often bilobed; lateral lobes of lip usually clasping column . . . **Lepanthes**
4. Column short and broad with short foot, or longer with well-developed foot; petals not bilobed; lateral lobes of lip not clasping column 5

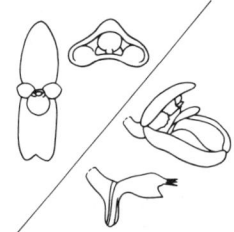

5(4). Column short; stigma apical and bilobed . . . **Lepanthopsis**
5. Column short or long; stigma ventral, not bilobed . . . . **Trichosalpinx**

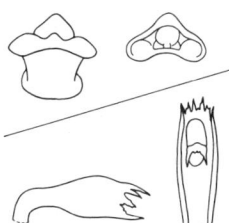

6(3). Column short and broad; stigma apical and more or less 2-lobed 7
6. Column elongate or cylindrical; stigma usually ventral, not 2-lobed 10

7(6). Petals and lip subequal to column, petals truncate and thickened apically.............................***Stelis***
7. Petals or lip usually much longer than column, petals not truncate or thickened apically 8

8(7). Stems elongate, from half as long as leaves to much longer; leaf base broad and notched (leaf more or less cordate)................................***Pleurothallis***
8. Stems short or leaves tapering basally 9

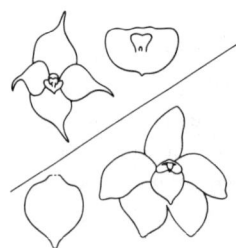
9(8). Lip short, wider than long, subequal to column; lateral sepals united, dorsal sepal, synsepal, and petals similar, usually tapering to narrow tails; pollinia 6 or 8; plants creeping or like tiny shrubs.............***Brachionidium***
9. Lip longer than wide; lateral sepals free; pollinia 2...... ....***Platystele***

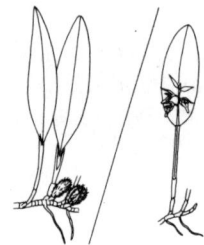
10(6). Single flowers borne on rhizome or lower stem, densely hairy or bristly........................***Myoxanthus***
10. Flowers borne near stem tip, *or* racemose, not densely hairy 11

11(10). Stem usually much shorter than leaf; leaves tapering basally and widest near apices (basal part of leaf may be no thicker than stem) 12
11. Stem elongate, usually half as long as, to much longer than, leaf; leaf base narrow to very wide (not gradually tapering)... 22

12(11). Flower with lip uppermost; lateral sepals each with thick cushion and slender apical tail, dorsal sepal may be thick and finger-like .... *Scaphosepalum*
12. Flower with lip lowermost; lateral sepals not as above 13

13(12). Column expanded at apex (trumpet-shaped); stigma apical; petals rudimentary; lip clasping column ... *Salpistele*
13. Column not trumpet-shaped; stigma usually ventral; petals larger; lip not clasping column.................. 14

14(13). Petals subequal to and parallel with column; lip jointed and motile; lateral sepals united and flat, dorsal sepal hooded........................ *Acostaea costaricensis*
14. Lip not jointed; sepals not as above 15

15(14). Tips of all 3 sepals united, leaving "windows" at sides 16
15. Tip of dorsal sepal free 17

16(15). Inflorescence short; flowers orange or brick red .... *Pleurothallis tribuloides*
16. Inflorescence elongate; flowers greenish streaked with purple .... **Ophidion pleurothallopsis**

17(15). Dorsal sepal usually markedly united with lateral sepals basally 18
17. Dorsal sepal usually free nearly or quite to base...... 20

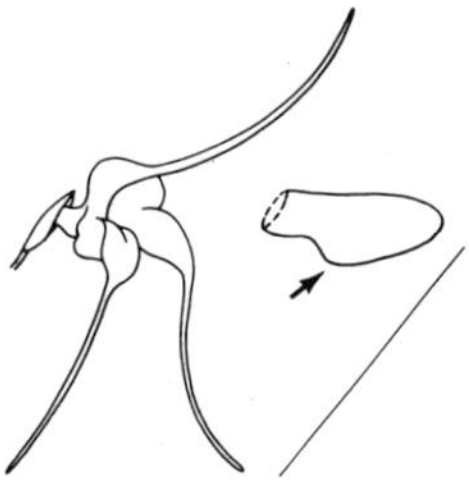

18(17). Petals thickened along lower side (near lip), sepals usually ending in slender tails....................*Masdevallia*
18. Petals not thickened along lower side.............. 19

19(18). Leaves thinly leathery; lip divided into hypochile and broader epichile, usually with fleshy ridges ..... ***Dracula***
19. Leaves fleshy; lip not divided into hypochile and epichile ....*Trisetella*

20(17). Lateral sepals bent sharply downward, with fleshy callus at bend; diminutive plants with fleshy leaves and more or less triangular flowers; lip hinged, with long claw .... ***Dryadella***
20. Lateral sepals not as above...................... 21

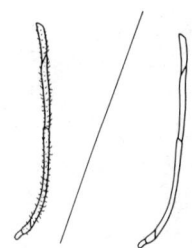

21(20). Plants creeping, rhizome usually longer than stem; usually with fleshy, erect leaves; flowers solitary, usually borne above leaves; pollinia 4 .... ***Barbosella***
21. Plants various, not as above; flowers usually not solitary; pollinia 2 .... ***Pleurothallis***

22(11). Stem sheaths hairy, scurfy, or glandular............. 23
22. Stem sheaths smooth 24

Key to Genera of Pleurothallidinae

23(22). Leaves fringed or hairy; pollinia 4 (2 large, 2 small) .... *Dresslerella*
23. Leaves smooth; pollinia 2 .... *Myoxanthus*

24(22). Apex of dorsal sepal united with lateral sepals, leaving a window on each side of flower, sepals often with keels or ridges; ovary with fleshy ridges .... *Zootrophion*
24. Apex of dorsal sepal free.......................... 25

25(24). Pollinia 8; flowers simple, usually yellow, fascicled; leaves may be narrow and fleshy .... *Octomeria*
25. Pollinia 2 or 4 26

26(25). Pollinia 2.............................. *Pleurothallis*
26. Pollinia 4 27

27(25). Dorsal sepal and petals with slender tails, these thickened and clublike at tips .... *Restrepia*
27. Dorsal sepal and petals broad and rounded.......... 28

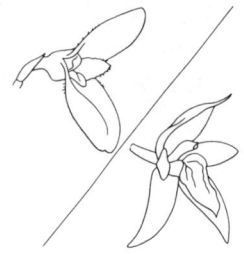

28(27). Plant robust, 15–25 cm tall; sepals fleshy, finely hairy
.... ***Restrepiella ophiocephala***
28. Plants smaller, usually < 10 cm tall; sepals thin, not hairy
.... ***Restrepiopsis***

Acostaea. Small epiphyte, 2–3 cm tall; stems short, clustered, each with 1 oblanceolate leaf; leaves tapering below to narrow stalks; inflorescence terminal, erect, producing 1 flower at a time; dorsal sepal hooded, lateral sepals united, together flat and longer than dorsal; petals attached to long column foot and parallel with column; column broadly winged, lip hinged, motile; 1 variable species in our area; **C:** yellow, yellow flushed with purple, or wine red; **D:** CR, Pma, SA (Mt); **M:** dS 2.5–5 mm, synS 4–6 mm .... ***Acostaea costaricensis***
Plate 20(1)

The shape of the flower is quite distinctive. The lip normally lies on the flat synsepal, but if disturbed it flips up against the column, catching any small insect that steps on the lip and briefly trapping it against the column.

Barbosella. Creeping epiphytes; leaves fleshy and usually erect, usually oblanceolate; solitary flowers generally brownish and held well above leaves; dorsal sepal narrow, tapering to finger-like apex, petals also quite narrow; pollinia 4.

Key to *Barbosella*

1. Leaves about as wide as long, borne flat on bark; **C:** pale pink with darker spots within; **D:** Pma, SA (Cb, Pc); **M:** lvs 3–7 mm, S 6–7 mm .... ***Barbosella orbicularis***
1. Leaves oblanceolate, erect................................. 2
2(1). Ovary and pedicel forming a loop so that synsepal is borne uppermost; **C:** pale yellow slightly suffused with brown; **D:** wPma (Mt); **M:** pl 3 cm, dS 13 mm .... ***Barbosella circinata***
2. Ovary and pedicel curving slightly so that synsepal and lip are borne on lower side of flower....................................... 3

3(2). Rhizome short, shoots clustered; **C:** usually brownish or purplish; **D:** No, CR, Pma (Mt); **M:** pl 3 cm, dS 11 mm .... *Barbosella anaristella*
Plate 20(2)
3. Rhizome creeping, shoots scattered; **C:** usually greenish or straw, may be lined with red; **D:** No, CR, wPma (Mt); **M:** pl 4 cm, dS 1.2–1.7 cm .... *Barbosella prorepens*

Brachionidium. Plants may creep or hang down; some species grow upward, forming miniature bushes; petals smaller than dorsal sepal or synsepal, but all 4 similar; lip short, wide, and fleshy, often hemicircular or crescent-shaped in front view; pollinia 6 or 8. Most species occur on mountaintops, and the flowers are short-lived. They are hard to find in flower and not well studied. Several of these species were described in the late 1980s, and there are at least 6 others yet to be named.

Key to *Brachionidium*

1. Plants creeping or pendent, mostly very small 2
1. Shoots clustered, or usually forming erect shrublets 4
2(1). Plant pendent; sides of lip rounded, apex shallowly 3-lobed; with hairpin-shaped pollinia (members of 2 pairs united); **C:** dS and P translucent greenish yellow, synsepal maroon-red; **D:** Pma, SA (Mt, Pk); **M:** lvs 8.5–14 mm, S 11.5–12 mm .... *Brachionidium kuhniarum*
2. Plant creeping; lip with pointed side lobes or not 3-lobed at apex; pollinia all club-shaped 3
3(2). Tails much longer than sepal blades; margin of lip toothed; **D:** CR (Pk); **M:** lvs 1–2 cm, S 1.5–2.4 cm .... *Brachionidium valerioi*
3. Tails shorter than sepal blades; margin of lip smooth; **D:** CR, Pma (Mt, Pk); **M:** lvs 7 mm, S 5.2–6 mm ..... *Brachionidium pusillum*
4(1). Tails much shorter than sepal blades; **C:** greenish yellow; **D:** CR (Pk); **M:** lvs 1.5–2 cm, S 10–12 mm .... *Brachionidium cruzae*
4. Tails much longer than sepal blades; **C:** pale green flushed with purple to red-purple; **D:** Pma (Mt, Pk); **M:** lvs 1.3–2 cm, S 2.6–3.3 cm .... *Brachionidium folsomii*
Plate 20(3)

Dracula. Epiphytic; shoots clustered, stems short, leaves oblanceolate, relatively thin, tapering below to narrow stalks; inflorescence pendent, with a single flower or producing 1 flower at a time; sepals united, each with a slender terminal tail, petals very small, usually thickened apically; base of lip (hypochile) narrow, apical portion (epichile) wider and usually concave with fleshy keels, looking strikingly like the spore-producing body of a fungus; pollinia 2. These flowers apparently mimic fungi and attract fungus gnats that normally lay their eggs on fungi.

Key to *Dracula*

1. Sepal blades < 10 mm long, tails subequal to blades, spreading or parallel; C: yellow-cream flushed and speckled with red; D: No, CR, Pma, SA (Mt); M: pl to 18 cm, S 1.5–2 cm ....*Dracula pusilla*
1. Sepal blades > 10 mm long; tails much longer than blades...... 2
2(1). Sides of epichile recurved so that opening is narrow, smooth within; C: S pale yellow spotted and blotched with red-purple, lip white; D: No, CR, SA (Mt); M: pl 10–20 cm, S 3.5–8.5 cm ....*Dracula vespertilio*
2. Sides of epichile not recurved, opening wide, with fleshy ridges within 3
3(2). Apex of lip inflated, incurved, with a minutely warty blunt tip; C: S white, basal halves blotched and stained with purple, tails red-purple, lip white; D: CR, SA (Mt); M: pl 14–25 cm, S 4–8.5 cm .... *Dracula ripleyana*
3. Apex of lip rounded, smooth; C: S white more or less flushed and speckled with red-brown, tails red, lip white; D: No, CR, wPma (Mt, Wt); M: pl 8–23 cm, S 5–5.5 cm ..... *Dracula erythrochaete*
Plate 21(1)

Dresslerella. Epiphytic; shoots clustered, stems subequal to leaves; inflorescence terminal, short, usually 1 flower at a time; plants hairy or minutely scurfy throughout, and leaves with a fringe of hairs; pollinia 4 (2 large and 2 small). In some species the flowers are nearly closed, with an apical opening between the sepals. *Dresslerella pilosissima*, however, has an open flower much like a shaggy *Restrepia* flower.

Key to *Dresslerella*

1. Sepals free to base; plant and flower hairy, with spreading hairs; C: pale green, dS and P striped with purple, synS speckled with purple; D: CR (Mt, Wt); M: lvs 2.5–4 cm, S 10 mm...................
....*Dresslerella pilosissima*
Plate 20(4)
1. Sepals at least partially united; hairs not long and spreading 2
2(1). Opening of flower less than one-fourth length of flower 3
2. Opening of flower one-third or more length of flower 4
3(2). Flower widest near base; covered with branched (starlike) hairs; C: S purplish red; D: CR, SA (Wt); M: lvs 8–12 cm, S 1.5–3 cm
....*Dresslerella stellaris*
3. Flower widest near apex; hairs not branched; C: dull red; D: Pma (Mt); M: lvs 4–10 cm, S 10–14 mm .... *Dresslerella pertusa*
Known only from Cerro Jefe in central Panama.
4(2). Dorsal sepal as long as lateral sepals, petals linear; C: dark red; D: Pma (Mt); M: lvs 3–7 cm, S 9 mm ....*Dresslerella elvallensis*
4. Dorsal sepal shorter than lateral sepals; petals obovate or oblanceolate 5
5(4). Apices of lateral sepals saccate; C: dark red; D: Pma (Cb, Pc); M: lvs 1.3–3.5 cm, S 6–9 mm................ *Dresslerella powellii*
5. Apices of lateral sepals cupped but not saccate (margins not recurved above); C: dark red; D: No, CR, Pma, SA (Mt); M: lvs 1.5–6.5 cm, S 5–10 mm....................... *Dresslerella hispida*

Dryadella. Dwarf, epiphytic; shoots clustered, stems short; inflorescences short, few-flowered; flowers rather triangular, sepals generally with definite tails, lateral sepals have thickened transverse callus and are sharply recurved at the callus; petals short, broad, and angled; lip with a slender stalk; column winged and hooded over anther; pollinia 2.

Key to *Dryadella*

1. Column wings long and narrow, at least twice as long as wide
.... 2
1. Column wings short and broad, obtuse...................... 4

2(1). Column wings curved forward; blade of lip widest across midlobe; C: pale green with dark red spots; D: Pma (Pc, Mt); M: pl 2–2.5 cm, S 3.5 mm .... *Dryadella butcheri*
Plate 21(2)
2. Column wings curved back; midlobe subequal to or slightly wider than lateral lobes 3

3(2). Leaves oblanceolate, about 2 cm long; sepals abruptly apiculate; C: pale green; D: CR, Pma, SA (Cb, Mt); M: pl 1.5–2 cm, S 3–3.5 mm .... *Dryadella odontostele*
3. Leaves elliptic, very fleshy, about 10 mm long; sepals tapering gradually; C: S yellow with red-purple spots, P white with red-purple spots, lip red-purple; D: CR, SA (Mt); M: pl 1 cm, S 4.5 mm .... *Dryadella minuscula*
The Costa Rican plants differ in several details from the Colombian material on which the species is based and may be a distinct species.

4(1). Leaves linear, > 15 times longer than wide; C: pale green with red-purple spots; D: No, CR, SA; M: pl 8–12 cm, S 8.5–9 mm....... .... *Dryadella guatemalensis*
4. Leaves oblanceolate, 4–10 times longer than wide 5

5(4). Sepals about 3 times longer than wide, narrowly tapering; C: lvs purplish, fls pale green heavily barred with purple; D: Pma (Cb); M: pl 1.5–2 cm, S 3.5–4 mm.................... *Dryadella dressleri*
Frequent in wet lowland forests of central Panama.
5. Sepals about twice as long as wide, tapering abruptly; leaves green; flowers green or spotted 6

6(5). Blade of lip strongly recurved, widest near apex; C: spotted; D: CR, Pma (Cb); M: pl 0.8–1.5 cm, S 3–4 mm ..... *Dryadella sororcula*
6. Blade of lip not strongly recurved, sides parallel; C: pale green; D: CR, Pma (Cb, Pc); M: pl 2–3 cm, S 3.5–5 mm.................. .... *Dryadella gnoma*

Lepanthes. Stems usually longer than leaves, always with distinctive oblique, funnel-like sheaths. The leaves are usually wide, and the inflorescences may be borne above or beneath the leaf. The inflorescence produces 1 flower at a time over a long period, causing a distinctive fishbone pattern on the old inflorescence. The petals are usually 2- or 3-lobed, and the lip is small, with the lateral lobes clasping the column.

There are 2 pollinia. Even the tiniest *Lepanthes* flower is attractive, and a few species are relatively large. More than 50 species have been described or recorded from Costa Rica and Panama (see Appendix B), and there are undoubtedly others still to be described. These complex little flowers are fascinating, but I have not attempted to write a key for the species of our area. Plate 19(1).

Lepanthopsis. Epiphytic; shoots clustered, stems usually longer than leaves, with funnel-shaped sheaths; leaves elliptic; inflorescence racemose, flowers usually arranged in 2 rows; sepals spreading, petals small, column very short, with 2-lobed stigma; pollinia 2. The plants of *Lepanthopsis* are much like those of *Lepanthes*, but the flowers resemble those of *Platystele*.

Key to *Lepanthopsis*

1. Sepals all united for about half their length, flowers bell-shaped; C: yellow-green; D: CR, SA (Mt); M: pl 3–6 cm, S 2.5–4.5 mm .... ***Lepanthopsis obliquipetala*** (Syn. *Pleurothallis connata*)
1. Dorsal sepal free from laterals, flowers open widely 2

2(1). Synsepal oblong, shallowly notched at apex; all flowers open at once; C: translucent yellowish or purplish green, lip yellow; D: No, CR, Pma, SA (Mt); M: pl 2–5 cm, S 3.5–4 mm .... *Lepanthopsis floripecten* Plate 19(2)
2. Lateral sepals free for more than half their length, together rounded; only 1 or 2 flowers open at a time; C: S red-brown; D: CR (Mt); M: pl 2 cm, S 1.6 mm .... ***Lepanthopsis comet-halleyi***

Masdevallia. Epiphytic; shoots clustered, stems short; leaves generally oblanceolate, tapering below, often with distinct petioles; inflorescence of 1 to many flowers; flowers small to quite large, sepals usually united to form a tube, usually with long, slender tails at their tips; petals small, with definite thickenings on lower margins; lip usually oblong; pollinia 2. Many of the *Masdevallia* species are quite attractive. All are much sought by growers of miniatures, though some of the flowers are a bit too large to fit that category.

The peculiar architecture of the *Masdevallia* flower does not lend itself well either to clear keys or to standard botanical descriptions, and so I have supplied brief descriptions of each species after the key. These descriptions emphasize the structure and form of the sepaline tube that is so characteristic of *Masdevallia*. The phrase *free blade* refers to the flattened portion of a sepal that is not part of the tube. The depth of the tube is its vertical dimension, or the distance from ceiling to floor. Color and distribution notes are given with the descriptions. Plate 21(3–5).

Key to *Masdevallia*

1. Lateral sepals without tails, or tails wide and flattened 2
1. Lateral sepals with slender or club-shaped tails 8
2(1). Dorsal sepal free from lateral sepals nearly or quite to base 3
2. Dorsal sepal clearly united to lateral sepals for at least a quarter of its length. 4
3(2). Lateral sepals free nearly to base..... *Masdevallia pleurothalloides*
3. Lateral sepals united to above middle, sepals fleshy, ridged within .... *Masdevallia pelecaniceps*
4(2). Lateral sepals broad, abruptly narrowing to acute angles.......... .... *Masdevallia utriculata*
4. Lateral sepals narrower, free portions longer than wide......... 5
5(4). Lateral sepals united for less than half their length .... *Masdevallia livingstoneana*
5. Lateral sepals united for half their length 6
6(5). Lateral sepals tapering abruptly to short tails................... .... *Masdevallia striatella*
6. Lateral sepals tapering evenly to acute apices................. 7
7(6). Lateral sepals abruptly expanded in front of tube .... *Masdevallia nicaraguae*
Plate 21(3)
7. Lateral sepals gradually expanding in front of tube.............. .... *Masdevallia tubuliflora*
8(1). Diminutive, creeping plant; flowers < 5 mm long............... .... *Masdevallia pygmaea*
8. Plant erect; flowers > 10 mm long......................... 9
9(8). Tails of lateral sepals thick and fleshy, often clublike at apices 10
9. Tails of lateral sepals slender throughout 16

10(9). Inflorescence pendent or creeping on substrate.................
.... *Masdevallia zahlbruckneri*
Plate 21(5)
10. Inflorescence erect, not creeping on substrate................ 11
11(10). Tails very short, widest at bases............ *Masdevallia striatella*
11. Tails long and narrow or constricted at bases................ 12
12(11). Sepals smooth on outside................................. 13
12. Sepals warty or hairy externally........................... 15
13(12). Floral tube notched beneath in profile .... *Masdevallia flaveola*
13. Floral tube only slightly curved beneath.................... 14
14(13). Flowers solitary........................ *Masdevallia cupularis*
14. Flowers normally paired .... *Masdevallia chontalensis*
15(12). Ovary and flower covered with conspicuous spinelike hairs........
.... *Masdevallia erinacea*
15. Ovary and flower merely warty......... *Masdevallia molossoides*
16(9). Tails less than half as long as blades of lateral sepals, blades may taper gradually 17
16. Tails subequal to lateral sepals or much longer; blades usually tapering abruptly... 21
17(16). Flowers several on inflorescence; sepals free nearly to base ....
.... *Masdevallia rafaeliana*
17. Flowers solitary, or produced 1 at a time; sepals united into definite tube... 18
18(17). Blade of synsepal much longer than blade of dorsal sepal..........
.... *Masdevallia floribunda*
18. Blade of dorsal sepal nearly as long as blade of synsepal 19
19(18). Tube strongly arched .... *Masdevallia collina*
19. Tube straight or nearly so............................... 20
20(19). Lateral sepals united for more than half their length (including tails), tapering abruptly to tails..................... *Masdevallia chasei*
20. Lateral sepals united for less than one quarter their length, tapering gradually to tails.................... *Masdevallia tokachiorum*
21(16). Sepals free nearly to base................ *Masdevallia picturata*
21. Sepals united for at least one quarter their length 22
22(21). Inflorescence pendent or creeping along substrate................
.... *Masdevallia zahlbruckneri*
Plate 21(5)
22. Inflorescence erect or ascending........................... 23
23(22). Sepaline tube about as wide as long, bulging beneath; dorsal sepal about as wide as synsepal .... *Masdevallia nidifica*

23. Sepaline tube longer than wide, dorsal sepal smaller than synsepal, or sepals subequal. 24

24(23). Sepals subequal 25
24. Synsepal forming much larger surface than dorsal sepal 27

25(24). Sepals tapering gradually to tails, tails not spreading widely . . . .
. . . . *Masdevallia schizopetala*
25. Sepals tapering abruptly to tails, tails spreading widely 26

26(25). Sepaline tube about twice as long as wide . . . .
. . . . *Masdevallia scabrilinguis*
26. Sepaline tube less than twice as long as wide . . . .
. . . . *Masdevallia cupularis*

27(24). Synsepal abruptly expanded and cuplike above sepaline tube. . . . . . .
. . . . *Masdevallia lata*
27. Synsepal relatively flat, not abruptly expanded. 28

28(27). Flowers basically white, cream, or yellow, may have narrow purple stripes on tube, but no red or purple on sepal blades 29
28. Flowers with pink, red, or purple on sepal blades 33

29(28). Sepaline tube narrow, only slightly expanded above 30
29. Sepaline tube much wider above, gaping 31

30(29). Lip as wide apically as basally . . . . *Masdevallia attenuata*
30. Lip much wider basally than apically . . . . *Masdevallia walteri*

31(29). Blade of dorsal sepal subequal in length to blades of lateral sepals, hairy within . . . . *Masdevallia tonduzii*
31. Blades of lateral sepals much longer than blade of dorsal sepal, not hairy within 32

32(31). Tail of dorsal sepal recurved; lip not constricted . . . .
. . . . *Masdevallia laucheana*
32. Tail of dorsal sepal straight; lip constricted above middle . . . .
. . . . *Masdevallia marginella*

33(28). Flowers red-purple or red-brown, with yellow tails 34
33. Flowers with both white and rose-purple on blades of sepals . . . 36

34(33). Inflorescence 1-flowered. 35
34. Inflorescence eventually several-flowered 37

35(34). Tube narrow, scarcely expanded above; tails of lateral sepals shorter than blades (including tube); lip narrowly triangular. . . . .
. . . . *Masdevallia demissa*
35. Tube markedly expanded above; tails of lateral sepals longer than blades; lip ovate-lanceolate . . . . *Masdevallia rolfeana*

36(33). Sepaline tube laterally compressed; inflorescence stiff . . . .
. . . . *Masdevallia calura*

36. Sepaline tube not markedly compressed; inflorescence arched......
.... *Masdevallia thienii*

37(34). Blades of lateral sepals, together, < 2 cm wide; tube narrow.......
.... *Masdevallia fulvescens*

37. Blades of lateral sepals, together, > 2.5 cm wide; tube expanded above.. 38

38(37). Lateral tails longer than blades of lateral sepals ....
.... *Masdevallia schroederiana*
Plate 21(4)

38. Lateral tails subequal to or shorter than blades ....
.... *Masdevallia reichenbachiana*

Masdevallia attenuata. Leaves 12–13 cm; inflorescence subequal to leaves, 1-flowered; sepaline tube 5–6 mm long, relatively narrow; free blade of dorsal sepal subtriangular, with slender tail 10–11 mm long, lateral sepals united for about 7 mm, free blades rounded, with tails about 9 mm long; **C:** cream with yellow tails, with 3 purple stripes on each side of tube; **D:** CR, Pma (Mt).

Masdevallia calura. Leaves 8–10 cm; inflorescence surpassing leaves, few-flowered, 1 at a time; sepaline tube about 10 mm long, somewhat compressed, distally a bit deeper than half the length; free blade of dorsal sepal triangular, with slender tail 3.5–4 cm long, lateral sepals united for 2–3 mm beyond tube, free blades ovate-triangular, **folded up and parallel** in front of tube, each with slender tail 3.5–4 cm long; **C:** purple with yellow-green tails; **D:** CR (Mt).

Masdevallia chasei. Leaves 5–15 cm; inflorescence about half as tall as leaves, spreading, with 1 or 2 flowers, 1 at a time; sepaline tube about 15 mm long, laterally compressed, about half as deep as long; dorsal sepal abruptly narrowed in front of tube to slender tail about 3 cm long, tail scarcely curving, lateral sepals united for 2–3 mm in front of tube, free blade broadly triangular, notched, tails about 8 mm long; **C:** S yellowish white streaked with purple near base; **D:** CR (Mt).

Masdevallia chontalensis. Leaves 3–7.5 cm; inflorescences slightly surpassing leaves, usually 2-flowered; sepaline tube 5–7 mm long, width about half the length; free blade of dorsal sepal triangular, with fleshy, clublike tail 3–7 mm long, lateral sepals similar; **C:** white with yellow tails; **D:** No, CR, Pma, SA (Mt).

Masdevallia collina. Leaves 5–11 cm; inflorescence surpassing leaves, few-flowered, usually 1 at a time; sepaline tube 14–15 mm long, arched, markedly indented below near base; dorsal sepal abruptly narrowed to slender tail about 16 mm long, lateral sepals united 5–6 mm beyond tube, free blade broadly rounded-triangular, with 2 slender tails about 8 mm long; **C:** S dark purple with yellow tails; **D:** Pma (Pc, Mt).

Masdevallia cupularis. Leaves 5–7 cm; inflorescence subequal to leaves, 1-flowered; sepaline tube about 11 mm long, nearly as wide as long; free blades of sepals triangular with slender tails about 15 mm long, thickened apically; apex of lip toothed or fringed; **C:** greenish yellow flushed and diffused with red-brown; **D:** CR (Pk).

Masdevallia demissa. Leaves 8–15 cm; inflorescence about half as long as leaves, 1-flowered; sepaline tube about 10 mm long, longer than wide; free blade of dorsal sepal shortly triangular, tapering to slender tail 2–2.2 cm long, lateral sepals united for about 5 mm in front of tube, free blade of synsepal ovate, notched, each half with slender tail 10–15 mm long; lip narrowly triangular; **C:** red-brown with green-yellow tails; **D:** CR (Mt).

Masdevallia erinacea. Leaves 2–5 cm; inflorescence surpassing leaves, few-flowered, 1 flower at a time, flowers covered with spinelike hairs; sepaline tube 4–5 mm long, cuplike, above as wide as long, free blades subtriangular, abruptly narrowed to tails 3–5 mm long, apically thickened; **C:** S greenish yellow flushed and speckled with red-brown, tails yellow; **D:** CR, Pma, SA (Cb, Mt).

Masdevallia flaveola. Inflorescence surpassing leaves, 1 flower at a time; sepaline tube about 5 mm long, thickest basally and then notched below; free blade of dorsal sepal triangular-rounded, with clublike tail about 11 mm long, blades of lateral sepals oblong, largely free, each with clublike tail 8–9 mm long; **C:** yellow, tails orange, brown on bases of lateral sepals; **D:** CR.

Masdevallia floribunda. Leaves 5–15 cm; inflorescence subequal to leaves, 1 or 2 flowers, 1 at a time; sepaline tube 6–7 mm long, about as wide as long; dorsal sepal abruptly narrowing to slender tail 10–15 mm long, lateral sepals united for 5–6 mm in front of tube, together broadly obovate, each with slender tail 5–10 mm long; **C:** either light yellow or heavily dotted with purple (apparently color forms of a single species); **D:** No, CR (Mt).

Masdevallia fulvescens. Leaves 6–11 cm; inflorescence surpassing leaves, few flowers, 1 at a time; sepaline tube 1.5–1.8 cm long, less than half as deep as long; free blade of dorsal sepal triangular with slender tail 5–7 cm long, this ascending, lateral sepals united for 3–5 mm beyond tube, free blade 14–16 mm wide, each half asymmetrically elliptic with tail 4–5.5 cm long; C: S light yellow suffused with dull purple-brown, tails yellow-green; D: CR (Mt).

Masdevallia lata. Leaves 8–14 cm; inflorescence about equal to leaves, few-flowered, 1 at a time; sepaline tube about 10 mm long, markedly bulging below beyond middle and then constricted below mouth of tube; dorsal sepal with tail 5–6 cm long, lateral sepals much expanded beyond tube, free blades transversely oblong if flattened, with tails 3–3.5 cm long: C: sepaline tube dark red, tails ocher-yellow, P pale creamy yellow, lip heavily spotted with maroon; D: CR, wPma (Pc).

Masdevallia laucheana. Leaves 10–13 cm; inflorescence about half length of leaves, 1 flower (at a time?); sepaline tube about 5–10 mm, tube wider distally; free blade of dorsal sepal broad and rounded with narrow tail about 2.5 cm long, lateral sepals united beyond tube, about a third longer than dorsal sepal, oblong, each with tail about 2.5 cm long, tails all curved backward; C: white with yellow tails; D: CR (Mt).

Masdevallia livingstoneana. Leaves 6–12 cm; inflorescence subequal to leaves, 1- or 2-flowered, 1 at a time; sepals united into tube for about 9 mm, tube narrow, about half as wide as long, slightly curved; blade of dorsal sepal slightly expanded above tube, then tapering to narrow tail with margins curled back, tail recurved, free blades of lateral sepals obliquely lanceolate, spreading; C: greenish cream to yellow marked with brown or purple on tails and in throat; D: Pma (Cb).

Masdevallia marginella. Leaves 5.5–10 cm; inflorescence surpassing leaves, 2-flowered; sepaline tube 12–14 mm long, distal width less than length; dorsal sepal tapering abruptly to slender tail 3–3.5 cm long; lateral sepals united for about 3 mm in front of tube, blades ovate tapering to tails 2.5–2.7 cm long; C: greenish white with yellow tails; D: CR (Mt).

Masdevallia molossoides. Leaves to 8 cm; inflorescence surpassing leaves, 1-flowered; sepaline tube about 8 mm long and about as wide, base saclike beneath, tube then indented ventrally in middle, with bumpy keels on outside; free portions of sepals rounded-triangular, abruptly narrowing to apically thickened tails, 4–5 mm long; C: S pale green flushed with red-brown, tails yellow; D: No, CR, Pma (Mt).

Masdevallia nicaraguae. Leaves 6–11 cm; inflorescence subequal to leaves, 1-flowered; sepaline tube 11–12 mm long, bell-shaped; dorsal sepal abruptly narrowed to slender tail about 12 mm long, lateral sepals united for 6–7 mm beyond tube, blades obliquely ovate tapering to narrowly triangular tips 3–4 mm long; C: white with faint to distinct purple streaks in throat, tails pale yellow; D: No, CR (Cb).

Masdevallia nidifica. Leaves 3–6 cm; inflorescence longer than leaves, 1-flowered; ovary with toothed wings; sepaline tube about 5 mm long, wide and **bulging below**; free blade of dorsal sepal 6–7 mm long, concave, rounded, with slender tail 0.8–3 cm long, lateral sepals united beyond tube, blades ovate tapering to slender tails 1–2 cm long; lip 3-lobed; C: S translucent white, yellow, or dull rose, suffused or striped with red-purple; D: No, CR, Pma (Mt).

Masdevallia pelecaniceps. Stems surrounded by loose, inflated sheaths; leaves 2–7 cm; inflorescence 1-flowered, 6–11 cm long; dorsal sepal free, 2 cm long, short-stalked, elliptic, acute, ridged within, lateral sepals more than half united, deeply concave, about 2 cm long, ridged within; C: S red or reddish externally, yellow with purple spots within; D: Pma (Pc, Mt), known only from the region of Cerro Jefe. This is an anomalous species that does not fit well into any genus. As the name suggests, the flower resembles a pelican's beak.

Masdevallia picturata. Leaves 2–7.5 cm; inflorescence subequal to leaves, 2-flowered; dorsal sepal free to base, 8–12 mm long, oblong, tapering to slender tail 3–7 cm long, lateral sepals narrowly oblong, incurved, 8–15 mm long, with slender tails; C: S white with purple spots; D: CR, Pma, SA (Pk).

Masdevallia pleurothalloides. Leaves 1.5–3.5 cm; inflorescence surpassing leaves, 1-flowered; sepals lanceolate, not forming a tube, 9–10 mm long; dorsal sepal acute, about 3.5 mm long, lateral sepals narrower; C: S cream with brownish spots, P white, lip bright orange; D: Pma (Pc, Mt), known only from the region of Cerro Jefe. This species has the appearance of a *Pleurothallis*.

Masdevallia pygmaea. Rhizome creeping, leaves narrowly oblanceolate, obtuse, to 2.5 cm long; inflorescence shorter than leaves, 1-flowered; sepals united for about 1.8 mm, about 4 mm long, bases (about 1 mm) triangular-ovate, rest narrow, fleshy tails; lip oblong; **C:** white with yellowish tails; **D:** CR (Mt).

Masdevallia rafaeliana. Leaves 11–18 cm; inflorescence surpassing leaves, simultaneously several-flowered; sepals 13–14.5 mm long; dorsal sepal free to base, ovate, with slender tail about 3.5 mm long, lateral sepals united and concave basally, obovate, tapering to slender tails about 4 mm long; **C:** dS light rose, lS white dotted with purple, P yellow, lip purple; **D:** CR (Pk).

Masdevallia reichenbachiana. Leaves 5–12 cm; inflorescence surpassing leaves, with few flowers, 1 at a time; sepaline tube about 2.4 cm long, distally about half as deep as long; free blade of dorsal sepal triangular-ovate with slender tail about 4 cm long, this turned back over tube, lateral sepals united for about 6 mm beyond tube, together subcircular, notched, each with slender reflexed or spreading tail about 2.5 cm long; **C:** S orange within, purple externally, rose to yellow-white above middle, tails green; **D:** CR (Mt).

Masdevallia rolfeana. Leaves about 10 cm; inflorescence shorter than leaves, with 1 flower; sepals united for 10 mm, tube narrow basally, wider above, bulging below near middle; free blade of dorsal sepal about 5 mm long, broadly triangular, ending in a tail about 3.5 cm long, lateral sepals united beyond tube, blades ovate, tapering, tails about 3 cm long; **C:** dark purple with yellowish tails; **D:** CR (Mt).

Masdevallia scabrilinguis. Leaves 6–10 cm; inflorescence a bit shorter than leaves, 1 flower; sepaline tube about 6 mm long, about half as wide, slightly arched; free blades triangular or triangular-ovate tapering to tails 8–9 mm long: **C:** S white with yellow tails, P white, lip yellow-white with yellow apex; **D:** wPma, CR (Mt).

Masdevallia schizopetala. Leaves 2–4.5 cm; inflorescense may be more than twice length of leaves, 1 flower at a time; sepaline tube about 4 mm long, wider than long; free portion of dorsal sepal triangular tapering to slender tail 5–8 mm long, blades of lateral sepals triangular-ovate tapering to slender tails, sepaline tails **spreading little** and gradually; **C:** green with pale purple streaks; **D:** CR, Pma, SA (Mt).

Masdevallia schroederiana. Leaves 7–12 cm; inflorescence subequal to or slightly surpassing leaves, 1-flowered; floral tube about 15 mm long, distally about as deep as long, upper profile arching; dorsal sepal abruptly tapering to slender tail about 8 cm long; lateral sepals united for about 10 mm beyond tube, together forming broadly ovate, notched blade, each sepal with slender tail about 6 cm long; C: S basally purple, inner halves of lS white above middle, tails yellow; D: CR (Mt).

Masdevallia striatella. Leaves 5–10 cm; inflorescence subequal to leaves, 1-flowered, sometimes 2 consecutive flowers; sepaline tube 7–10 mm, bell-shaped; free blades triangular-ovate narrowed to fleshy tails 2–3 mm long; C: straw with purple streaks; D: CR, SA (Mt).

Masdevallia thienii. Leaves 10–13 cm; inflorescence pendent, much longer than leaves, producing several flowers 1 at a time; sepaline tube about 15 mm long, nearly as wide as long, free blade of dorsal sepal triangular-semicircular tapering to tail 1.8–2 cm long, lateral sepals united for about 10 mm in front of tube, blades ovate tapering to tails 2–2.2 cm long; C: red with yellow base and tails; D: CR, Pma, SA (Cb, Mt).

Masdevallia tokachiorum. Leaves 6–7 cm; inflorescence subequal to leaves, 1 flower or 2 consecutive flowers; sepaline tube 3–4 mm long, very wide; dorsal sepal united with lateral sepals for about 2 mm in front of tube, free portion narrowly triangular with slender tail about 8 mm long, this not spreading, lateral sepals asymmetrically lanceolate tapering to tails about 8 mm long; C: dS yellow, lS yellow suffused with brown, P white, lip brown; D: wPma (Mt).

Masdevallia tonduzii. Leaves 4.5–9 cm; inflorescence subequal to leaves, 1-flowered; sepaline tube 9–11 mm long, bell-shaped, distally about as wide as long; free blades of sepals subtriangular, abruptly narrowed to slender tails 1.9–2.5 cm long, lateral sepals united for about 5 mm in front of tube; C: white with some yellow within, tails yellow, tube may be yellow externally; D: CR, Pma (Cb).

Masdevallia tubuliflora. Leaves 5–10 cm; inflorescence subequal to leaves, 1-flowered; sepaline tube 7.5–8 mm long, narrow, curved; free blade of dorsal sepal elliptic-lanceolate, acute, lateral sepals united for 2–3 mm beyond tube, free blades comma-shaped; C: white with yellow tails and purple markings in throat; D: No, CR (Cb).

Masdevallia utriculata. Leaves 6–8.5 cm; inflorescence 2–3 cm long, ascending or horizontal, few-flowered, 1 at a time; sepaline tube about 5 mm long, and 6–7 mm wide, free blade of dorsal sepal subtriangular with tail about 11 mm long, lateral sepals united for about 10 mm in front of tube, blades ovate, abruptly short-acuminate; C: S yellow, lateral sepals studded with red warts within, lip yellow intensely marked with red-purple; D: wPma (Mt), to be expected in adjacent Costa Rica.

Masdevallia walteri. Leaves 7–13 cm; inflorescence subequal to leaves, 1-flowered; sepaline tube about 7 mm long, curved, about half as wide as long; free blades subtriangular with slender tails 12–14 mm long; C: yellow-orange, tails orange; D: CR (Pk).

Masdevallia zahlbruckneri. Leaves 5–8 cm; inflorescence to 6 cm long, creeping or pendent, 1 flower at a time; sepaline tube wide, 4–5 mm long; free blade of dorsal sepal subtriangular with slender tail 8–15 mm long, lateral sepals united 6–8 mm beyond tube, blades broadly ovate, notched, each with slender tail 7–10 mm long; C: lS cream heavily speckled with red-purple, dS and tails yellow; D: CR, Pma, SA (Cb, Mt). One of the commoner species and quite variable in flower size; the tails may be long and slender or short and thick, and may even be curled.

Myoxanthus. Epiphytic; shoots clustered, rarely shrubby or viny with superposed shoots; stems long, sheaths loose, more or less scabrous; inflorescence terminal or often from rhizome or lower stem, dense fascicles or solitary flowers; petals often tapering to slender or thickened, clublike tips; pollinia 2.

Key to *Myoxanthus*

1. Inflorescence apical, not markedly hairy or bristly.............. 2
1. Flowers borne from rhizome or lower stem, densely hairy or bristly 10
2(1). Leaf sheaths tight on stem, distinctly bristly 3
2. Leaf sheaths loose, only slightly bristly 8
3(2). Lateral sepals united for at least half their length, petals arched, apical halves thick and finger-like; C: pale green flushed and speckled with red-purple, petal tips dark red-purple; D: CR, Pma, SA (Mt); M: pl to 24 cm, S 7–9 mm .... *Myoxanthus hirsuticaulis*
Plate 22(2)
3. Lateral sepals free or united only near base 4

4(3). Lip not 3-lobed; petals not or only gradually tapering above; C: white to creamy white; D: No, CR, Pma, SA (Cb); M: pl to 30 cm, S 6.5–8 mm .... *Myoxanthus octomeriae*

4. Lip distinctly 3-lobed; petals tapering to narrow or linear apices 5

5(4). Petals forming slender, narrow tails about twice length of petal blades; apex of lip rounded; C: yellow-white; D: CR, Pma, SA (Mt, Pc); M: pl to 30 cm, S 7–10 mm .... *Myoxanthus colothrix*

5. Tails of petals flattened, subequal to petal blades, or less than twice as long 6

6(5). Plants usually proliferating, forming new stems from upper nodes of old stems, becoming much-branched shrubs; C: dull yellowish flecked or suffused with purple-brown; D: CR, Pma, SA (Cb, Mt, Pc); M: pl to 50 cm, S 4–5 mm .... *Myoxanthus scandens*

6. Plants not proliferating, new stems arising from base of old stem 7

7(6). Plant about 10 cm tall, creeping; lip with toothlike lateral lobes; C: yellow-green; D: No, CR, Pma, SA (Cb, Mt, Pc); M: pl 10 cm, S 5–7.5 mm........................... *Myoxanthus trachychlamys*
   This species forms extensive mats on large branches at relatively low elevations.

7. Plant 20–40 cm tall, shoots clustered; lateral lobes of lip rounded; C: pale yellow with red-purple stripes; D: CR, Pma, SA, (Mt, Pk); M: pl 11–30 cm, S 9.5–12 mm .... *Myoxanthus speciosus*

8(2). Lip obtuse; flowers self-pollinating without opening; C: S dull green suffused and mottled with dark purple-brown, P yellow dotted with purple, lip red; D: CR, Pma (Mt); M: pl 18–30 cm, S 12–15 mm .... *Myoxanthus sempergemmatus*

8. Lip acute; flowers opening normally....................... 9

9(8). Lip with narrow U-shaped callus at base; C: S dull greenish marked with dull purple, P and lip yellow spotted with purple or red; D: CR, Pma, SA (Pc, Mt); M: pl 10–25 cm, S 1.7–2.5 mm .... *Myoxanthus aspacicensis*
(Syn. *Pleurothallis alexandrae*)

9. Callus at base of lip hemispheric, rounded; C: S cream to yellow, marked with purple, P and lip yellow marked with purple; D: No, CR, Pma (Cb, Mt, Pc); M: pl 10–25 cm, S 12–15 mm .... *Myoxanthus uncinatus*
Plate 22(4)

10(1). Sepals > 3 cm long, with clustered hairs; **C:** dark red; **D:** Pma (Mt); **M:** pl 25 cm, S 3.5–3.6 cm .... *Myoxanthus balaeniceps*
Plate 22(3)
10. Sepals < 2 cm long, with stiff simple hairs 11
11(10). Dorsal sepal lanceolate tapering evenly to an acute apex; **C:** dark red; **D:** No, CR, Pma, SA (Cb, Mt, Pc); **M:** pl 15–18 cm, S 1.1–1.9 cm .... *Myoxanthus lappiformis*
11. Dorsal sepal wider, obtuse 12
12(11). Synsepal relatively flat; no warts under sepaline hairs; petals with triangular lobes in middle; **C:** dull green spotted with purple; **D:** CR, Pma (Mt); **M:** pl 15–35 cm, S 1.5–2 cm .... *Myoxanthus pan*
12. Synsepal deeply concave; sepaline hairs borne on warts; petals spatulate; **C:** dark red-purple with white hairs; **D:** CR (Cb); **M:** pl 10–25 cm, S 2–2.2 cm .... *Myoxanthus stonei*

Octomeria. Epiphytic; shoots clustered, or rhizome creeping, stems usually longer than leaves; flowers fascicled at leaf base, usually yellow, rather bell-shaped, with free sepals and petals; pollinia 8. Most species of *Octomeria* are monotonously similar in flower structure, but the Central American species are easily distinguished by vegetative features.

Key to *Octomeria*

1. Leaves cylindrical, not or scarcely wider than stems, which are much longer; **C:** yellow; **D:** No, CR, Pma (Cb); **M:** pl 10 cm, S 4.5 mm .... *Octomeria hondurensis*
1. Leaves flattened, much wider than stems, which are subequal in length 2
2(1). Leaves 3–4 cm long; **C:** pale yellow with maroon flush in center; **D:** CR, SA; **M:** pl to 45 cm, S 3.3–4.3 mm ..... *Octomeria apiculata*
O. graminifolia has been reported from Nicaragua and is very similar to *O. apiculata*. *O. graminifolia* has narrower and more narrowly acute sepals and petals, and the lip is shallowly and indistinctly lobed. Either species could easily be confused with *Myoxanthus trachychlamys*.
2. Leaves 7–25 cm long. 3
3(2). Leaves 7–10 cm long; **C:** cream; **D:** CR, Pma (Cb); **M:** pl 13–17

cm, S 7 mm........................... ***Octomeria costaricensis***
   O. surinamensis, reported from Nicaragua, is about the same size as *O. costaricensis* but has larger white flowers with dark red calli.
3. Leaves larger, 20–25 cm long; **C:** pale yellow; **D:** No, CR, Pma (Cb, Mt); **M:** pl to 40 cm, S 4.5–6 mm ....*Octomeria valerioi*
Plate 19(3)

Ophidion. Small epiphyte; stems clustered, longer than leaves, covered with loose sheaths; leaves elliptic, narrowed basally; inflorescence short, with few flowers, 12–13 mm long; sepals united at tips, with "windows" on sides; lip smooth, lateral lobes rounded; column slender, arched, somewhat winged above; 1 species in Panama; **C:** greenish streaked with purple; **D:** ePma, SA (Mt); **M:** pl to 10 cm, S 11 mm.................
.... ***Ophidion pleurothallopsis***
Plate 22(1)
This species resembles a miniature *Zootrophion*, but the leaf, lip, and column, and especially the long, loose raceme, are distinctive.

Platystele. Plants small or tiny; creeping or shoots clustered, stems short; leaves oblanceolate to obovate; inflorescences racemose with several to many flowers; sepals and petals free; lip simple; column short and broad with 2 stigmatic lobes; pollinia 2.

Key to *Platystele*

1. Plants creeping 2
1. Plants forming clumps, not creeping......................... 4
2(1). Leaves > 7 mm long; inflorescence less than twice length of leaves; **C:** pale yellow-green; **D:** No, CR, Pma (Mt); **M:** pl 4–9 mm, S 1–1.2 mm..............................*Platystele minimiflora*
   2. Leaves < 5 mm long; inflorescence more than twice length of leaves
   3
3(2). Leaves about 2.5 mm long, subequal to spaces between leaves, suberect; **C:** translucent yellow-green, lip light tan; **D:** No, CR, Pma (Cb, Mt, Pc); **M:** pl 2–2.4 mm, S 1–1.2 mm ....
   ....*Platystele jungermannioides*
   This may be the smallest of all orchids.

3. Leaves about 5 mm long, much longer than spaces between leaves, wide and flattened against substrate; C: sepals and petals translucent pale yellow, lip yellow or red; D: No, Pma, SA (Mt, Pc); M: pl 2–6 mm, S 0.8–1.2 mm .... *Platystele ovalifolia*
Plate 19(4)

4(1). Flowers umbellate; lip constricted above middle; C: sepals and petals translucent purple, lip purple; D: Pma, SA (Cb); M: pl 3–4 cm, S 1–1.2 mm, infl < lvs .... *Platystele ortiziana*

4. Flowers racemose, scattered along inflorescence, or produced few at a time over a long time 5

5(4). Inflorescence usually shorter than leaves (very old inflorescences may be longer than leaves) 6

5. Inflorescence longer than leaves 7

6(5). Stem subequal to leaf; lip densely papillose; C: sepals and petals greenish yellow to orange, lip orange to dark purple; D: No, CR, Pma, SA (Mt, Pk); M: pl 5–10 cm, raceme becoming > lvs, S 1.3–1.8 mm .... *Platystele stenostachya*
This is one of the commonest and most distinctive species of the genus. Lip color is quite variable

6. Stem much shorter than leaf; lip smooth (minutely hairy); C: sepals and petals pale translucent greenish yellow, lip purple-black; D: Pma (Cb, Mt); M: pl 1.3–2.5 cm, raceme < lvs, S 1.3–1.5 mm .... *Platystele dressleri*

7(5). Inflorescence elongate with several scattered flowers at once 8

7. Inflorescence with with 1–3 flowers at a time, or elongating slowly with several densely clustered flowers at tip 11

8(7). Lateral sepals held together (not united), petals arching downward and widest near apices; C: pale translucent green; D: Pma (Pc); M: pl 2–3.5 cm, S 1.8–2 mm .... *Platystele calymma*

8. Lateral sepals spreading, petals straight or slightly curved, widest below middle 9

9(8). Apex of lip curved upward; C: translucent greenish yellow; D: CR, Pma (Mt); M: pl 1.5–3 cm, S 1.2 mm .... *Platystele perpusilla*

9. Apex of lip not markedly curved 10

10(9). Lip obtuse, without basal cavity, petals widest near middle; C: S, P pale translucent yellow, lip yellow to dark purple; D: No, CR, Pma (Mt); M: pl 2–4.5 cm, S 1–1.5 mm .... *Platystele ovatilabia*

10. Lip acute, with definite basal cavity, petals widest near base; C: S, P translucent, pale yellow-green, lip yellow-green; D: CR (Mt); M: pl 1.5–3 cm, S 1.7–2 mm .... *Platystele lancilabris*

11(7). Sepals wide and obtuse or abruptly acute 12

11. Sepals extended into long, narrow tails 13

12(11). Tiny plant, about 2 cm tall; petals narrow and acute, lip with thick basal callus; C: S, P pale translucent green or suffused with rose, lip yellow; D: CR (Mt, Pk); M: pl 0.9–1.7 cm, S 1–1.2 mm .... *Platystele microtatantha*
Frequent in the area of Monteverde, this species is easily recognized by the thick callus.

12. Plant 3–4 cm tall; petals broad and obtuse, lip without obvious callus at base; C: flowers yellow; D: No, CR, wPma (Mt, Pk); M: pl 2–5 cm, S 2–3 mm .... *Platystele compacta*
Common at higher elevations in central Costa Rica.

13(11). Apex of lip upturned; C: S, P translucent light purple, rose, tan, or yellow, lip red, red-brown, or orange; D: Pma, SA (Mt); M: pl 1–4.5 cm, S 2.5–4 mm........................ *Platystele resimula*

13. Apex of lip straight or slightly recurved...................... 14

14(13). Sepals with long, slender tails about twice as long as petals..... 15

14. Sepals acuminate but only slightly longer than petals 16

15(14). Dorsal sepal > 10 mm long, lip ovate, acute; C: S translucent green to yellow with red along midveins, P translucent yellow, lip red, orange or yellow; D: No, CR, Pma, SA (Cb, Mt); M: pl 1.2–2.5 cm, S 9–12 mm........................... *Platystele caudatisepala*

15. Dorsal sepal about 6 mm long, lip acuminate; C: S translucent light yellow with red midveins, P translucent yellow, lip red; D: Pma, SA (Cb); M: pl 6–8 mm, S 4–6 mm.............. *Platystele brenneri*

16(14). Leaves < 8 mm long; C: S translucent light yellow, lip red-purple; D: Pma (Pc); M: pl 4–13 mm, S 1.3–1.6 mm ....*Platystele taylori*

16. Leaves > 10 mm long 17

17(16). Inflorescence becoming more than twice as long as leaves; lip much shorter than lateral sepals; C: S, P translucent yellow-green, S may be suffused with purple, lip red-purple; D: No, CR, Pma, SA (Mt, Pk); M: pl 0.8–2.5 cm, S 2–5 mm .... *Platystele oxyglossa*

17. Inflorescence much less than twice as long as leaves; lip subequal to lateral sepals; C: S, P translucent purple, lip deep purple; D: CR (Mt, Pk); M: pl 2–2.2 cm, S 4.8–5.5 mm.... *Platystele propinqua*

Pleurothallis. This group is diverse and difficult to characterize. Usually epiphytic. The number of pollinia is always 2, but in other features *Pleurothallis* may be confused with nearly every other genus in the subtribe Pleurothallidinae. It is quite likely that *Pleurothallis* will eventually be divided into a few more uniform genera.

For purposes of identification I divide *Pleurothallis* into 3 major groups and 1 rather minor one. These are not natural groups, but the features that separate them are usually not ambiguous, and I have not needed to key out many species in different keys. Group 3 corresponds very closely to the *Specklinia* group, which may eventually be treated as a separate genus, while group 2 corresponds fairly well to the *Acianthera* group. The round-leaved species of Key 4 are probably variants of the *Specklinia* pattern. Group 1, unfortunately, is rather more diverse, and I doubt that all or most species would remain in *Pleurothallis* if the genus were eventually divided into several genera. Still, it is fairly easy to assign most *Pleurothallis* to one of these artificial categories, even without flowers. *Pleurothallis ruscifolia* is the type species of the genus, but it is not very representative of the large heart-leaved group with fascicled flowers. Appendix B indicates the major groups for the species I have not been able to key.

General Key to *Pleurothallis*

1. Leaves coinlike, nearly as wide as long, procumbent or flat on substrate; stems short; inflorescence never much longer than leaves .... **Key 4**
1. Leaves not coinlike *or* erect or on long stems; inflorescence short or long... 2

2(1). Stems short, usually less than one-fourth length of leaves; leaves tapering basally (may be narrow and stemlike at base, but this seeming "stem" is above the inflorescence) .... **Key 3**
2. Stems long, usually more than half length of leaves; leaves usually rounded or abruptly tapering basally 3

3(2). Flowers fascicled at leaf base, each on a separate stalk (or appearing so) .... **Key 1**
3. Flowers racemose on a definite rachis .... **Key 2**

Key 1
Flowers fascicled at leaf base; stems relatively long; *P. cardiothallis* group, *P. ruscifolia* group.

1. Stems flattened and 2-edged, at least near leaf; peduncles very long, mostly more than half leaf length 2
1. Stems not flattened; peduncles usually much shorter............. 6
2(1). Lip 3-lobed 3
2. Lip not 3-lobed... 4
3(2). Lateral lobes of lip long and narrow, acute; leaves wide and folded basally; **C**: S, P white to yellowish, with or without purple spots, lip dark purple; **D**: No, CR, Pma, SA (Mt); **M**: pl 15–30 cm, pedicels 3–6.5 cm, S 8–14 mm .... *Pleurothallis crocodiliceps*
(Syn. *P. arietina*)
3. Lateral lobes of lip short and triangular, obtuse; leaves tapering basally; **C**: S rose-white with rose spots, P white, lip rose dotted and suffused with red-purple; **D**: CR (Mt); **M**: pl 7–12 cm, pedicels 2–2.3 cm, S 10 mm .... *Pleurothallis caniceps*
4(2). Lip hooked, tip curved up and back; **C**: S, P white with purple flecks or hairs, lip white marked with purple; **D**: ePma (Mt); **M**: pl 10–15 cm, pedicels 2.7–3.3 cm, S 4 mm. ... *Pleurothallis harpago*
4. Lip straight 5
5(4). Sepals 6–8 mm long; lip with low callus above middle; **C**: white, S may be dotted with purple, lip marked with purple; **D**: CR, Pma (Mt); **M**: pl 8–20 cm, pedicels 2–3.5 cm, S 6–7 mm............. .... *Pleurothallis eumecocaulon*
5. Sepals 4–5 mm long; lip without callus above middle; **C**: S and lip lavender spotted with purple, P white spotted with purple; **D**: wPma (Mt); **M**: pl 6–15 cm, pedicels 2–2.5 cm, S 4.5–5 mm ....*Pleurothallis instar*
6(1). Lip deeply 3-lobed, lateral lobes equal to or surpassing midlobe; leaves long-tapering; **C**: greenish white, midlobe of lip red-purple; **D**: CR (Cb, Mt); **M**: pl 15–30 cm, S 6–8 mm................... .... *Pleurothallis fantastica*
6. Lip entire or 3-lobed, but midlobe always longer than lateral lobes 7
7(6). Lip fleshy, toothed, both sides and apex curled under, rigid, surface warty; **C**: S yellowish with brownish veins and suffused with brown or dull purple, P red-brown, lip deep purple; **D**: CR, Pma (Mt); **M**: pl 15–30 cm, S 10–13 mm .... *Pleurothallis tonduzii*
7. Lip never with both sides and apex curled under............... 8

8(7). Base of leaf tapering to stem 9
  8. Base of leaf rounded or notched 23
9(8). Leaves broadly elliptic to ovate, abruptly narrowed to a narrow base above inflorescence 10
  9. Leaves elliptic to obovate, gradually narrowed basally 13
10(9). Sepals united for more than half their length; tapering abruptly; C: pale orange-yellow; D: CR (Mt); M: pl 6–8 cm, S 5 mm .... *Pleurothallis lentiginosa*
  10. Dorsal sepal free to base, sepals and petals tapering and acute 11
11(10). Leaves elliptic; synsepal lanceolate; C: yellow-green to white; D: No, CR, Pma, SA (Cb, Mt); M: pl 10–40 cm, S 7 mm............ .... *Pleurothallis ruscifolia*
  11. Leaves ovate; synsepal ovate............................... 12
12(11). Lip uppermost, shaped like horse collar (or toilet seat), clasping column; petals narrowly lanceolate, > 8 mm long; C: S translucent with purple stripes, P purple; D: CR, Pma (Mt); M: pl 25–30 cm, S 11–12 mm .... **Pleurothallis crescentilabia**
Plate 23(1)
  12. Lip lowermost, ovate, not clasping column; petals < 6 mm long, widest above middle; C: yellow; D: CR, Pma, SA (Mt); M: pl 30–80 cm, S 7 mm ....*Pleurothallis ventricosa*
13(9). Lateral sepals free for most of length; leaves obovate; C: white; D: No, SA; M: pl 15 cm, S 4 mm............. **Pleurothallis obovata**
  13. Lateral sepals united to apex or nearly so; leaves elliptic, ovate, or lanceolate 14
14(13). Lip distinctly 3-lobed, midlobe ovate to triangular............. 15
  14. Lip oblong to semicircular, with or without small basal lobules clasping column, but not distinctly 3-lobed.................. 22
15(14). Lateral lobes of lip narrow, more or less acute 16
  15. Lateral lobes of lip broad, rounded, midlobe triangular 19
16(15). Lateral lobes of lip longer than midlobe, apex of lip acute or obtuse; C: yellow; D: CR (Pc); M: pl 10 cm, S 3–4 mm ....*Pleurothallis aurita*
  16. Lateral lobes shorter than midlobe, apex of lip acuminate, upturned 17
17(16). Petals obovate; apex of lip short; lateral lobes triangular, midlobe elliptic, shallowly concave; C: violet; D: No, rCR; M: pl 10 cm, S 4.2–4.5 mm .... **Pleurothallis saccatilabia**
  17. Petals oblong or oblong-spatulate, apex of lip slender, tapering, lateral lobes finger-like, midlobe ovate, deeply concave........... 18

18(17). Leaves elliptic or lance-elliptic, acute; synsepal shallowly concave, sepals strongly papillose within; C: deep purple; D: CR, Pma (Mt); M: pl 5–10 cm, S 4–5 mm .... *Pleurothallis excavata*
18. Leaves broadly ovate, blunt or notched; synsepal deeply concave, hoodlike, sepals smooth within or nearly so; C: yellowish flecked with purple or brown; D: Pma (Mt); M: pl 5–10 cm, S 4.5–6 mm .... *Pleurothallis imago*
19(15). Petals markedly toothed on both margins 20
19. Petals smooth or slightly toothed on 1 margin 21
20(19). Dorsal sepal and synsepal tapering to narrow points; lip distinctly toothed; C: maroon; D: Pma (Mt); M: pl 5–10 cm, S 7 mm .... *Pleurothallis simulans*
20. Dorsal sepal and synsepal ovate, acute, but not abruptly narrowed apically; lip only minutely toothed; C: S yellow-green marked with purple, P and lip maroon; D: Pma (Mt); M: pl 10–15 cm, S 7–8 mm .... *Pleurothallis cobraeformis*
21(19). Petals elliptic, narrowly acute, slightly S-shaped; midlobe of lip warty; C: dark red-purple; D: Pma (Mt); M: pl 6–15 cm, S 14–15 mm .... *Pleurothallis allenii*
21. Petals oblong, abruptly acute, curved; midlobe smooth; C: yellow striped or blotched with red; D: Pma (Cb, Mt); M: pl 6–15 cm, S 4–5 mm .... *Pleurothallis archicolonae*
22(14). Leaves broadly elliptic; synsepal wider than long; lip with small basal lobules; C: S light green with few purple specks, P purple, lip yellow marked with purple; D: CR; M: pl 20–25 cm, S 4 mm .... *Pleurothallis dorotheae*
22. Leaves narrowly elliptic or lance-elliptic; synsepal longer than wide; lip without basal lobules; C: greenish yellow; D: No, rCR; M: pl 7–14 cm, S 3.3 mm .... *Pleurothallis leucantha*
23(8). Leaf base rounded to slightly notched, leaves ovate or elliptic to lanceolate 24
23. Leaf base deeply notched and lobed, lobes usually overlapping, leaves usually broadly ovate 30
24(23). Lip deeply saccate with finger-like basal lobes clasping column and upcurved, acuminate apex 25
24. Lip not as above 26
25(24). Leaves elliptic or lance-elliptic, acute; synsepal shallowly concave, sepals strongly papillose within; C: deep purple; D: CR, Pma (Mt); M: pl 5–10 cm, S 4–5 mm .... *Pleurothallis excavata*
25. Leaves broadly ovate, blunt or notched; synsepal deeply concave, hoodlike, sepals smooth within or nearly so; C: yellowish flecked

with purple or brown; **D:** Pma (Mt); **M:** pl 5–10 cm, S 4.5–6 mm
.... *Pleurothallis imago*

26(24). Petals fringed; leaves shiny; **C:** deep purple; **D:** wPma (Mt); **M:** pl 5–15 cm, pedicels 7–10 mm, S 5–6 mm ...... *Pleurothallis nitida*
26. Petals smooth or minutely toothed 27

27(26). Lip shallowly 3-lobed, lateral lobes rounded, midlobe subtriangular; pedicels about half leaf length; **C:** reddish with dark red flecks; **D:** Pma (Mt); **M:** pl 10–15 cm, S 4–5 mm....*Pleurothallis annectans*
27. Lip semicircular to oblong, may have basal lobules clasping column but not clearly 3-lobed.................................... 28

28(27). Lip about as wide as long, base clasping column; **C:** dS yellow faintly mottled with purple, synS yellow-green, P purple, lip purple with white center; **D:** No, CR, Pma (Mt); **M:** pl 20–30 cm, S 4.5–5 mm..................................*Pleurothallis sanchoi*
28. Lip oblong or ovate, basal lobules very small................. 29

29(28). Petals lanceolate tapering to very narrow tips; synsepal also tapering to narrow tip, dorsal sepal oblong; **C:** reddish greenish brown with red lip; **D:** CR (Mt); **M:** pl 15–20 cm, S 9–10 mm...............
.... *Pleurothallis nervosa*
29. Petals linear-oblong, tapering abruptly at tips; sepal and synsepal ovate 30

30(29). Lip ovate, widest near base; cavity at base of lip long and narrow; **C:** S, P light yellow-green to pale yellow, lip orange to yellow-orange; **D:** CR, Pma (Mt); **M:** pl 7–20 cm, S 5.5–8 mm..........
....................................*Pleurothallis isthmica*
30. Lip oblong, widest near middle, cavity at base of lip subcircular
.... 31

31(30). Sepals 8–10 mm long; lip subquadrate-oblong, more than half as wide as long; **C:** dS, P rose, synS white, lip maroon; **D:** wPma (Mt); **M:** pl 10–20 cm, S 8–10 mm.............*Pleurothallis hemileuca*
31. Sepals 13–15 mm long; lip oblong, about a third as wide as long; **C:** S, P yellow, lip red; **D:** Pma (Mt); **M:** pl 20–35 cm, S 13.5–15 mm
....................................*Pleurothallis telamon*

32(23). Lip 3-lobed, lateral lobes triangular; sepals and petals hairy within; leaves long tapering; **C:** dark red; **D:** Pma (Mt); **M:** pl 20–40 cm, S 8–9 mm .... *Pleurothallis peculiaris*
An aptly named species.
32. Lip circular to oblong or shallowly lobed 33

33(32). Dorsal sepal deeply concave, wider than synsepal; flowers relatively large.. 34
33. Dorsal sepal longer than wide or narrower than synsepal, usually not deeply concave; flowers relatively small 35

34(33). Petals falcate-oblong, about as wide near apex as near base, not markedly toothed; C: S yellowish, dorsal tinged with dark red, lip deep orange; D: Pma (Mt); M: pl 20–25 cm, S 1.8–2 cm .... *Pleurothallis cardiochila*
34. Petals strongly falcate tapering to narrow apices, distinctly saw-toothed on both margins; C: dS brown or yellow streaked with brown, lS pale yellow or streaked with brown, P and lip red or pinkish red; D: CR, Pma (Mt, Wt); M: pl 20–25 cm, S 1.5–2 cm .... *Pleurothallis palliolata*
Plate 23(3)

35(33). Lip rough, coarsely warty; petals toothed; spathe prominent, erect; C: yellow; D: CR (Mt); M: pl 15–35 cm, S 14 mm .... *Pleurothallis radula*
35. Lip smooth or minutely warty; spathe usually small, procumbent 36

36(35). Lip subquadrate-oblong, squarish 37
36. Lip ovate or oblong, rounded.............................. 41

37(36). Petals linear-oblong, curved; lip minutely warty; C: reddish brown; D: CR, Pma (Mt); M: pl 12–20 cm, S 7 mm ....*Pleurothallis homalanthoides*
37. Petals lanceolate to linear-lanceolate, straight or nearly so...... 38

38(37). Dorsal sepal lanceolate, narrower than synsepal; C: yellow to reddish brown; D: CR, SA; M: pl 12–20 cm, S 5–6 mm............ .... *Pleurothallis undulata*
38. Dorsal sepal oblong to ovate, subequal to synsepal............ 39

39(38). Sepals < 6 mm long; C: yellow; D: No, CR, Pma, SA (Mt); M: pl 10–15 cm, S 3 mm................*Pleurothallis phyllocardioides*
39. Sepals > 10 mm long.. 40

40(39). Dorsal sepal 10–18 mm long; lip widest basally; C: yellow to orange or red-orange; D: No, CR, Pma, SA (Cb, Mt); M: pl 40–50 cm, S 1.1–1.8 cm ....*Pleurothallis cardiothallis*
Plate 23(2)
40. Dorsal sepal 3–3.5 cm long; lip widest above middle; C: S, P pale rose-brown, lip pale yellow; D: wPma (Mt); M: pl 20–35 cm, S 3–3.5 cm...................................... *Pleurothallis titan*

41(36). Lip triangular-ovate, widest at base, with shallow concavity at each side of small basal cavity; spathe erect, prominent; C: deep purple; D: CR, Pma (Mt); M: pl 15–40 cm, S 11–15 mm.. *Pleurothallis phyllocardia*
41. Lip oblong or ovate, without paired concavities.............. 42
42(41). Sepals > 10 mm long; with prominent, erect spathe; leaves concave; C: yellow striped with purple; D: wPma (Mt); M: pl 15–35 cm, S 11–13 mm .... *Pleurothallis scitula*
42. Sepals < 10 mm long; spathe inconspicuous or procumbent.... 43
43(42). Dorsal sepal 5-veined 44
43. Dorsal sepal 3–veined 45
44(43). Sepals 8–9 mm long; C: red-brown; D: No, CR, Pma (Mt); M: pl 15–30 cm, S 8–9 mm .... *Pleurothallis homalantha*
44. Sepals 5–6 mm long; C: brownish yellow with reddish veins; D: No, CR, Pma (Mt); M: pl 12–15 cm, S 5–8 mm .... *Pleurothallis nemorum*
45(43). Lip triangular-ovate; petals minutely serrate; C: S yellow, P orange to red-brown, lip yellow suffused with red-brown; D: No, Pma (Mt); M: pl 15–30 cm, S 6–7 mm .... *Pleurothallis antonensis*
45. Lip ovate; petal margins smooth 46
46(45). Dorsal sepal fleshy, 3-keeled externally; C: dS yellow, synS pale brown, P red, lip yellow flecked with brown; D: Pma (Mt); M: pl 10–15 cm, S 4.5–5.5 mm ....*Pleurothallis veraguacensis*
46. Dorsal sepal membranous, without keels; C: S, P translucent white, lip red-purple; D: CR, Pma (Pc); M: pl 12–20 cm, S 4.5–6 mm .... *Pleurothallis rhodoglossa*

Key 2

Plants with long stems; inflorescences racemose or stalked; *Acianthera* group.

1. Inflorescence basal, racemose; C: deep red-purple, lip yellow marked with purple; D: No, CR, Pma (Mt); M: pl 10–25 cm, S 10–14 mm .... ***Pleurothallis johnsonii***

    (Syn. *Brenesia costaricensis*)

    The inflorescence may be borne either at the stem tip or from the rhizome; both types of inflorescence are sometimes found on the same plant.
1. Inflorescence terminal 2
2(1). Leaves ovate, base broadly rounded, notched or lobed.......... 3

2. Leaves usually elliptic (or tapering basally above inflorescence) 13

3(2). Large plants, 0.6–1 m tall; with multiple erect inflorescences; dorsal sepal markedly concave.. 4
3. Plants < 0.5 m tall, with single arching or spreading inflorescence; dorsal sepal not markedly concave 6

4(3). Lip much wider than long (when flattened), semicircular; C: white dotted with dark purple; D: ePma (Mt); M: pl 1 m, raceme to 20 cm, S 2.5–3.5 mm .... *Pleurothallis praegrandis*
4. Lip longer than wide, shallowly 3-lobed 5

5(4). Petals acute, toothed; C: S pale rose-brown, P and lip yellow; D: CR (Mt); M: pl 30–35 cm, raceme sub=lvs, S 5.5 mm .... *Pleurothallis stevensii*
5. Petals blunt, not toothed; C: cream spotted with purple; D: CR (Mt); M: pl 1 m, raceme to 10 cm, S 4.5–5 mm................ .... *Pleurothallis saccata*

6(3). Stem thickened and 3-angled, especially above; inflorescence flat on leaf, shorter than leaf; C: S yellow or green spotted with dark purple, lip brown; D: No, CR, Pma, SA (Cb, Mt, Pc); M: pl 10–20 cm, raceme > lvs, S 6–12 mm ....*Pleurothallis cogniauxiana*
Plate 24(1)
6. Stem slender, cylindrical; inflorescence arching or erect, subequal to leaf or longer.. 7

7(6). Sepals hairy; leaf narrowed to a short petiole above leaf; lip with distinct claw; inflorescence normally borne behind leaf; C: yellow, yellow spotted with purple, or black-purple; D: No, CR, Pma, SA (Mt); M: pl 20–50 cm, S 6 mm..............*Pleurothallis imraei*
Common in some areas; the flowers may be nearly black and are normally borne beneath the leaves, where they are easily overlooked.
7. Sepals smooth; leaf without petiole; lip not clawed; inflorescence borne above leaf... 8

8(7). Petals obovate, toothed or fringed........................... 9
8. Petals narrow with smooth margins 11

9(8). Plant creeping, < 8 cm tall; few-flowered; petals toothed, acuminate; C: S dull yellow with red-brown veins, P translucent with red midvein, lip brown; D: wPma (Mt); M: pl 2–3.5 cm, raceme > lvs, S 13–14 mm ....*Pleurothallis juxtaposita*
9. Shoots clustered, > 10 cm tall; many-flowered; petals fringed or toothed.. 10

10(9). Petals fringed, blunt or subacute; C: purplish; D: CR, Pma (Mt); M: pl 15–35 cm, raceme > lvs, S 4–5 mm ....
.... *Pleurothallis dentipetala*
10. Petals toothed, narrowly acute; C: S pale rose-brown, P and lip yellow; D: CR (Mt); M: pl 30–35 cm, raceme sub=lvs, S 5.5 mm
.... *Pleurothallis stevensii*
11(8). Flowers > 1 cm long, dark purple; pedicels much shorter than flowers; C: S maroon, P yellow-green flecked with purple, lip orange-yellow; D: CR, Pma (Mt); M: pl 35–50 cm, raceme > lvs, S 1.7 cm..*Pleurothallis powellii*
11. Flowers much less than 1 cm long, pale green or yellow; pedicels longer than flowers... 12
12(11). Lip oblong-ovate; sepals 6–8 mm long; C: S dull red-purple, P orange-brown, lip yellow-brown; D: CR, Pma (Mt, Pk); M: pl 20–40 cm, raceme sub=lvs, pedicels 0.8–2 cm, S 6–8 mm ....
.... *Pleurothallis longipedicellata*
12. Lip shallowly 3-lobed, midlobe narrowly triangular; sepals 2.5–4 mm long; C: pale rose, lip pale yellow; D: Pma (Mt); M: pl 10–20 cm, raceme > lvs, S 3.5–4 mm .... *Pleurothallis volcanica*
13(2). Inflorescence subequal to leaves or much longer............... 14
13. Inflorescence distinctly shorter than leaves.................... 44
14(13). Dorsal sepal markedly united with lateral sepals, base of flower tubular or cuplike, sepals all similar 15
14. Dorsal sepal free to base 16
15(14). Sepals united to well above middle; petals widest apically; D: CR (Mt); M: pl 4–12 cm, 1-sided raceme > lvs, S 3.5 mm...........
.... *Lepanthopsis obliquipetala*
(Syn. *Pleurothallis connata*)
15. Sepals united only basally; petals ovate; C: greenish or yellowish white; D: No, CR, Pma, SA (Mt); M: pl 6–20 cm, raceme > lvs, S 2.5–3 mm .... *Pleurothallis deregularis*
16(14). Lateral sepals free for at least half their lengths 17
16. Lateral sepals united to apices or notched................... 22
17(16). Sepals < 10 mm long....................................... 18
17. Sepals > 10 mm long....................................... 19
18(17). Apex of lip squarish; sepals hairy within; C: translucent yellow-green; D: No, CR, Pma, SA (Mt); M: pl 30–50 cm, raceme sub=lvs, S 5–8 mm .... *Pleurothallis gelida*
18. Apex of lip acute; sepals minutely fringed; D: CR, Pma, SA (Mt); M: pl 40–50 cm, raceme < lvs, S 5–5.5 mm ....
.... *Pleurothallis floribunda*

19(17). Stems shorter than leaves; bracts prominent, about as long as pedicels; C: golden ocher, lip brown-ocher; D: No, CR, Pma, SA (Mt); M: pl to 15 cm, raceme > lvs, S 1.2–1.7 cm .... *Pleurothallis erinacea*
        Similar to P. endotrachys but floral bracts longer than pedicels.
19. Stems much longer than leaves............................ 20
20(19). Sepals elliptic-oblong, > 2 cm long; C: S greenish tan flushed with pink, P pale green; D: CR (Mt); M: pl to > 1 m, raceme to 50 cm, S 3 cm...................................... *Pleurothallis grandis*
        A giant among the pleurothallids.
20. Sepals lanceolate or linear-lanceolate, long-tapering, < 2 cm long 21
21(20). Lateral sepals united for about a third their length and then spreading; lip spoon-shaped; C: S dark purple, each with white stripe, P brown, lip red-brown; D: Pma (Cb, Mt); M: pl 17–23 cm, raceme > lvs, S 14 mm......................... *Pleurothallis mystax*
21. Lateral sepals free to base, finely hairy within; lip oblong-ovate, shallowly 3-lobed; C: pale yellow; D: No, CR, Pma, SA (Mt); M: pl 10–25 cm, raceme > lvs, S 2 cm........ *Pleurothallis sclerophylla*
        (Syn. P. dolichopus)
22(16). Sepals bristly-hairy, bristles long and obvious; inflorescence or plant pendent... 23
22. Sepals smooth or finely hairy, not bristly; plants erect 24
23(22). Petals and lip fringed; dorsal sepal oblong, abruptly tapering; stems ascending or spreading; inflorescence pendent; C: S green, P translucent white, lip pink, all heavily marked with purple; D: No, Pma (Mt); M: pl 12–25 cm, raceme > lvs, S 12–14 mm .... *Pleurothallis oscitans*
23. Petals and lip smooth or minutely warty; dorsal sepal linear-lanceolate; plant pendent; C: leaves purplish, S shiny yellow-green with purple spots, lip brown; D: wPma (Mt); M: pl 10–25 cm, raceme sub=lvs, S 15 mm...................... *Pleurothallis butcheri*
24(22). Petals widest near apices, blunt, square or abruptly acute....... 25
24. Petals lanceolate, acuminate 37
25(24). Synsepal deeply concave, much wider than dorsal sepal; flowers with abundant hair within; C: green or pale yellow; D: CR, Pma (Mt); M: pl 8–12 cm, raceme > lvs, S 7 mm ...*Pleurothallis amparoana*
25. Sepals not as above 26

26(25). Petals abruptly acute.................................... 27
26. Petals rounded to square or notched....................... 29
27(26). Petals oblanceolate, strongly curved and parallel with synsepal; C: greenish white or pale brownish green to deep purple; D: CR, Pma, SA (Mt); M: pl 10–30 cm, raceme > lvs, S 8–13 mm ....
....*Pleurothallis rowleei*
27. Petals rhombic, spreading................................ 28
28(27). Leaf ovate-elliptic; dorsal sepal and synsepal ovate or elliptic-ovate, smooth; petals stalked; C: S translucent yellow with or without reddish brown spots, P orange speckled with red-brown, lip orange; D: CR, Pma (Mt); M: pl 5–12 cm, raceme > lvs, S 11–12 mm.......
.... *Pleurothallis carpinterae*
28. Leaf oblanceolate; dorsal sepal lanceolate, synsepal narrowly elliptic, finely hairy externally; petals not stalked; C: maroon or semi-transparent with maroon stripes; D: CR, Pma (Mt); M: pl 10–15 cm, raceme > lvs, S 12–16 mm............*Pleurothallis pompalis*
The right-angled lateral lobes of the lip are distinctive.

29(26). Petal apices square or notched 30
29. Petal apices rounded..................................... 31
30(29). Petals and lip distinctly bilobed, lip tubular, shaped like double-bladed ax head if flattened, sepals white-hairy within; C: S reddish brown to purple, P and lip maroon; D: No, CR, Pma (Mt); M: pl 12–30 cm, 1-sided raceme > lvs, S 8– 10 mm..................
.... *Pleurothallis convallaria*
30. Petals and lip apices squarish, lip in-rolled to form tube, petals cuneate, sepals smooth; C: strongly mottled with red-brown to purple, dS and P 3-striped, lip orange suffused with purple; D: Pma (Mt, Pk); M: pl 10–17 cm, raceme > lvs, S 1.8–2.1 cm...............
.... *Pleurothallis macrantha*
31(29). Sepals < 5 mm long; lip obovate, rounded; petals oblong, curved inward, fleshy, smooth; C: pale green; D: Pma (Mt, Pc); M: pl 5–10 cm, raceme > lvs, S 3–4.5 mm ....*Pleurothallis mammillata*
31. Sepals > 5 mm long; lip lanceolate, fleshy; petals obovate, warty
.... 32
32(31). Inflorescence with prominent sheath up to 7 cm long; C: S dark red externally, synS flecked with purple within, P white marked with purple; D: No, CR, Pma (Mt); M: pl 15–40 cm, raceme > lvs, S 2.5–2.6 cm....................... *Pleurothallis tuerckheimii*
32. Spathe small and inconspicuous............................ 33

33(32). Dorsal sepal narrowly lanceolate, tubular above, synsepal in-rolled; C: S purple-brown, P yellow with purple veins, lip yellow with red-purple suffusion; D: Pma (Mt); M: pl 7–12 cm, zigzag raceme > lvs, S 2.1–2.2 cm .... *Pleurothallis fortunae*

33. Dorsal sepal broadly lanceolate or ovate-lanceolate, not tubular 34

34(33). Lip curved downward, lower surface smooth or nearly so; C: deep purple, P yellow marked with purple; D: CR, rPma (Mt); M: pl 14–20 cm, raceme > lvs, S 1.5–2 cm .... *Pleurothallis ramonensis*

34. Apex of lip curved upward, lower surface distinctly warty 35

35(34). Midlobe of lip ovate or elliptic-ovate in outline; C: maroon, lateral lobes of lip white; D: Pma (Pc); M: pl 8–15 cm, raceme > lvs, S 11 mm................................ *Pleurothallis thymochila*

35. Midlobe of lip lanceolate, apex long and tapering............ 36

36(35). Sepals smooth within; upper surface of lip with warty ridges; C: S dark violet-purple, P white with few purple streaks; D: No, CR (Wt); M: pl 20–30 cm, raceme > lvs, S 13–15 mm ....*Pleurothallis pachyglossa*

36. Sepals hairy within; upper surface of lip smooth or nearly so; C: S dark purple, P and lip white marked with purple; D: CR (Mt); M: pl 10–20 cm, raceme > lvs, S 2.1–2.2 cm .... *Pleurothallis dracontea*

37(24). Lip oblong, apex rounded 38

37. Lip narrowed apically, acute to long-tailed 39

38(37). Leaf leathery, tapering below; dorsal sepal elliptic-ovate, similar to synsepal; C: yellow to greenish yellow, more or less translucent; D: No, CR, Pma, SA; M: pl 25 cm, raceme > lvs, S 9–11 mm .... *Pleurothallis racemiflora*
(Syns. *P. ghiesbreghtiana*, *P. quadrifida*)
Plate 24(3)

A common species north of our area; some self-pollinating forms are rather drab.

38. Leaf fleshy, elliptic; dorsal sepal oblanceolate, longer and much narrower than synsepal; C: yellow with reddish brown spots; D: CR, Pma, SA (Cb?); M: pl 8–15 cm, raceme > lvs, S 12–13 mm .... *Pleurothallis lanceana*

39(37). Inflorescence from prominent spathe that is about a third as long as leaf and diverges from leaf blade; petals narrowly lanceolate; lip shallowly 3-lobed; C: S, P translucent pale yellow-green to pale pink, spotted with purple; D: CR, Pma, SA (Cb, Mt); M: pl 10–20 cm, raceme > lvs, S 8 mm............ *Pleurothallis loranthophylla*

39. Sheath much smaller, or appressed to leaf blade 40
40(39). Leaves elliptic to ovate, about half as wide as long 41
40. Leaves lanceolate to narrowly elliptic, much less than half as wide as long... 42
41(40). Leaves elliptic; lip ovate, not lobed; C: pale green or S flushed with rose; D: Pma (Mt); M: pl 6–12 cm, raceme > lvs, S 7–8 mm, pedicels 0.8–1.7 cm .... *Pleurothallis pallida*
41. Leaves ovate; lip triangular, shallowly 3-lobed; C: pale rose, lip pale yellow; D: Pma (Mt); M: pl 10–20 cm, raceme > lvs, S 3.5–4 mm .... *Pleurothallis volcanica*
42(40). Dorsal sepal and synsepal narrowly lanceolate, width about one-fourth length; C: purple; D: No, CR, Pma, SA (Mt); M: pl 10–18 cm, raceme > lvs, S 1.6–2 cm .... *Pleurothallis luctuosa*
42. Dorsal sepal and synsepal ovate or triangular-ovate, about half as wide as long... 43
43(42). Lip with basal lobules clasping column; C: greenish white with light purple flecks; D: ePma (Mt); M: pl 7–12 cm, raceme sub=lvs, S 6 mm................................... *Pleurothallis polysticta*
43. Lip without basal lobules; C: pale yellowish green; D: No, CR, Pma, SA (Mt); M: pl 3–10 cm, raceme > lvs, S 2.5–4 mm........ .... *Pleurothallis pruinosa*
44(13). Leaves ovate or broadly elliptic, blade narrowed above base (where inflorescence is attached); usually with multiple inflorescences 45
44. Leaves elliptic or lance-elliptic, not narrowed above base; inflorescences solitary or 1 at a time 47
45(44). Dorsal sepal and synsepal lanceolate, smooth; petals linear-lanceolate; C: greenish yellow with red dots on lip; D: No, SA; M: pl to 50 cm, raceme 10 cm, S 6–9 mm .... *Pleurothallis helleri*
45. Dorsal sepal ovate 46
46(45). Leaves broadly ovate; petals much broader apically; lip not 3-lobed; flowers strongly hairy within; C: yellow, yellow spotted with purple, or black-purple; D: No, CR, Pma, SA (Mt); M: pl 20–50 cm, short raceme, S 6 mm......................... *Pleurothallis imraei*
> Common in some areas; the flowers may be nearly black, and they are normally borne beneath the leaves, where they are easily overlooked.

46. Leaves oblong or oblong-ovate; petals oblong; lip 3-lobed; flowers sparsely hairy within; D: CR, Pma, SA (Mt); M: pl 40–50 cm, raceme < lvs, S 5–5.5 mm .... *Pleurothallis floribunda*

47(44). Plant pendent, purplish, stem flexible; leaves elliptic; petals linear, attached to base of column; synsepal much wider than dorsal sepal; **C:** green with maroon stripes; **D:** wPma (Mt); **M:** pl 15–25 cm, raceme < lvs; S 13–14 mm................ *Pleurothallis lepidota*
47. Plant erect or spreading, stem stiff; petals not attached to base of column; not with above combination of features.............. 48

48(47). Stems slender, cylindrical.................................... 49
48. Stems 3-cornered above, grooved, often fleshy................ 55

49(48). Dorsal sepal two-thirds united with lateral sepals, sepals warty externally; flowers < 5 mm long; **C:** pale orange-yellow; **D:** CR (Mt); **M:** pl 6–8 cm, fls fasc, S 5 mm .... *Pleurothallis lentiginosa*
49. Dorsal sepal free from lateral sepals........................ 50

50(49). Lip spoonlike, with erect basal lobes, midlobe obtuse, concave **C:** S maroon, lip white; **D:** No, Pma (Cb); **M:** pl 5–10 cm, raceme < lvs, S 8–10 mm......................... *Pleurothallis carnosilabia*
(Syn. *P. caligularis*)
50. Lip entire, or lobes well above base 51

51(50). Sepals < 4 mm long.. 52
51. Sepals > 4 mm long.. 53

52(51). Flowers about 2 mm long, subglobose; lip fleshy, rhomboid; **C:** yellow-green suffused or marked with purple; **D:** wPma, SA (Mt); **M:** pl 7–20 cm, raceme sub=lvs, S 2 mm .... *Pleurothallis divaricans*
52. Flowers 3–4 mm long, oblong, on short raceme; lip ovate, clawed; **D:** CR; **M:** pl 5–8 cm, S 3.5 mm .... *Pleurothallis listerophora*

53(51). Lip widest near base, warty; lateral sepals united only basally; **C:** deep red-purple, lip yellow marked with purple; **D:** No, CR, Pma (Mt); **M:** pl 10–25 cm, S 10–14 mm ...... *Pleurothallis johnsonii*
(Syn. *Brenesia costaricensis*)
The inflorescence may be borne either at the stem tip or from the rhizome; both types of inflorescence are sometimes found on the same plant.
53. Lip widest above base, not warty; lateral sepals at least half united 54

54(53). Sepals warty externally, smooth within; **C:** S light greenish, P translucent yellow, lip yellow; **D:** Pma (Mt); **M:** pl 10–18 cm, raceme < lvs, S 7.5–8 mm .... *Pleurothallis cucumeris*
54. Sepals smooth externally, hairy within; **C:** S yellow-green, P white, lip greenish white, all speckled with purple; **D:** wPma (Mt); **M:** pl 5–10 cm, raceme < lvs, S 5 mm .... *Pleurothallis campicola*

55(48). Base of inflorescence clasped by leaf; inflorescence appearing to arise near middle of leaf 56
55. Inflorescence arising at base of leaf......................... 57
56(55). Midlobe of lip about one-third length of lip; synsepal wider than long, much shorter than dorsal sepal; stem of about same thickness throughout; C: olive green; D: No, CR (Mt); M: pl to 25 cm, raceme < lvs, **from midleaf,** S 4–6 mm ....
.... *Pleurothallis circumplexa*
Plate 23(4)
56. Midlobe of lip about half length of lip; synsepal longer than wide, subequal to dorsal sepal; stem much thicker above than near base; C: greenish becoming orange with age; D: No, CR, SA; M: pl 8–12 cm, short raceme, **from midleaf,** S 5.5 mm ....
.... *Pleurothallis pacayana*
The name was first published as *gacayana*, a spelling that has been in use recently, but the name is clearly based on Volcán Pacaya, where the type was collected, so the *g* must be considered a misspelling.
57(55). Stem strongly winged above, wings continuous with leaf margins 58
57. Stem 3-angled, but not strongly winged above 60
58(57). Leaf blades lanceolate; sepals warty within; lip with rounded lateral lobes near base; C: dark red-brown; D: No, CR (Mt); M: pl 15–20 cm, short raceme, S 5–6.5 mm........... *Pleurothallis pantasmi*
58. Leaf blades elliptic; sepals smooth within; lip with acute lobes near middle... 59
59(58). Sepals concave throughout, not spreading widely; midlobe of lip distinctly toothed; C: S olive green striped with purple, lip yellow marked with purple; D: CR, Pma (Pc, Mt); M: pl 12–35 cm, raceme < lvs, S 6–7 mm......................... *Pleurothallis alpina*
59. Sepals concave basally, flattened and spreading apically; midlobe of lip without teeth; C: yellowish with purple stripes; D: CR, Pma, SA; M: pl 20–30 cm, **short** raceme, S 7–10 mm... *Pleurothallis sicaria*
60(57). Lip with prominent claw much longer than wide; blade squarish, widest near apex; petals cuneate, truncate, C: green; D: Pma (Mt); M: pl 10–15 cm, raceme 4 cm, S 9–10 mm ....
.... *Pleurothallis aberrans*
60. Lip rounded or acute, without a claw, or claw short and wide.....
.... 61
61(60). Petals triangular-ovate from narrow claw, margins deeply toothed or fringed; lip with large basal pit; C: dS white with 3 purple veins,

laterals veined externally, purple spotted within, P white with purple markings, lip purple; D: No, SA; M: pl 10–20 cm, raceme 3 cm, S 6 mm...*Pleurothallis pubescens*
61. Petals oblanceolate to obovate from a wide claw, entire or minutely toothed 62
62(61). Petals oblanceolate; C: yellow-green suffused and veined with red-brown; D: Pma (Mt); M: pl 6–12 cm, raceme < lvs, S 7–8 mm ....*Pleurothallis citrophila*
62. Petals obovate or elliptic 63
63(62). Petals finely sharp-toothed, narrowly acute to acuminate 64
63. Petals smooth or blunt-toothed, obtuse 65
64(63). Apex of lip rounded, lip widest above middle; petals obovate, acuminate; C: purplish; D: CR, Pma, SA (Cb, Pc); M: pl 8–15 cm, raceme < lvs, S 9–11 mm............ *Pleurothallis geminicaulina*
Plate 23(6)
64. Apex of lip acute, lip widest near base; petals elliptic-lanceolate, acuminate; C: light green; D: Pma (Cb, Pc); M: pl 10–15 cm, raceme < lvs, S 14 mm............... *Pleurothallis ellipsophylla*
65(63). Lateral lobes of lip low, triangular; lip rounded-subtruncate, widest near apex; C: S dull yellow suffused with red-purple or red-brown; D: CR, Pma, SA (Pc, Mt); M: pl 10–20 cm, raceme < lvs, S 5–7.5 mm................................. *Pleurothallis decipiens*
65. Lateral lobes narrow, acute; lip widest near middle; C: green mottled with purple; D: No, CR, Pma, SA (Cb, Pc); M: pl 15–30 cm, raceme < lvs, S 7–8 mm.................*Pleurothallis verecunda*

Key 3

Plants with short stems, leaves tapering basally; leaves erect if subcircular; *Specklinia* group.

1. Floral bracts conspicuous, keeled, flattened, 5–15 mm long; flowers produced successively; plants relatively large, leaves 10–12 cm long 2
1. Floral bracts inconspicuous, not markedly keeled or flattened; mostly small plants.. 3
2(1). Ovary and fruit warty or spiny; peduncle cylindrical; sepals smooth within; C: golden ocher, lip brown-ocher; D: No, CR, Pma, SA (Mt); M: pl to 15 cm, raceme > lvs, S 1.2–1.7 cm............... .... *Pleurothallis erinacea*
2. Ovary and fruit smooth; peduncle flattened, 2-edged; sepals warty within; C: red to red-orange; D: No, CR, Pma, Sa (Mt); M: pl 10–

25 cm, raceme > lvs, S 1.2–2 cm....... *Pleurothallis endotrachys*
(Syn. *P. pfavii*)

3(1). Flowers solitary or usually fascicled, usually produced 1 at a time; old pedicels crowded together 4
3. Flowers usually several at a time, not clustered together; if 1 at a time, pedicels well separated on rachis........................ 26

4(3). Sepals hairy or bristly 5
4. Sepals smooth or warty, without prominent hairs or bristles..... 7

5(4). Sepals hairy, connivent (looking like a *Zootrophion* flower, see Plate 24[6]); plant creeping or pendent; C: S gray-green with magenta veins, mottled magenta within, P and lip wine red; D: No, CR, SA (Mt); M: pl 30 cm, lvs 2.7 cm, 1 fl < lvs, S 5 mm .... *Pleurothallis testaefolia*
5. Sepals bristly, spreading; shoots clustered 6

6(5). Sepals long-tailed, with spinelike hairs; petals widest apically, notched; C: S translucent white faintly suffused with pink, P pale green flushed with rose, lip red; D: Pma (Mt); M: pl 7–10 mm, stalked fasc > lvs, S 10–11 mm .... *Pleurothallis cactantha*
6. Sepals acute; petals widest near middle, acute; C: orange-red or yellow; D: No, CR, Pma (Mt, Pc); M: pl 1–2 cm, 1–2 fls > lvs, S 5–5.5 mm.............................. *Pleurothallis glandulosa*

7(4). Lip with distinct fringe 8
7. Lip without fringe 10

8(7). Sepals long-tailed; petals fringed; lip widest across lateral lobes C: green with red nerves and spots; D: No, CR, (Mt); M: pl 2 cm, stalked fasc > lvs, S 7–8 mm............... *Pleurothallis hastata*
8. Sepals abruptly acute; petals not or only microscopically fringed... 9

9(8). Tips of petals thick, minutely warty; C: orange or red; D: CR, Pma (Mt); M: pl 2 cm, stalked fasc > lvs, S 3–5 mm .... *Pleurothallis casualis*
9. Tips of petals acute or acuminate; C: S, P yellow with red nerves and dark red lip; D: No, CR, Pma, SA (Mt); M: pl 1–2 cm, stalked fasc sub=lvs, S 3–4 mm................. *Pleurothallis barbulata*
(Syn. *P. abjecta*)
Tiny plant with movable, fringed lip.

10(7). Lateral sepals long-tailed; ovary spiny; C: translucent yellow, lip purple; D: CR (Mt); M: pl 1–1.5 cm, 1 fl sub=lvs, S 10–11 mm .... *Pleurothallis turrialbae*
10. Lateral sepals acute or obtuse; ovary not spiny 11

11(10). Lateral sepals completely united and deeply concave, synsepal more than twice as wide as dorsal sepal; lip uppermost, widest near apex, tapering to base; C: cream flecked with red-purple; D: CR, SA; M: pl 3–5 cm, pendent raceme sub=lvs, S 9–11 mm ....
.... *Pleurothallis aryter*
11. Lateral sepals free or united but not deeply concave or twice as wide as dorsal sepal.. 12

12(11). Lip widest near apex 13
12. Lip widest near base or middle, or oblong.................... 16

13(12). Apex of lip with a pair of knobby, knuckle-like calluses; C: S, P golden yellow dotted with purple-brown, lip maroon; D: wPma (Pc); M: pl 5–8 cm, raceme sub=lvs, S 13 mm......................
.... *Pleurothallis condylata*
13. Apex of lip rounded or squarish 14

14(13). Petal apices thick and fleshy; surface of lip minutely warty and hairy; flower pendent; C: S, P wine red, bases of S with white stripes, lip yellow with red callus; D: No, Pma; M: pl 5–6 cm, 1–2 fls on stalk > lvs, S 13–15 mm .... *Pleurothallis alexii*
14. Petal apices not thick and fleshy 15

15(14). Lip half as wide as long, with a circular depression at base of fleshy blade; blade thickest near base; leaves acute; D: CR; M: pl 3–6 cm, stalked fasc (or condensed raceme) sub=lvs, S 8–8.5 mm ....
.... *Pleurothallis pyrsodes*
15. Lip much longer than twice width, thickest near middle; leaves obtuse or notched; C: orange; D: No, CR (Pc); M: pl 5–8 cm, stalked fasc < lvs, S 9–10 mm .... *Pleurothallis guanacastensis*

16(12). Lateral lobes of lip distinctly pointed forward or backward..... 17
16. Lateral lobes of lip triangular or rounded, not pointed forward or backward 18

17(16). Lateral lobes of lip pointing forward, midlobe somewhat warty; C: brown, orange, or orange-red; D: CR, Pma (Mt); M: pl 5–8 cm, stalked fasc or raceme > lvs, S 12–13 mm ... *Pleurothallis fulgens*
17. Lateral lobes of lip pointed back toward base, whole lip warty and bristly; D: CR (Wt); M: pl 1.6–2 cm, stalked fasc > lvs, S 5–6 mm
.... *Pleurothallis acicularis*

18(16). Leaves sublinear; D: Pma (Mt); M: pl 3 cm, loose fasc > lvs, S 5.5 mm.................................... *Pleurothallis vitariifolia*
18. Leaves elliptic to elliptic-oblanceolate....................... 19

19(18). Usually with a single long-pediceled flower, pedicel about twice flower length; dorsal sepal and synsepal obtuse; C: yellow-green to

greenish orange; D: No, Pma (Mt, Pc); M: pl 2–5 cm, 1 fl > lvs, S 8–10 mm........................ *Pleurothallis barboselloides*
Similar to *P. brighamii* but with 1-flowered inflorescence, pedicel much longer than flower.
19. Usually with several flowers, pedicels less than twice flower length 20

20(19). Inflorescence subequal to leaves; petals long-tailed; dwarf creeping plant, < 15 mm tall; C: S orangish, more or less transparent, P and lip dark red; D: Pma, SA (Pc); M: pl to 15 mm, 1 fl sub=lvs, S 4 mm................................. *Pleurothallis polygonoides*
20. Inflorescence much longer than leaves 21

21(20). Lateral sepals lanceolate, narrowly acute 22
21. Lateral sepals oblong-ovate, abruptly acute................... 24

22(21). Lip equally wide across lateral lobes and midlobe; petals elliptic, lower margin only slightly more convex than upper; C: S, P yellow with purple stripes; D: No, CR; M: pl 2–4 cm, raceme to 7 cm, S 8–8.5 mm............................... *Pleurothallis periodica*
22. Lip widest across lateral lobes; petals asymmetrical, lower margin much more convex than upper........................... 23

23(22). Leaves obtuse; sepals somewhat acuminate; C: S, P green with spots and blotches of purple, P dark purple, lip yellow to purple; D: CR, Pma (Cb, Pc); M: pl 8 cm, stalked fasc > lvs, S 7–10 mm .... *Pleurothallis acrisepala*
A common lowland species that differs from *P. brighamii* in the narrow, tapering sepals.
23. Leaves acute; sepals acute to blunt; C: green-yellow with orange lip; D: No, CR, Pma; M: pl to 25 cm, raceme sub=lvs, S 5.5–6 mm .... *Pleurothallis corniculata*
Similar to *P. brighamii*.

24(21). Leaves acute; C: green-yellow with orange lip; D: No, CR, Pma; M: pl to 25 cm, raceme sub=lvs, S 5.5–6 mm .... *Pleurothallis corniculata*
Similar to *P. brighamii*.
24. Leaves obtuse or notched 25

25(24). Petals elliptic-obovate, lower edge rounded; C: S yellow striped with brown or purple, laterals purple-brown below middle, P yellow striped with brown; D: No, CR, Pma (Cb, Pc); M: pl 5–10 cm, stalked fasc > lvs, S 8 mm................ *Pleurothallis brighamii*
Common species in Central American lowlands.

25. Petals rhombic-obovate, lower half broadly triangular; D: Pma (Mt); M: pl 3 cm, stalked fasc > lvs, S 5–6 mm .... *Pleurothallis uniflora*
26(3). Sepals hairy within 27
26. Sepals without hair within (though margins may be fringed).... 34
27(26). Base of inflorescence clasped by leaf, inflorescence appearing to arise above middle of leaf; C: yellow-orange; D: No, CR, Pma, SA (Mt); M: pl 10–20 cm, raceme > lvs, S 11–12 mm.................. .... *Pleurothallis immersa*
   Unmistakable because of the way the leaf clasps the peduncle, which is easily exposed by spreading the leaf.
27. Inflorescence arising from base of leaf 28
28(27). Petals with club-shaped tips; lateral sepals not deeply concave, less than twice as wide as dorsal sepal; C: dark purple-red; D: CR, Pma (Mt); M: pl 3 cm, zigzag raceme becoming > lvs, S 6–7 mm .... *Pleurothallis fractiflexa*
28. Petal tips not thickened 29
29(28). Lateral sepals deeply concave (apparently united, but merely connivent), together more than twice as wide as dorsal sepal; white hairs conspicuous; lip with rounded lateral lobes; C: greenish cream; D: CR, Pma (Mt); M: pl 8–12 cm, raceme > lvs, S 7 mm............ .... *Pleurothallis amparoana*
29. Lateral sepals united nearly to tips, not deeply concave or much wider than dorsal; lateral lobes of lip erect, with angles or slender lobes pointing forward (*P. segoviensis* complex) 30
30(29). Lateral lobes of lip basally angular 31
30. Lateral lobes of lip basally rounded 32
31(30). Lateral lobes wider than long, with very short lobes pointing forward, callus of low ridges; C: purplish; D: Pma (Mt); M: pl to 8 cm, raceme sub=lvs, S 7–10 mm .... *Pleurothallis canae*
31. Lateral lobes as long as wide, lobule pointing forward, about 0.3 mm long, callus of high, rounded ridges; C: maroon or semitransparent with maroon stripes; D: CR, Pma (Mt); M: pl 10–15 cm, raceme > lvs, S 12–16 mm................ *Pleurothallis pompalis*
   The right-angled lateral lobes of the lip are distinctive.
32(30). Sepals 15–16 mm long; lateral lobes of lip hooklike, acute, midlobe ovate; C: brownish striped; D: CR (Mt); M: pl 5–10 cm, raceme sub=lvs, S 8–14 mm...................... *Pleurothallis wercklei*

32. Sepals 5–10 mm long; lateral lobes finger-like, obtuse, midlobe oblong 33

33(32). Lateral lobes of lip slender and curved; D: CR (Mt); M: pl 15–20 cm, raceme sub=lvs, S 11 mm .... *Pleurothallis falcatiloba*
33. Lateral lobes of lip tapering and straight or nearly so; C: yellow green with purple spots or stripes to nearly black; D: No, CR, Pma (Mt); M: pl 5–10 cm, raceme > lvs, S 5.5–10 mm .... *Pleurothallis segoviensis*
This species is highly variable, and there may be a number of species lumped together here. The variation is bewildering, and it is not easy to sort them out.

34(26). Sepals, petals, or lip with prominent fringe 35
34. Sepals, petals, and lip without fringes, at most a few short bristles along margins 44

35(34). Bristle-like hairs on margins of sepals, petals, and lip, and on outer surface of flower; inflorescence pendent; C: purplish pink, P and upper S green, tails yellow; D: CR (Cb); M: pl 2 cm, creeping or pendent raceme > lvs, S 9–12 mm .... *Pleurothallis setosa*
35. Hairy fringes on sepals, petals, or lip, not on all 3 36

36(35). Sepals fringed... 37
36. Petals and/or lip fringed, but not sepals 38

37(36). Petals subequal to sepals, long-tailed; sepals with slender tails forming about a quarter of their length; C: translucent yellow-green marked with purple; D: No, CR? SA; M: pl 1.5–2 cm, raceme > lvs, S 5.5 mm .... *Pleurothallis samacensis*
37. Petals ovate-lanceolate, acute, less than half length of sepals; sepals with long, thickened tails about half of length; C: S, P translucent spotted with maroon; D: CR (Mt); M: pl 15 cm, stalked fasc > lvs, S 6.5–7 mm .... *Pleurothallis strumosa*

38(36). Petals without fringes, narrowed to slender tails 39
38. Petals distinctly fringed at tip or at base (if they have slender tails) 40

39(38). Tails of petals not more than half length of petals; sepals broadly acute; leaves very long-petioled; C: dark red, P tips and lip maroon; D: wPma (Mt); M: pl 8–20 cm, zigzag raceme > lvs, S 3–3.3 mm .... *Pleurothallis rubella*

39. Tails of petals about two-thirds length of petals; sepals acuminate; C: maroon; D: Pma (Mt); M: pl 3–4 cm, zigzag raceme > lvs, S 9 mm .... *Pleurothallis cuspidata*

40(38). Midlobe of lip fringed, circular, with hollow cavity in base of lip; sepals elliptic-ovate, acute; C: dark red-purple; D: No, SA; M: pl 2–2.5 cm, zigzag raceme > lvs, S 4–5 mm........................ .... *Pleurothallis exesilabia*

40. Midlobe of lip not circular, lip without hollow cavity; sepals long-tailed 41

41(40). Petals lanceolate, acute, evenly fringed; C: maroon purple; D: CR, Pma; M: pl 1.5–2.4 cm, raceme > lvs, S 8–10 mm.............. .... *Pleurothallis aristata*

41. Petals fringed basally, ending in long tails 42

42(41). Lip long-fringed, especially in front, with narrow, curved lateral lobes; tails of petals about half of length; C: S translucent greenish white spotted with dark purple, P white with purple midvein, lip white suffused and spotted with dark purple; D: CR (Mt); M: pl 3–4 cm, raceme sub=lvs, S 7 mm............ *Pleurothallis herpestes*

42. Apex of lip not fringed; lateral lobes triangular 43

43(42). Tails of sepals and petals much longer than blades; D: CR (Mt); M: pl 2 cm, raceme 4–6 cm, S 9–11 mm.... *Pleurothallis quinqueseta*
Possibly an extreme form of *P. setigera*.

43. Tails of sepals and petals subequal to blades; D: CR, Pma, So (Mt); M: pl 2–3 cm, raceme > lvs, S 6 mm .... *Pleurothallis setigera*
Floral details are quite variable in this species.

44(34). Lateral sepals completely united, obtuse; flowers relatively large, produced 1 at a time 45

44. Lateral sepals at least partly free, acute, *or* flowers produced simultaneously 47

45(44). Synsepal more than twice as wide as dorsal sepal; lip uppermost, widest near apex, tapering to base; C: cream flecked with red-purple; D: CR, SA; M: pl 3–5 cm, pendent raceme sub=lvs, S 9–11 mm................................... *Pleurothallis aryter*

45. Synsepal and dorsal sepal similar; lip widest below middle 46

46(45). Inflorescence pendent; lip elliptic, toothed near apex; C: S translucent pale yellow-green with blotches of purple, P and lip red-purple;

D: Pma (Mt, Pc); **M:** pl 5–8 cm, raceme sub=lvs, S 8–10 mm
.... *Pleurothallis guttata*

46. Inflorescence ascending; lip widest near base, weakly 3-lobed, lateral lobes toothed; C: translucent yellow suffused and spotted with rose, P and lip yellow marked with brown; D: CR (Mt); **M:** pl 6–10 cm, raceme sub=lvs, S 1.3–1.8 cm .... *Pleurothallis janetiae*

47(44). Petals ending in long tails 48
47. Petals obtuse to acute but without long tails 49

48(47). Tails of petals shorter than blades; C: dS rosy red-brown, P white with red-brown tails, lip maroon; D: No, SA; **M:** pl 10–15 cm, zigzag raceme > lvs, S 3.5–4 mm .... *Pleurothallis abbreviata*
48. Tails of petals much longer than blades; C: maroon; D: Pma (Mt); **M:** pl 3–4 cm, zigzag raceme > lvs, S 9 mm ....
.... *Pleurothallis cuspidata*

49(47). Ovary and fruit with spine-like warts; sepals warty, tips sticking together; inflorescence very short, flowers borne near base of leaves; C: brick red or orange-red; D: No, CR, Pma; **M:** pl 3–6 cm, S 7 mm
.... *Pleurothallis tribuloides*
Plate 24(2)
A common and distinctive species in Central America.
49. Ovary smooth or warty; sepals not markedly warty 50

50(49). Each sepal markedly keeled, flower thus triangular in cross section; inflorescence pendent or creeping; leaves long-petiolate, petioles about a third length of leaf; C: greenish with purple tints; D: No, CR, Pma, SA (Mt); **M:** pl 12 cm, raceme sub=lvs, S 12–13 mm
.... *Pleurothallis tripterantha*
A widespread and distinctive species.
50. Sepals not or weakly keeled; flower not triangular in cross section; leaves short-petiolate; inflorescence erect or arching 51

51(50). Petals narrowly oblanceolate, narrow-stalked, longer than lip; sepals narrowly acute; dwarf creeping plant with 2–3 flowers; C: greenish white; D: CR, Pma (Mt); **M:** pl 10–15 mm, 2–3 fls > lvs, S 5 mm, P 3.5 mm .... *Pleurothallis calyptrostele*
Similar to *P. geminiflora* but with narrower leaves and longer petals.
51. Petals lanceolate or oblanceolate to obovate, not longer than lip
.... 52

52(47). Petals narrowly acute to acuminate 53
52. Petals blunt to broadly acute 56

53(52). Petals acuminate; dwarf creeping plant; C: S orangish, more or less transparent, P and lip dark red; D: Pma, SA (Pc); M: pl to 15 mm, 1 fl sub=lvs, S 4 mm................... *Pleurothallis polygonoides*
53. Petals acute 54

54(53). Dwarf creeping plant, < 3 cm tall; sepals and petals lanceolate; C: yellow with orange-yellow tips; D: No, SA; M: pl 1–3 cm, 1–2 fls sub=lvs, S 2.5 mm................... *Pleurothallis sertularioides*
54. Stems clustered, plants larger 55

55(54). Lateral sepals united to near apex; dorsal sepal long-tailed; C: yellow; D: Pma.............................. *Pleurothallis picta*
Similar to *P. grobyi* but with long, tapering, acute sepals. The typical *P. picta* seems quite different from the typical *P. grobyi*, but there is a bewildering array of intermediates and it is not clear that these are distinct species.
55. Lateral sepals free to near bases, dorsal sepal lanceolate, acute; C: S dull yellow spotted with purple, P and lip maroon; D: Pma (Mt); M: pl 8–13 cm, raceme sub=lvs, S 15 mm .... *Pleurothallis areldii*

56(52). Lateral lobes of lip linear, about one-third length of lip; sepals fuzzy within; C: yellow-green with purple spots or stripes to nearly black; D: No, CR, Pma (Mt); M: pl 5–10 cm, raceme > lvs, S 5.5–10 mm .... *Pleurothallis segoviensis*
This species is highly variable. There may be a number of species lumped together here, but the variation is bewildering and it is not easy to sort them out.
56. Lateral lobes of lip (if present) rounded or triangular 57

57(56). Lip with distinct rounded or toothlike lateral lobes 58
57. Lip oblong or elliptic-oblong, without distinct lateral lobes 59

58(57). Sepals with narrow tails; lateral lobes of lip toothlike; C: yellow or cream yellow with or without purple spots; D: No, Pma; M: pl 2 cm, raceme > lvs, S 6–7 mm.................. *Pleurothallis fuegii*
P. fuegii var. *echinata*, which has spinelike hairs on the ovary, may prove to be a distinct species.

58. Sepals acute, without tails; lateral lobes rounded; **C:** pink with deep wine red veins; **D:** No, CR (Mt); **M:** pl 10–12 cm, raceme > lvs, S 7–8 mm .... *Pleurothallis cobanensis*

59(57). Dorsal sepal ending in thick knob; **C:** dark red; **D:** No, CR, Pma (Mt); **M:** pl to 1.5 cm, raceme > lvs, S 4 mm................... .... *Pleurothallis segregatifolia*
(Syn. *P. calyptrosepala*)
59. Dorsal sepal apex flat, not fleshy........................... 60

60(59). Lateral sepals united to apex............................. 61
60. Lateral sepals largely free 63

61(60). Synsepal broadly rounded (may be notched); stems clustered, not ascending 62
61. Synsepal tapering; plants ascending (new shoots arising above bases of older stems); both dorsal sepal and synsepal tend to be acute or tapering; **C:** light yellow-green; **D:** CR, Pma, SA (Mt); **M:** pl 3–9 cm, raceme > lvs, S 4.5–7 mm ....*Pleurothallis costaricensis*

62(61). Leaves oblanceolate, tapering basally; **C:** creamy white suffused with pink or dark red-maroon; **D:** No, CR, Pma, SA (Cb); **M:** 4–5 cm, raceme > lvs, S 4–5.5 mm...................*Pleurothallis grobyi*
A common lowland species.
62. Leaves subcircular; **C:** yellow; **D:** No, Pma, (Pc) .... *Pleurothallis microphylla*
Very similar to *P. grobyi* and possibly not distinct.

63(60). Leaves broadly elliptic-obovate, nearly as wide as long; flowers borne well above leaves; plant creeping; **C:** greenish white; **D:** CR, Pma (Mt); **M:** pl 5–8 mm, 2 fls > lvs, S 4 mm, P 1.3 mm......... .... *Pleurothallis geminiflora*
Similar to *P. calyptrostele* but with wider leaves and shorter petals.
63. Leaves oblanceolate 64

64(63). Lip oblong-straplike, sides parallel; **D:** CR (Mt); **M:** pl 3 cm, 2–3 fls > lvs, S 4–4.8 mm .... *Pleurothallis minuta*
64. Lip narrowly obovate, tapering at each end; **D:** No, CR, Pma; **M:** pl 3–3.5 cm, raceme > lvs, S 7 mm .... *Pleurothallis lanceola*
(Syn. *P. lateritia*)

Key 4

Dwarf plants with subcircular or very wide leaves and short inflorescences, often creeping.

    1. Flowers bristly or hairy externally............................ 2
    1. Flowers smooth or warty externally 3

    2(1). Leaves subcircular; dorsal sepal at least one-third united with lateral sepals, lateral sepals united to apex; plant creeping; 1 short-stalked flower; C: S translucent pale yellowish stained with purple, P pale yellowish, lS and lip blotched with deep maroon; D: CR; M: lvs 6 mm, S 5–6 mm.................... *Pleurothallis peperomioides*
    2. Leaves broadly elliptic; sepals free (but tips may cling together); plant creeping or pendent; C: S gray-green with magenta veins, mottled magenta within, P and lip wine red; D: No, CR, SA (Mt); M: pl 30 cm, lvs 2.7 cm, 1 fl < lvs, S 5 mm..... *Pleurothallis testaefolia*
        A distinctive species with synonyms under several generic names.

    3(1). Leaves subcircular, light green with dark veins; lip about as wide as long, minutely toothed; C: lvs light green with dark veins, dS yellow-brown with purple stripes, lS purple, P yellow, lip red-brown; D: Pma (Mt); M: lvs 2–4 mm, 1 fl > lvs, S 7.5 mm..............
    .... *Pleurothallis dressleri*
        Found in cloud forests of central Panama; with watermelon-striped leaves and proportionately huge flowers.
    3. Leaves broadly elliptic, not veined; plant creeping; lip much longer than wide; C: wine red; D: No, CR, Pma (Cb); M: lvs 0.6–1.8 cm, 1 fl, stalk sub=lvs, S 5–6 mm.............. *Pleurothallis lewisae*
        Plate 23(5)

Restrepia. Epiphytic; shoots clustered, stems relatively long, covered by loose sheaths; leaves ovate, subacute; flowers fascicled, produced 1 or a few at a time; lateral sepals united, petals and dorsal sepal narrow, attenuate, with clublike apical thickenings; column arched; pollinia 4. The 2 species of *Restrepia* in Central America are quite variable and have many synonyms.

Key to *Restrepia*

1. Flowers on short stalks, borne near base of leaf; C: dS and P cream flushed with pink, lip red, lS striped with red; D: No, CR, Pma, SA (Cb, Mt); M: pl to 20 cm, S 10–12 mm ...... ***Restrepia muscifera***
(Syn. *R. xanthophthalma*)
1. Flowers on long stalks, held at level of leaf tips; C: P red, dS cream streaked and tipped with red, lip and lS yellow basally densely spotted with red, spots becoming sparse distally; D: CR, Pma (Mt); M: pl to 15 cm, S 1.6–2 cm .... ***Restrepia subserrata***
(Syn. *R. lankesteri*)
Plate 20(5)

Restrepiella. Epiphytic; stems clustered, stout, subequal to leaf; leaves narrow basally, then abruptly wider, tapering gradually to a narrow apex; inflorescence short, flowers fascicled, produced 1 at a time from a papery sheath; lateral sepals united, together similar to dorsal sepal, so that they form the lower jaw of an imagined snake's head; C: greenish cream heavily spotted or flushed with red; D: No, CR (Cb); M: pl 20–30 cm, S 1.5–2 cm .... ***Restrepiella ophiocephala***
Plate 24(4)

Restrepiopsis. Epiphytic; shoots clustered, stems subequal to leaves or longer; inflorescence a fascicle, flowers usually short-stalked; sepals and petals similar, lip simple; pollinia 4.

Key to *Restrepiopsis*

1. Flowers on stalks longer than leaves; lateral sepals united to near apex; C: S, P translucent green, dS sometimes with red veins, lip green, lateral lobes suffused with purple: D: No, CR (Cb, Mt); M: pl 2–3 cm, S 10–11 mm. . . . . ***Restrepiopsis reichenbachiana***
   This species is rather unlike the other species of *Restrepiopsis*, though it fits better here than it did in 3 other genera.
1. Peduncles short, flowers borne near bases of leaves; lateral sepals free or nearly so 2
   2(1). Midlobe of lip much larger than lateral lobes; plant > 5 cm tall; C: S, P white, greenish white, or green, lip light green, may be suffused

with brown or purple; **D:** CR, Pma (Cb, Mt); **M:** pl 5–15 cm, S 5.5–8 mm .... *Restrepiopsis tubulosa*
2. Lobes of lip subequal; plant < 4 cm tall; **C:** light green; **D:** No, CR, Pma, SA (Mt, Pk); **M:** pl 2–4 cm, S 3–3.5 mm ....
.... *Restrepiopsis ujarrensis*
Plate 19(5)

Salpistele. Epiphytic; shoots clustered; leaves elliptic, more or less stalked; inflorescence racemose, usually creeping or pendent, producing 1 flower at a time over a long period; lateral sepals united, petals linear or lance-linear; lip shorter than column, 3-lobed, lateral lobes clasping column; apex of column expanded and flat, anther terminal, surrounded by stigma; pollinia 2. The flowers of this genus tend to hide near the base of the plant and are easily overlooked.

Key to *Salpistele*

1. Dorsal sepal and synsepal ovate, at least 2 mm wide; lip about as long as column 2
1. Sepals narrowly ovate, about 1.5 mm wide; lip distinctly shorter than column; apex of column ovate 3
2(1). Apex of column triangular; midlobe recurved; **C:** S red-brown marked with yellow; **D:** CR, Pma (Mt); **M:** pl 1.5–3 cm, S 4.5–6.5 mm .... *Salpistele brunnea*
Plate 19(6)
2. Apex of column ovate to subcircular; midlobe of lip pointing forward; **C:** S, P yellow with brown spots and flecks, lip brown; **D:** Pma (Mt); **M:** pl 1.3–2.5 cm, S 5–8 mm .... *Salpistele lutea*
3(1). Sepals about 5 mm long; lip about half as long as column; **C:** yellow-brown, lip yellow; **D:** wPma (Mt); **M:** pl 2–3.5 cm, S 5 mm
.... *Salpistele dressleri*
3. Sepals about 3 mm long; lip nearly as long as column; **C:** S red-brown with yellow streaks and margins; **D:** Pma (Mt); **M:** pl 1–1.5 cm, S 3.8–4 mm .... *Salpistele parvula*

Scaphosepalum. Epiphytic; shoots clustered, stems short; leaves petiolate; inflorescence elongate, often sprawling on surface, producing 1 flower at a time over a long period; flowers with lip uppermost; dorsal sepal often fleshy near apex, lateral sepals more or less united, each with a fleshy callus on blade and ending in a slender or fleshy tail; pollinia 2. The 3 common species in our area are all quite variable in form and color, and each has several synonyms.

Key to *Scaphosepalum*

1. Tails of lateral sepals as long as cushions or longer, spreading; flowers usually pink or purplish 2
1. Tails of lateral sepals very short, not spreading; flowers usually brownish 3
2(1). Sepals and tails with teeth, cushions narrower than bases of lateral sepals; **C:** S greenish white suffused with purple basally within and spotted with purple, P and lip purple; **D:** Pma (Mt); **M:** pl 7 cm, dS 9 mm............................. *Scaphosepalum viviparum*
2. Sepals and tails without teeth, cushions (together) wider than bases of lateral sepals; **C:** S yellow-green suffused and spotted with dark purple, to solid purple or red-violet, P yellow marked with purple, lip yellow or orange suffused or spotted with red or purple; **D:** CR, wPma (Mt); **M:** pl 20 cm, dS 10–15 mm .... *Scaphosepalum anchoriferum*
Plate 21(6)
3(1). Dorsal sepal much longer than laterals; **C:** S yellow-brown or orange to red, P yellow, each with 2 red veins, lip yellow to red; **D:** CR, Pma, SA (Cb, Pc); **M:** pl 11 cm, dS 11–13 mm.............. .... *Scaphosepalum clavellatum*
3. Dorsal sepal shorter than laterals or subequal; **C:** S light yellow or yellow-green suffused, dotted, or spotted with brown to red, P translucent yellow, sometimes marked with red, lip light yellow marked with red; **D:** No, CR, Pma, SA (Mt, Pc); **M:** pl to 18 cm, dS 4.5–7.5 mm.................... *Scaphosepalum microdactylum*

Stelis. Shoots clustered, or rhizome creeping, stems short or relatively long; leaves oblanceolate, fleshy; inflorescence usually elongate, borne near leaf base, with several to many flowers; sepals similar, or lateral sepals united; petals subequal to column, truncate and fleshy at apices; lip short, truncate, and fleshy; column short; stigma divided into 2 lobes which may be connected under rostellum; pollinia 2. The flowers of *Stelis* are easily recognized by the very short petals, lip, and column. However, they are difficult to identify even with a microscope. The genus is badly in need of revision, and I cannot write a key for the 80 or so species recorded from our area. Most of the names that have been used in Central America are listed in Appendix B. Plate 22(5, 6).

Trichosalpinx. Epiphytic; shoots clustered or superposed; stems relatively long with bristly, oblique, funnel-like sheaths; leaves usually wide; inflorescence terminal, short or elongate; flower structure relatively simple; pollinia 2.

Key to *Trichosalpinx*

1. Stems usually shorter than leaves; leaves obovate, widest above middle; small, creeping plant; C: dark red-purple; D: CR (Mt); M: pl 2–2.5 cm, S 3.3 mm . . . . ***Trichosalpinx navarrensis***
1. Stems usually much longer than leaves *or* leaves lanceolate to subcircular, widest at or below middle. 2
2(1). Inflorescences much longer than leaves; plants frequently produce new stems on top of others, forming a chain of stems, each with 1 terminal leaf. 3
2. Inflorescence shorter than leaves; new stems arising from bases of older stems (plants not forming small bushes) 10
3(2). Sepals with long, narrow tails, tail of dorsal sepal at least half as long as blade 4
3. Sepals acute or short-acuminate, tails less than half length of blade 7
4(3). Lip 3-lobed or rhombic; column without wings, column foot short 5
4. Lip oblong or narrowly ovate; column with prominent wings, column foot prominent. 6
5(4). Lip distinctly 3-lobed; C: yellow-green to nearly white; D: Pma, SA (Mt); M: lvs 10–12 mm, S 6–6.5 mm . . . . . ***Trichosalpinx intricata***
5. Lip nearly rhombic, much wider near base, more or less triangular above; C: yellowish; D: No, CR, Pma, SA (Mt); M: lvs 8 mm, S 8 mm . . . . ***Trichosalpinx cedralensis***
6(4). Tails subequal to blades of sepals; column foot subequal to column; C: S yellow to yellow-orange, often suffused with brown or purple, lip yellow or suffused with red or brown; D: CR, Pma, So (Mt); M:

lvs 2–5.5 cm, S 9–15 mm .... *Trichosalpinx arbuscula*
(Syn. *T. moschata*)
6. Tails shorter than sepal blades; column foot shorter than column; C: lvs with 3 purple stripes, fls yellow or yellow-green; D: CR, Pma, SA (Mt); M: lvs 1.5–2 cm, S 7 mm .... *Trichosalpinx dura*
(Syns. *Pleurothallis broadwayi*, *P. foliata*)
7(3). Column wings lacking; lip distinctly 3-lobed; C: yellowish with purplish veins; D: Pma (Mt); M: lvs 7–15 mm, S 2.1–4 mm ....
.... *Trichosalpinx tantilla*
7. Column wings projecting forward (parallel with anther); lip oblong, subquadrate or obovate.................................. 8
8(7). Sepals elliptic-lanceolate, acute, more than twice as long as wide; column with lateral lobules below middle; C: yellowish white, lip yellow; D: CR, Pma (Mt, Wt); M: lvs 6–10 mm, S 2.3–2.5 mm
.... *Trichosalpinx membraniflora*
8. Dorsal sepal ovate or elliptic-ovate, acuminate, wider than half its length; column without lobules below middle 9
9(8). Lip with prominent median keel and 2 lower lateral keels; C: lvs purplish beneath, S, P pale yellow-green, lip pale green with dark green keel; D: CR, Pma (Mt); M: lvs 9–12 mm, S 2.5 mm........
.... *Trichosalpinx carinilabia*
9. Lip with single median keel; C: S pale yellow-green, P translucent yellow, lip green; D: Pma (Mt); M: lvs 9–12 mm, S 2–2.5 mm
.... *Trichosalpinx tropida*
10(2). Leaves broadly obovate to subcircular; inflorescence short and condensed... 11
10. Leaves ovate to lanceolate; inflorescences about half leaf length
....12
11(10). Sepals 1.8–2 cm long with long, whiplike tails; leaves ovate or obovate; C: maroon; D: CR, Pma, SA (Mt); M: pl 10–12 cm........
...................................*Trichosalpinx pergrata*
11. Sepals < 10 mm long; leaves often very wide, nearly circular; C: dark reddish purple; D: No, CR, Pma, SA (Mt); M: pl 12–14 cm, S 4.5–5.5 mm........................ *Trichosalpinx orbicularis*
The names *operculata* and *rotundata* are based on very similar plants, and 1 or both may be synonyms of *T. orbicularis*. There may be 2 distinct species in our area, but I have not seen enough material to convince me of this.
12(10). Leaves about 6 times as long as wide; stems usually zigzag; C: margins and lower surface of lvs purple, fls red-purple and cream; D:

No, CR, rPma; **M:** pl 4–8 cm, S 3 mm...... *Trichosalpinx ciliaris*
T. ciliaris is reported from a wide area, but many records may be based on plants of other species.
12. Leaves ovate or elliptic, at least one-third as wide as long...... 13
13(12). Sepals minutely hairy externally; apex of column fringed; lateral sepals markedly concave or saclike; **C:** S purple-red,P translucent, lip yellow with red lines; **D:** No, CR, Pma, SA (Mt); **M:** pl 6–10 cm, S 3–3.5 mm........................... *Trichosalpinx memor*
Plate 24(5)
13. Sepals smooth externally (though margins fringed); apex of column not fringed; lateral sepals concave or not; **C:** red-purple and translucent yellow and cream; **D:** No, CR, Pma, SA (Cb, Mt); **M:** pl 10 cm, S 4–6 mm......................... *Trichosalpinx blaisdellii*
A highly variable species.

Trisetella. Plants small or very small; shoots clustered, stems short; leaves narrow, fleshy; flower stalks longer than leaves, congested racemes producing 1 flower at a time; sepals basally united into cup or tube, with slender tails, lateral sepals united; pollinia 2; flowers brownish or yellowish, often striped or suffused with purple; plants of relatively wet forests.

Key to *Trisetella*

1. Tails of lateral sepals attached to sides of sepals, spreading; flowers about 10 mm long.. 2
1. Tails of lateral sepals attached at tips of sepals, parallel; flowers < 10 mm long.. 3
2(1). Synsepal tapering from near base, tails longer than synsepal; leaves slender; peduncle warty or scaly; **C:** dS yellow-orange often suffused with purple, synS purple, P translucent yellow suffused with purple along midvein, lip purple; **D:** CR, Pma, SA (Cb, Mt, Pc); **M:** pl 3–5 cm, dS 2–3 cm, synS 1.4–1.8 cm........... *Trisetella triaristella*
Plate 20(6)
2. Sides of synsepal parallel, tails shorter than synsepal; leaves fleshy; peduncle smooth; **C:** dS yellow, yellow-orange, orange-brown, red-brown, to purple, synS purple, red-brown, to orange-brown, P translucent yellow suffused with purple along midvein, lip purple;

D: CR, Pma, SA (Pc, Mt); **M:** pl 2–6 cm, dS 3–6 mm, synS 0.8–2.1 cm. . . . .*Trisetella triglochin*
A variable species with many synonyms.

3(1). Synsepal wider than dorsal sepal, deeply concave throughout; **C:** dS yellow suffused with red, lS red-purple, P translucent yellow suffused with purple medially, lip maroon; **D:** Pma (Mt, Pc); **M:** pl 1.5–2 cm, S 4–4.5 mm . . . . **Trisetella dressleri**

3. Synsepal narrower than dorsal sepal, shallowly concave at base; **C:** dS yellow with 3 purple veins, synS yellow-white with purple stripes along veins, P translucent white with purple midveins, lip red; **D:** Pma, SA (Pc, Mt); **M:** pl 10–15 mm, S 5–6 mm . . . .
. . . . *Trisetella tenuissima*

Zootrophion. Stems clustered, usually about as long as leaves, with prominent loose or funnel-shaped sheaths; leaves usually wide; flowers clustered near base of leaf; ovary with straight or wavy keels; sepals partially united at both base and apex, with window on each side; petals small; lip hinged to prominent column foot, usually 3-lobed; pollinia 2.

Key to *Zootrophion*

1. Stem much shorter than leaf; sepals warty and hairy externally; **C:** lvs purplish beneath, fls dark claret; **D:** No, CR (Mt); **M:** pl 5 cm, fl 14 mm. . . . . **Zootrophion moorei**
I have seen no specimens, but this species could be an odd *Pleurothallis* rather than a *Zootrophion*.
1. Stem subequal to leaf. 2
2(1). Flower smooth and rounded. 3
2. Flower with definite ridges, grooves, or keels 4
3(2). Leaves broad, purplish; apex of flower obtuse; windows above middle; **C:** dark purple; **D:** CR, Pma (Mt); **M:** pl 8–10 cm, fl 2 cm
. . . . **Zootrophion atropurpureus**
Plate 24(6)
3. Leaves narrow; flower tapering abruptly to a curved beak; windows small, well above middle; **C:** white, P and lip yellowish; **D:** CR (Mt); **M:** pl 12–20 cm, fl 3.3 cm . . . . **Zootrophion vulturiceps**

4(2). Flower with few keels, these somewhat toothed, especially near apex; windows small, mostly above middle; **C:** maroon; **D:** CR (Mt); **M:** fl 13 mm.................... ***Zootrophion gracilentus***
   4. Flower with many keels or ridges, these not markedly toothed... 5

5(4). Lateral lobes attached above middle of lip; flowers concave beneath, with little space within; windows a bit below middle; **C:** cream or pale yellow, may be spotted red; **D:** No, CR, Pma, SA (Mt); **M:** pl 10–15 cm, fl 2.7 cm .... ***Zootrophion endresianus***
   5. Lateral lobes attached near base of lip; synsepal not strongly concave beneath; windows in middle or below; **C:** large purple blotches on a pale background; **D:** Pma, SA (Mt); **M:** pl 6–15 cm, fl 1.5–2 cm .... ***Zootrophion hypodiscus***
(Syn. *Cryptophoranthus lepidotus*)

8 Maxillarieae with Four Pollinia

Usually epiphytic, sometimes terrestrial; pseudobulbs, when present, usually of a **single internode;** leaves pleated, conduplicate, or fleshy; sheathing leaves may enfold pseudobulbs; **inflorescence lateral,** usually basal, often 1-flowered; **pollinia** always **4,** with a distinct viscidium, usually with a stipe.

This diverse group is quite large enough to merit a separate chapter. The Maxillarieae are relatively highly evolved, in that the pollinia always have a distinct viscidium and there is usually a distinct strap, or stipe, between the viscidium and the pollinia. The genera treated in this chapter always have 4 pollinia. Only *Maxillaria* is likely to be confused with the Oncidiinae, and its 1-flowered inflorescence, with each flower on a separate peduncle, is usually enough to distinguish it, even when the inflorescences are old and dry.

The genera of the *Chondrorhyncha* complex are so easy to recognize from the plants alone that they may be discussed as a group. They usually have no pseudobulbs, though some species have very small pseudobulbs nearly hidden by the leaf bases. The leaves are thin and closer to conduplicate than to pleated, but the larger plants, especially, have several major veins in each leaf. Each growth has several leaves in a more or less fanlike arrangement. Each flower is always on its own peduncle; that is, plants have 1-flowered inflorescences (usually several at once). It is still not clear that *Chondrorhyncha* and *Cochleanthes* are distinct. Each is variable in some features, and future study may change the number of genera either way. This complex includes the genera *Chaubardiella, Chondrorhyncha, Cochleanthes, Huntleya, Kefersteinia,* and *Pescatorea.*

Key to the Genera of Maxillarieae with Four Pollinia

1. Leaves distinctly pleated or at least with several major veins; with distinct pseudobulbs 2
1. Leaves distinctly conduplicate *or* pseudobulbs lacking 10

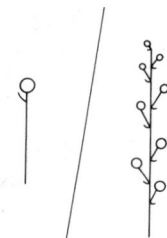

2(1). Inflorescence 1-flowered, each flower with peduncle ... 3
2. Inflorescence of several to many flowers on common peduncle 4

3(2). Each pseudobulb covered with a hard, spotted sheath; 1 stalked leaf per pseudobulb................. ***Teuscheria***
3. Pseudobulbs exposed or with thin, papery sheaths, often with 2 apical spines; leaves several, without stalks .... ***Lycaste***

4(2). Pseudobulbs very large, 7–15 cm long, strongly grooved; lip proportionately small, deeply 3-lobed, midlobe linear-lanceolate, narrowly acute.......... ***Neomoorea wallisii***
4. Pseudobulbs smooth or shriveled, but without rounded ridges or grooves................................... 5

5(4). Pseudobulbs distinctly flattened and 4-angled, rugose, each bearing a single leaf; lip deeply 3-lobed, with narrow basal stalk .... ***Bifrenaria* (*Rudolfiella*) *picta***
5. Pseudobulbs rounded, smooth, not 4-angled; lip not as above 6

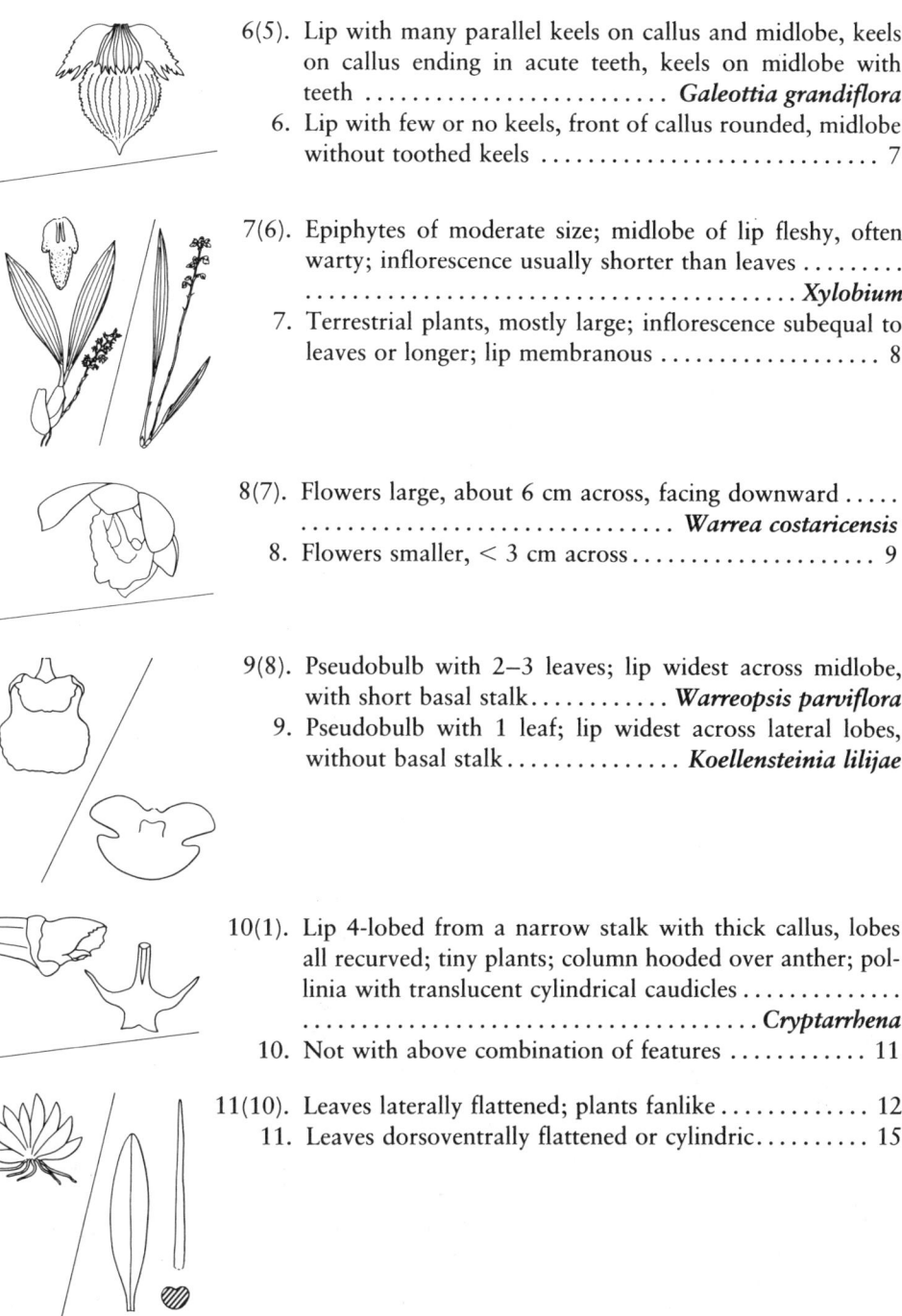

6(5). Lip with many parallel keels on callus and midlobe, keels on callus ending in acute teeth, keels on midlobe with teeth .... ***Galeottia grandiflora***
6. Lip with few or no keels, front of callus rounded, midlobe without toothed keels 7

7(6). Epiphytes of moderate size; midlobe of lip fleshy, often warty; inflorescence usually shorter than leaves .... ***Xylobium***
7. Terrestrial plants, mostly large; inflorescence subequal to leaves or longer; lip membranous 8

8(7). Flowers large, about 6 cm across, facing downward .... ***Warrea costaricensis***
8. Flowers smaller, < 3 cm across..................... 9

9(8). Pseudobulb with 2–3 leaves; lip widest across midlobe, with short basal stalk............ ***Warreopsis parviflora***
9. Pseudobulb with 1 leaf; lip widest across lateral lobes, without basal stalk............... ***Koellensteinia lilijae***

10(1). Lip 4-lobed from a narrow stalk with thick callus, lobes all recurved; tiny plants; column hooded over anther; pollinia with translucent cylindrical caudicles .... ***Cryptarrhena***
10. Not with above combination of features 11

11(10). Leaves laterally flattened; plants fanlike 12
11. Leaves dorsoventrally flattened or cylindric.......... 15

12(11). Plant > 15 cm long; flowers solitary.................... .... *Maxillaria valenzuelana*
12. Plants usually < 15 cm long; flowers racemose 13

13(12). Column without long beak .... *Phymatidium panamense*
13. Column with long beak like a bird's head or an elephant's trunk 14

14(13). Beak of column slender .... *Ornithocephalus*
14. Beak distinctly wider above base .... *Sphyrastylis cryptantha*

15(11). Plants without pseudobulbs....................... 16
15. Plants with definite pseudobulbs 26

16(15). Stems elongate, much longer than leaves 17
16. Stems short, leaves crowded 19

17(16). Column bristly; inflorescence of several flowers; petals and lip similar; viscidium hooked .... *Telipogon*
17. Column without bristles; inflorescence 1-flowered; petals not similar to lip; viscidium not hooked 18

Key to Genera

18(17). Lip anchor-shaped, lateral lobes recurved and acute; column with erect appendage below stigma; plants often pendent . . . . ***Dichaea***

18. Lip not anchor-shaped; column without erect appendage below stigma . . . . ***Maxillaria***

19(16). Plants minute; peduncle flattened, several-flowered, longer than leaves; column or lip usually bristly . . . . ***Stellilabium***

19. Plants larger; peduncle not flattened, inflorescence 1-flowered; column and lip without bristles 20

20(19). Leaves leathery or fleshy; flower yellow-brown, starfish-like, with distinct spur that may be hidden by bracts . . . . ***Cryptocentrum***

20. Leaves thin; flower not as above, may have a chin, but not a spur (*Chondrorhyncha* complex) 21

21(20). Lip convex, with thick lunate callus with many longitudinal keels at base; generally large plants 21

21. Lip concave, or at least lateral margins incurved, callus diverse; small or medium plants 23

22(21). Keels of lip terminating in acute angles or bristles, lip distinctly constricted above callus; flowers very flat, sepals and petals in single plane . . . . ***Huntleya***

22. Keels of callus rounded in front, without prominent angles, lip not constricted above callus; sepals and petals somewhat cupped . . . . ***Pescatorea***

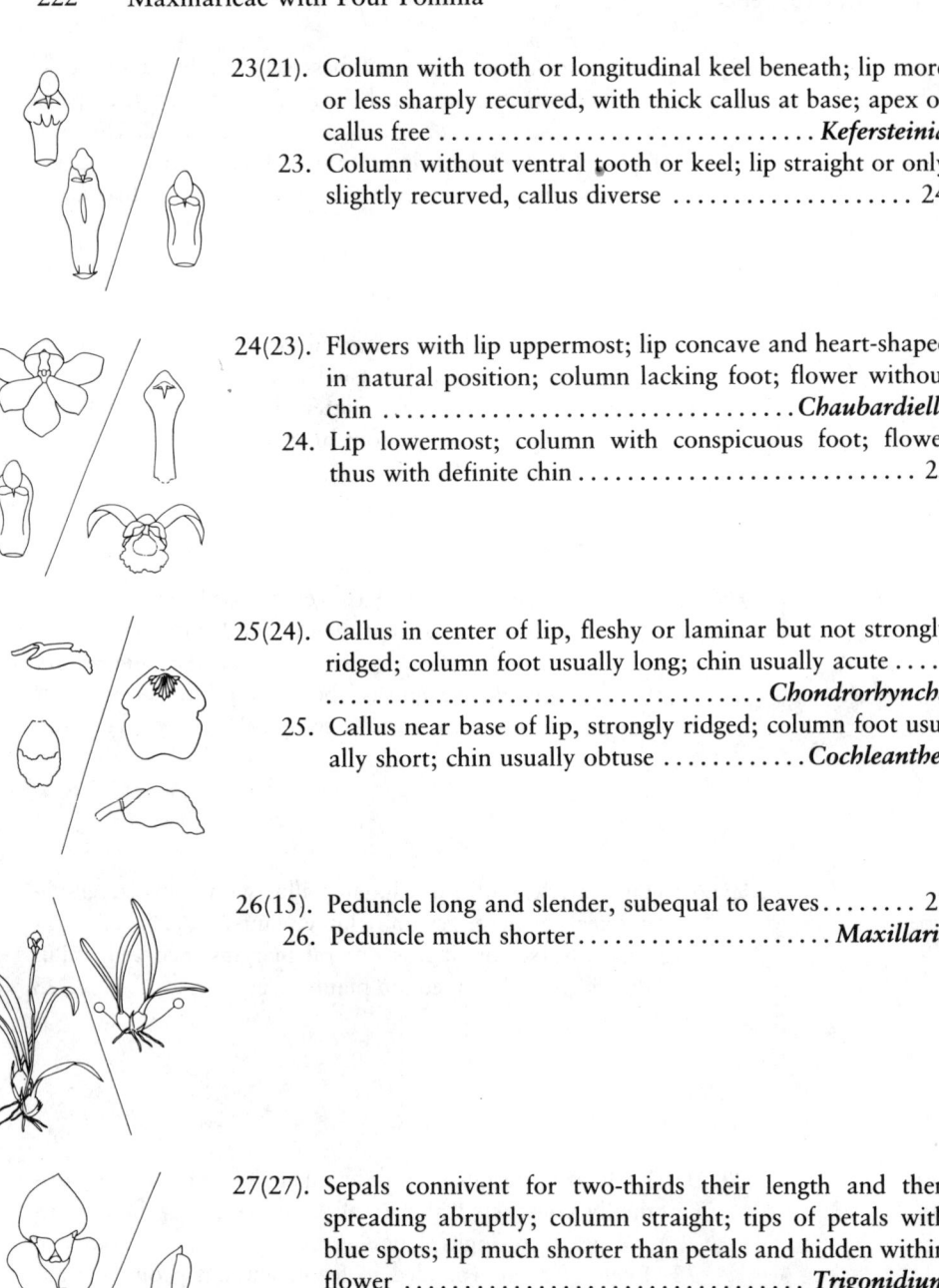

23(21). Column with tooth or longitudinal keel beneath; lip more or less sharply recurved, with thick callus at base; apex of callus free .... *Kefersteinia*
23. Column without ventral tooth or keel; lip straight or only slightly recurved, callus diverse 24

24(23). Flowers with lip uppermost; lip concave and heart-shaped in natural position; column lacking foot; flower without chin .... *Chaubardiella*
24. Lip lowermost; column with conspicuous foot; flower thus with definite chin 25

25(24). Callus in center of lip, fleshy or laminar but not strongly ridged; column foot usually long; chin usually acute .....
.... *Chondrorhyncha*
25. Callus near base of lip, strongly ridged; column foot usually short; chin usually obtuse .... *Cochleanthes*

26(15). Peduncle long and slender, subequal to leaves 27
26. Peduncle much shorter .... *Maxillaria*

27(27). Sepals connivent for two-thirds their length and then spreading abruptly; column straight; tips of petals with blue spots; lip much shorter than petals and hidden within flower .... *Trigonidium*
27. Sepals spreading from bases; column arched, without foot .... *Mormolyca ringens*

Bifrenaria. Epiphytic; pseudobulbs clustered, flattened, 4-angled, subparallel with substrate, overlapping, each with 1 stalked, pleated leaf; inflorescence basal, arching, of several to many flowers; lip distinctly stalked, deeply 3-lobed, lateral lobes prominent; 1 species in Central America; **C:** yellow with brown spots and blotches; **D:** ePma, SA (Cb, Pc); **M:** pl 30–40 cm, S 10–14 mm . . . . . . ***Bifrenaria* (*Rudolfiella*) *picta***
Plate 25(1)

This species belongs to a group often treated as a separate genus, *Rudolfiella*, but the genera in the *Bifrenaria* complex are poorly delimited. This is our only outlier of a South American, and primarily Brazilian, group.

Chaubardiella. Plants similar to *Chondrorhyncha*; leaves 3–5, markedly narrowed basally, acute; flowers borne near base of plant, with **lip uppermost**; sepals and petals similar, lip concave, somewhat heart-shaped. Leaf width varies greatly, even on a single plant.

Key to *Chaubardiella*

1. Flowers yellowish cream barred and spotted with red-purple; **D:** CR, SA (Wt); **M:** pl 12–17 cm, S 15–16 mm . . . . *Chaubardiella subquadrata*
   Described from the area of San Ramón, Costa Rica; I have seen no specimens.
1. Flowers without spots or bars 2

2(1). Flowers yellow; **D:** CR, Pma, SA (Mt); **M:** pl 12–16 cm, S 2–2.4 cm . . . . *Chaubardiella chasmatochila*
2. Flowers greenish cream; **D:** CR (Cb); **M:** pl 12–15 cm, S 1.8–2.2 cm . . . . *Chaubardiella pacuarensis*
Plate 26(1)

Chondrorhyncha. Leaves thin, conduplicate, few to several, arranged in a fan, small to medium in size; flowers with lip lowermost and enfolding column, with prominent, acute chin at base.

Key to *Chondrorhyncha*

1. Lip with single, laminar or fleshy, usually bilobed callus; stipe of pollinia short 2
1. Lip with second thickening or callus in front of basal callus; pollinia with prominent stipe 6
2(1). Callus broad and laminar, toothed 3
2. Callus fleshy, 2-parted 4
3(2). Lip about 4 cm long; chin short and rounded; leaves light green, large; C: S, P greenish cream to yellow-green, lip cream to yellow, often spotted with red or red-brown; D: Pma (Cb, Mt); M: pl 15–36 cm, dS 2.7–3.2 cm lS 3.4–4.2 cm....*Chondrorhyncha anatona*
Plate 26(4)

This species has the callus of *Chondrorhyncha*, but the short chin and pollinia of *Cochleanthes*; it is probably a *Chondrorhyncha* that is pollinated the same way as *Cochleanthes* (with the pollinia placed behind the head rather than on the back of the pollinator).

3. Lip < 3 cm long; chin long and pointed; leaves dark gray-green, usually narrow; C: white or creamy yellow, yellow in throat, with red specks or stains on lip; D: CR, Pma, SA (Cb, Mt); M: pl 15–30 cm, dS 13–15 mm, lS 1.8–2 cm ....
.... *Chondrorhyncha reichenbachiana*
Plate 26(3)
4(2). Callus a single thick, concave plate; C: S, P pale yellow, lip cream or pale yellow spotted with purple, spots more numerous laterally, callus flushed with purple above, darkest in front; D: Pma (Mt); M: pl 7.5–13 cm, dS 12–13 mm, lS 1.6–1.8 cm ....
......................................*Chondrorhyncha crassa*
4. Callus thin, or of 2 distinct elements 5
5(4). Callus narrow, 2-toothed; C: cream, yellowish within throat; D: No, rCR; M: pl 8–30 cm, S 2–3 cm ...*Chondrorhyncha lendyana*
5. Callus broad with 2 rounded lobes; C: white, may have maroon blotches in throat; D: CR, Pma (Cb); M: pl 7.5–10 cm, S 10–12 mm .... *Chondrorhyncha albicans*
6(1). Thickening in front of basal callus short; viscidium < 1 mm wide; C: cream with purple spots in throat; D: CR, Pma (Mt, Cb); M: pl 35–60 cm, S 2.5 cm....................*Chondrorhyncha bicolor*
6. Thickening in front of basal callus extending to apex of lip; viscidium about 2 mm wide; C: S greenish cream, P cream or yellowish

cream, lip cream or yellow-cream, callus yellow, calli and throat marked with red-purple spots; **D**: Pma (Mt); **M**: pl 15–37 cm, dS 2.5–2.7 cm, lS 3.5–3.8 cm.............*Chondrorhyncha eburnea*
Very close to *C. bicolor*, but the midlobe of the lip has a thickening that extends nearly to the apex, and the details of the pollinia are quite different.

Chrysocycnis. See *Maxillaria tigrina*.

Cochleanthes. Similar to *Chondrorhyncha*; 20–40 cm tall; flowers with short, blunt chin; callus usually basal and keeled or toothed.

Key to *Cochleanthes*

1. Lip spreading, not at all enfolding relatively short column; **C**: S, P green, lip white stained with blue to red-lavender; **D**: CR, wPma (Mt); **M**: pl 15–30 cm, S 2.8–3.5 cm .... ***Cochleanthes aromatica***
Plate 26(5)
1. Lip more or less enfolding column, at least basally, column 4–5 times as long as wide..................................... 2
2(1). Callus flat, front margin straight or slightly curved, irregularly toothed, at about middle of lip; margins of lip ruffled and slightly toothed........................ See *Chondrorhyncha anatona*
2. Callus thick and fleshy with distinct and regular teeth on front margin, which is near base of lip 3
3(2). Lip enfolding column basally and then spreading out and relatively flat; **C**: white with narrow rose-purple veins in center of lip; **D**: Pma (Cb); **M**: pl 15–30 cm, S 2.2–3 cm .....*Cochleanthes lipscombiae*
3. Lip covering column to apex and somewhat recurved beyond column.. 4
4(3). Flower small (about 3 cm long); **C**: P pale green, lip yellowish with heavy purple veins to margin; **D**: CR, Pma (Mt); **M**: pl to 30 cm, S 2.5–2.8 cm .... *Cochleanthes picta*
4. Flower larger (3.5–4 cm long); **C**: S, P greenish yellow or buff, lip largely blue-violet with yellow callus; **D**: CR, wPma (Mt); **M**: pl 15–35 cm, lS 3.2–3.5 cm, dS 2.5–3.7 cm... *Cochleanthes discolor*
Plate 26(6)

Cryptarrhena. Epiphytic; with or without pseudobulbs; leaves conduplicate; inflorescence lateral, of several or many flowers; column hooded over anther; lip stalked, with thick callus on stalk, 4-lobed, lobes recurved, rather anchor-shaped.

Key to *Cryptarrhena*

1. Plant without pseudobulbs; leaves several; C: pale green, or lip yellowish; D: No, CR, Pma, SA (Cb, Pc); M: pl to 25 cm, S 3–3.2 mm .... *Cryptarrhena lunata*
1. Plant with definite pseudobulbs, sheathing leaves and 1 or 2 terminal leaves on pseudobulb; C: pale green; D: No, CR, Pma (Mt, Pc); M: pl to 6.5 cm, S 4.5–5 mm .... *Cryptarrhena guatemalensis*
(Syn. *C. quadricornu*)
Plate 30(6)

Cryptocentrum. Epiphytic; without pseudobulbs (in our area), stems short; leaves 2-ranked or spiral, conduplicate or fleshy; inflorescences axillary, 1-flowered, with long peduncles clothed in tubular bracts; flowers brown or yellowish or greenish brown, sepals, petals, and lip narrow and spreading; lip with prominent spur at base (may be hidden by a bract); pollinia 4.

Key to *Cryptocentrum*

1. Leaves spiral, subcylindrical; inflorescences much longer than leaves; < than 3 cm tall; C: dull red to yellow or tan; D: CR, Pma (Cb, Mt); M: pl 1.5–2.2 cm, S 7–8 mm .... *Cryptocentrum standleyi*
1. Leaves 2-ranked, conduplicate; inflorescences subequal to leaves or shorter; plants > 6 cm tall 2
2(1). Plants > 12 cm tall; spur 2–3 cm long 3
2. Plants 6–12 cm tall; spur 1–2 cm long 5
3(2). Leaves 1.5–2 cm wide, obtuse; C: brown or tan; D: CR, Pma, SA (Cb, Mt); M: pl to 20 cm, S 3 cm ...... *Cryptocentrum latifolium*
3. Leaves 3–10 mm wide 4
4(3). Leaves 4–6 mm wide, with obliquely 2-lobed apices; C: brown or tan; D: CR, Pma (Mt); M: pl 6–12 cm, S 1.5–2.2 cm .... *Cryptocentrum calcaratum*

4. Leaves < 3.5 mm wide, abruptly obtuse; C: brown or tan; D: CR, Pma, SA (Mt); M: pl to 20 cm, S 1.5–1.9 cm .... *Cryptocentrum inaequisepalum*

5(2). Leaves flat, 4–10 mm wide; plants 20–25 cm tall; C: brown-olivaceous; D: CR, Pma (Mt, Wt); M: pl 5–10 cm, S 13 mm .... *Cryptocentrum gracilipes*

5. Leaves fleshy, to 4 mm wide; cross section a semicircle; 9–10 cm tall; C: dull olivaceous; D: CR, Pma, SA (Cb, Mt); M: pl 3–9 cm, S 8 mm .... *Cryptocentrum gracillimum*

Dichaea. Epiphytic; without pseudobulbs, stems slender, branching or not; leaves 2-ranked, conduplicate, jointed basally or not; flowers lateral on separate peduncles; lip usually anchor-shaped.

In Central America, the name *Dichaea muricata* has been used more than any other, though not always for the same kind of plant. As originally described, *D. muricata* was a much larger plant, probably what botanists now call *D. morrisii*. That being the case, the name *D. muricata* is conspicuously missing from this key and should be forgotten by all concerned. James Folsom's detailed treatment of the largest group of dichaeas, those with nonjointed leaves and spiny ovaries, should be published soon.

Key to *Dichaea*

1. Bases of leaves not jointed, dead leaves rotting in place 2
1. Bases of leaves distinctly jointed, dead leaves falling off with a clean break. 19

2(1). Ovary and fruit smooth. 3
2. Ovary and fruit with soft spines or sharp bumps. 4

3(2). Tiny plant with leaf blades < 6 mm long; C: S, P greenish spotted red-violet, lip white with few spots; D: No, CR, Pma (Mt); M: lvs 3.5–6.5 mm, S 4.8–6.5 mm................ *Dichaea tuerckheimii*

3. Plant larger, leaves 10–13 mm long; C: pale yellow-green; D: CR, Pma, SA; M: S 10 mm, peduncle long and slender, about 13 mm .... *Dichaea tenuifolia*

4(2). Leaf margins with tiny hairs or bristles; stems clustered, suberect; C: S, P pale green with purple spots and lines, lip white with pinkish lateral lobes; D: No, CR, Pma (Mt); M: lvs to 8 mm, S 5–6.5 mm .... *Dichaea hystricina*

4. Leaf margins smooth or nearly so; stems creeping or pendent ... 5

5(4). Leaf blades without apparent cross-veins..................... 6
5. Leaf blades with definite cross-veins......................... 17
6(5). Leaf blades > 2 cm long...................................... 7
6. Leaf blades < 2 cm long...................................... 8
7(6). Lower leaf surface with large bundles of crystals under the surface, glistening white in living material and forming spots in dried material; C: yellow-green to tan-yellow, lip white with purple spots; D: No, CR, Pma, SA (Mt); M: lvs 2.2–2.5 cm, S 8–10 mm..........
....*Dichaea dammeriana*
7. Bundles of crystals not visible on surface; C: S, P pale yellow, lip violet and white; D: CR, SA (Mt); M: lvs 1.8–2.9 cm, S 6.7–8.3 mm .... *Dichaea pendula*
8(6). Leaf sheaths tightly clasping stem; stem with sheaths cylindrical, not flattened; leaves succulent and corrugated in life, linear when dry
.... 9
8. Leaf sheaths loose, flattened, leaves thin in texture, smooth when dry... 10
9(8). Base of lip disk-shaped; front pollinia united; C: S, P yellowish white, callus of lip with purple spots; D: No, CR, Pma (Mt); M: lvs 6–10 mm, S 5.5–7.1 mm .... *Dichaea trichocarpa*
9. Base of lip rounded to tapering; pollinia free; C: S, P orange with or without purple flecks, lip white with purple spots; D: No, wPma (Mt); M: lvs 4–12 mm, S 6.5–8.2 mm......... *Dichaea squarrosa*
10(8). Petals obovate or oblanceolate, widest near apices............. 11
10. Petals widest basally....................................... 12
11(10). Plants creeping, much branched; C: S, P white heavily spotted with violet, lip white with dark purple-violet markings; D: CR, Pma (Cb, Mt); M: lvs 6–7 mm, S 6.5–9.8 mm ...... *Dichaea obovatipetala*
11. Stems clustered, loosely pendent; C: S, P cream with pale lavender bars; D: CR (Cb); M: lvs 6–9.5 mm, S 6.8–8 mm ....
.... *Dichaea sarapiquiensis*
12(10). Column with distinct ligule projecting below stigma 13
12. Ligule reflexed or lacking 16
13(12). Ligule oblong, more or less broad, apex square 14
13. Ligule narrowly lanceolate or nearly linear 15
14(13). Ligule widest apically; apex of lip long and narrow; C: S, P pale yellow-orange with light violet blotches, lip white with purple blotches; D: CR, wPma (Mt); M: lvs 4.5–7.1 mm, S 7.7–12 mm
.... *Dichaea oxyglossa*
14. Ligule oblong; apex of lip short, from rounded blade; C: S, P orange

with violet spots, lip white and purple; **D:** CR (Mt); **M:** lvs 7–15 mm, S 7.3–9 mm .... *Dichaea poicillantha*

15(13). Leaf blades usually > 8 mm long; ligule arching, 1.5–2 mm long; **C:** S, P green-yellow with purple spots, lip white with dark violet spots; **D:** No, CR, wPma (Mt); **M:** lvs 9–15 mm, S 8.2–13.7 mm ....*Dichaea cryptarrhena*

15. Leaf blades usually < 6 mm long; ligule straight, 1.3–1.8 mm long; **C:** S, P green to yellow-green with purple to violet spots; **D:** CR (Cb); **M:** lvs 4.5–5.5 mm, S 7.4–8.8 mm ..... *Dichaea schlechteri*

16(12). Ligule lacking; lower rim of stigma smooth; **D:** CR, wPma (Mt); **M:** lvs 6–8 mm, S 7.2–8.7 mm .... *Dichaea eligulata*

16. Ligule narrow, reflexed; **C:** S, P orange with violet bars, lip white with violet spots; **D:** Pma (Mt); **M:** lvs 4.5–4.8 mm, S 6.4–7.8 mm ....*Dichaea retroflexiligula*

17(5). Leaf blade with 2–3 lateral veins and a few evident cross-veins; **C:** cream marked with red-purple; **D:** CR (Cb, Mt, Pc); **M:** lvs 4–9 mm, S 8–9.7 mm ....*Dichaea costaricensis*

17. Leaf blade with > 5 lateral veins, numerous conspicuous cross-veins 18

18(17). Sepals acuminate; ligule short, broad, notched; **C:** S, P violet or mottled with violet, lip violet; **D:** Pma (Mt); **M:** lvs 11–12 mm, S 11–15 mm.. *Dichaea violacea*

18. Sepals acute; ligule oblong, tip rounded or square; **C:** cream with lavender spots; **D:** Pma, SA (Mt); **M:** lvs 9.5–14 mm, S 8–10 mm ....*Dichaea tuberculilabris*

19(1). Plants pendent; leaves 3–4 cm long; **C:** S white with violet spots, P brown-violet; **D:** No, CR, Pma, SA (Mt); **M:** lvs 3–4 cm, S 8 mm .... *Dichaea brachypoda*

19. Plants erect or spreading, rarely pendent 20

20(19). Leaves > 1 cm wide; **C:** S, P green with violet lines, lip violet, or fl ivory; **D:** No, CR, Pma, SA (Mt); **M:** lvs 5–7.5 cm, S 12 mm .... *Dichaea morrisii*

Costa Rican plants have ivory flowers; it is not clear whether this represents a distinct species or a color form of *D. morrisii*.

20. Leaves much less than 1 cm wide 21

21(20). Leaves > 8 cm long; **C:** S, P yellow-green, lip deep purple, often spotted; **D:** No, CR, Pma, SA (Cb, Mt); **M:** lvs to 14 cm, S 9 mm .... *Dichaea trulla*
Plate 30(2)

21. Leaves < 5 cm long 22
22(21). Plant usually > 20 cm tall; leaves oblong, tapering abruptly, strongly whitish, especially beneath; C: white; D: No, Pma (Mt); M: pl to 40 cm, lvs 2.7 cm, S 11–12 mm .... *Dichaea glauca*
22. Plants usually < 15 cm tall; leaves narrow, tapering gradually, usually not markedly whitish 23
23(22). Sepals warty externally; lateral lobes of lip rounded; lip with linear basal callus; C: S, P white with purple spots on dorsal sepal and petals, lip cream; D: No, CR, Pma, SA (Mt); M: lvs 3.5 cm, S 6 mm
.... *Dichaea graminoides*
23. Sepals smooth externally; lateral lobes of lip pointed; lip without linear callus 24
24(23). Tip of lip thick and turned downward; C: greenish white; D: No, CR, Pma; M: lvs to 4.5 cm, S 8 mm .... *Dichaea lankesteri*
D. acroblephara, D. amparoana, and *D. standleyi* are all very similar to *D. lankesteri*; there may be 2 or more species here (but surely not 4).
24. Tip of lip not thick; C: pale green or pinkish spotted with violet; D: No, CR, Pma (Cb, Mt); M: lvs 3 cm, S 6–8.5 mm ....
.... *Dichaea panamensis*
Plate 30(1)

Galeottia. Large epiphyte; pseudobulbs clustered, ovoid, somewhat 4-angled, smooth when young, each with 2–3 pleated leaves; inflorescence basal, with 1 or few large flowers; sepals and petals similar, lip 3-lobed, margins strongly toothed throughout, with grooved callus that ends in teeth, column thick, winged; 1 species in Central America; C: sepals and petals green with red-brown stripes, lip white with red markings; D: No, CR, Pma, SA (Cb, Pc); M: pl 40–50 cm, S 3–4.5 cm .....
.... *Galeottia grandiflora*
(Syn. *Mendoncella grandiflora*)
Plate 28(4)

Huntleya. Plants similar to *Chondrorhyncha*; quite large, to 45 cm tall; shoots may be separated on a creeping or climbing rhizome; straplike leaves with several major veins, but not pleated; flowers borne on erect peduncles, quite open and flat; callus with many ridges that end in acute angles in front, lip strongly constricted in front of callus.

Key to *Huntleya*

1. Flower about 8–9 cm across, usually 1 or 2 at a time; sepals and petals with blister-like bumps; **C**: S, P basally white shading through yellow to orange-red, petals with maroon marks basally, lip basally white, then orange-red; **D**: No, CR, Pma, SA (Cb, Mt) ....*Huntleya burtii*
Plate 29(1)
1. Flower 6–7 cm across, often several at once; sepals and petals smooth; **C**: S, P basally white, then yellow blotched or barred with red-brown, lip basally white, apically red-brown; **D**: CR, Pma, SA (Mt); **M**: pl 20–25 cm, S 3.5–4 cm ....*Huntleya fasciata*
Plate 29(2)

Kefersteinia. Plants similar to *Chondrorhyncha* but generally smaller, 10–15 cm tall. This delightful and variable little genus is distinguished by several floral details. There is almost always some sort of keel under the column, the callus at the base of the lip is relatively tall, and the viscidium curls up when removed. The pollinia are normally deposited on the bases of the bees' antennae. There are several species yet to be described.

Key to *Kefersteinia*

1. Lip deeply fringed; **C**: white; **D**: ePma, SA (Mt); **M**: pl to 16 cm, S 1.6–1.8 cm .... *Kefersteinia mystacina*
1. Lip at most ruffled or toothed, never fringed 2

2(1). Blade of lip thin, wide basally and curved abruptly downward near middle, sides upswept, more or less concealing column, margins of lip ruffled and toothed; flowers translucent white more or less speckled with red dots...................................... 3
2. Blade of lip relatively fleshy, narrow basally, not abruptly recurving in middle, margins usually not ruffled (except *K. maculosa*) 5

3(2). Sides of callus parallel or apex wider than midportion; **C**: white with some red specks in throat; **D**: CR, Pma (Mt); **M**: pl to 15 cm, S 10–14 mm............................... *Kefersteinia lactea*
Plate 28(5)
There are thought to be 3 similar white-flowered species in

our area. Until the original specimens are carefully studied, the application of the names is uncertain.
3. Callus rhombic or lyre-shaped 4

4(3). Lip broadly ovate; column without wings; callus rhombic; C: white spotted with brown within throat; D: CR, Pma (Mt) .... *Kefersteinia microcharis*
  I have seen no authentic material of this species.
4. Lip rhombic (when flattened); column with triangular wings above middle; callus lyre-shaped; flowers white or with red-brown specks; D: CR, Pma (Mt, Wt); M: pl 5–15 cm, S 14–15 mm .... *Kefersteinia alba*

5(2). Lip with distinct, rounded basal lobules, these separated from rest of lip by acute sinuses; C: white with red streaks and specks; D: Pma (Mt); M: pl 10–25 cm, S 9–13 mm .... *Kefersteinia auriculata*
  Similar to *K. costaricensis*, but the lateral lobules of the lip are easily seen from the front of the flower.
5. Lip without basal lobules, or these indistinct 6

6(5). Lateral outline of lip distinctly concave (lip constricted above base) 7
6. Lateral outline of lip straight or convex...................... 10

7(6). Callus oblong, rounded, not bilobed, without clear folded margins; lip fleshy, isthmus less than half width of midlobe; column without distinct plate below stigma; C: sepals and petals green with purple spots or blotches, lip yellow marked with purple or purple throughout; D: CR, Pma (Mt); M: pl 6–9 cm, S 1.4–2.3 cm............. .... *Kefersteinia parvilabris*
  (Syn. *K. deflexipetala*)
  Plate 28(6)
  K. deflexipetala is said to differ by the presence of a wart on the callus, but the flowers of this species are so variable in all their features that one can scarcely accept a single wart as the basis for a distinct species.
7. Callus bilobed, stalked or with margins folded under; lip membranous, isthmus more than half width of midlobe; column with distinct raised plate below stigma 8

8(7). Callus short, height less than length, tooth at rear of plate small, scarcely longer than corners of plate; lip usually tipped to 1 side; C:

S, P greenish cream speckled and flushed with red-purple, midlobe of lip purplish red; **D**: CR, Pma (Mt, Wt); **M**: pl to 15 cm, S 12–16 mm .... ***Kefersteinia* species**
8. Callus taller, prominent keel or tooth at rear of plate; lip not tipped to 1 side 9

9(8). Callus parallel with blade of lip, not much expanded apically; plate below stigma subquadrate, with distinct median keel; **D**: CR .... **"*Zygopetalum umbonatum*"**
   This old name was never validly published, even under *Zygopetalum*. Günter Gerlach plans to publish a valid name for this species.
9. Callus projecting upward from axis of lip, the apex expanded, keel forming prominent tooth at rear of plate, sides of plate rounded; **D**: CR; **M**: pl 10–12 cm, S 12 mm .... ***Kefersteinia wercklei***

10(6). Lip fleshy, margin not at all ruffled; keel under column a short tooth well back of stigma; **C**: S white, P and lip speckled with red; **D**: No, CR, Pma (Mt); **M**: pl to 14 cm long, S 10–15 mm .... ***Kefersteinia costaricensis***
10. Lip somewhat ruffled; keel elongate, starting next to stigma; **C**: S, P brownish yellow, lip cream, all more or less strongly spotted with purple; **D**: Pma (Mt); **M**: pl 4.5–7 cm, S 1–1.7 cm .... ***Kefersteinia maculosa***
   The few plants that have been found are rather variable, even within a single population.

Koellensteinia*.** Terrestrial; pseudobulbs clustered, small, narrowly conical, hidden by sheaths, each with a narrow, pleated leaf; inflorescence lateral, erect, with several to many flowers; sepals and petals similar, spreading, but not widely; lip 3-lobed, lateral lobes small, with callus near base of lip; column short, without wings; pollinia 4, with short stipe; 1 species in Central America; **C**: S, P greenish yellow, lip cream, all with red-brown spots or streaks; **D**: Pma, SA (Mt); **M**: pl 50–70 cm, S 10–12 mm .... ***Koellensteinia lilijae

In our area this species is known only from open brushland between Cerro Azul and Cerro Jefe, northeast of Panama City, an area now largely devoted to chicken farms. The name *K. kellneriana* has also been used for the Panamanian plants.

Lycaste. Epiphytic; pseudobulbs clustered, stout, ovoid, often with terminal spines (after leaves have fallen), each with 2–4 terminal leaves and 2–3 sheathing leaves; leaves pleated, wide, without distinct basal stalks, often deciduous; usually flowers with new growth, inflorescence lateral, each flower on separate stalk; flower with definite chin; sepals and petals similar, sepals usually spreading, petals generally hooded over column, with apices spreading; lip 3-lobed, lateral lobes enfolding column, midlobe spreading or recurved; column long, without wings; pollinia 4, with stipe.

Key to *Lycaste*

1. Old pseudobulbs normally leafless at flowering; pseudobulbs usually with distinct spines at apices. 2
1. Plants flowering while old leaves are still present; pseudobulbs usually without spines 5
2(1). Flowers pink or white with pink or rose spots 3
2. Flowers yellow, orange, or greenish, lip always yellow to orange. 4
3(2). Floral bract much shorter than ovary; spines prominent; lip shallowly lobed; C: S greenish tan, P white usually suffused with pink, lip white spotted with pink; D: No, CR, Pma (Mt); M: ps to 6.5 cm, S 2.5 cm . . . . *Lycaste brevispatha*
3. Floral bract as long as ovary or longer; spines small; lip deeply lobed; C: S pale greenish brown, P and lip paler, P with faint rose spots, lip spotted and suffused with pink; D: No, CR, Pma (Mt); M: ps to 8 cm, S 4.5 cm . . . . *Lycaste tricolor*
Plate 25(3)
4(2). Callus narrow, much longer than wide, obtuse; flowers < 3 cm wide; C: S, P greenish yellow, lip yellow; D: Pma, SA (Pc); M: ps 4–5 cm, S 1.4–2 cm. . . . . *Lycaste campbellii*
The smallest of all *Lycaste* species.
4. Callus wider than long, truncate; flowers 4–4.5 cm wide; C: S yellowish green, P and lip yellowish orange; D: No, CR; M: ps to 7 cm, S 2.2–2.8 cm. . . . . *Lycaste bradeorum*
Plate 25(2)
The only yellow-orange-flowered species known in Costa Rica.

5(1). Sepals more than twice as long as petals; C: S pale green suffused with tan, P and lip white with pink suffusion; D: wPma, SA (Wt); M: ps 8–9 cm, S 9–9.6 cm .... *Lycaste schilleriana*
5. Sepals only slightly longer than petals 6
6(5). Lateral lobes rounded, abruptly wider than base of lip.......... 7
6. Base of lip gradually expanding to lateral lobes................ 8
7(6). When lip is flattened, lateral lobes overlap with midlobe; callus tongue-like; C: S light apple green, lip and P creamy white; D: No, CR, Pma (Mt); M: ps to 6 cm, S 4–4.5 cm .... *Lycaste leucantha*
7. When lip is flattened, lateral lobes do not overlap with midlobe; callus with free ovoid apex; C: S light green suffused with tan, petals and lip creamy white; D: ePma, SA (Pc/Mt); M: ps 7–8 cm, S 3–4 cm .... *Lycaste xytriophora*
8(6). Sepals and petals blotched; sinuses narrow; apex of callus broadly rounded, grooved; C: S green blotched with chestnut brown, P and lip pinkish white with pink spots; D: Pma (Cb, Mt); M: ps to 7 cm, S 4.4–4.8 cm.................................. *Lycaste powellii*
8. Sepals and petals not blotched; sinuses rounded next to isthmus; callus tongue-like, not grooved 9
9(8). Peduncles much longer than pseudobulbs; sepals and petals 4.5–5 cm long; C: S usually red or red-brown, P and lip cream or yellowish, often marked with rose or red; D: No, CR, Pma, SA (Mt); M: ps to 12 cm, S 4.5–5 cm.................. *Lycaste macrophylla*
This species is unusually variable in color, and several geographic subspecies have been named.
9. Peduncles subequal to pseudobulbs; sepals and petals about 30 mm long; C: S dark brownish green, P cream white, lip white with yellow suffusion; D: No, CR (Mt); M: ps to 8 cm, S 3 cm ....
.... *Lycaste dowiana*

Maxillaria. Epiphytic or terrestrial; pseudobulbs, when present, either clustered or scattered on long stems, always of a **single internode**, slender stems may be long or short; leaves conduplicate or fleshy; flowers always on separate peduncles (solitary); lip usually hinged to column foot.

I have tried to use plant habit as a major feature in the keys, but the larger plants are sometimes difficult to identify. Botanical collectors often pass these plants by, as they are too big to fit easily in a plant press. Museum specimens are often fragmentary and without descriptive notes.

The "groups" into which I divide *Maxillaria* in the keys are based on features that are easy to see, rather than close relationships. At present, the "big horsey" maxillarias in Keys 3 and 4 are the most difficult to identify. John Atwood identified *M. ampliflora* (Plate 28[3]) for me barely in time to include it here. I have good photographs of several other distinctive big, bushy maxillarias from high elevations, but I cannot name any of them with confidence. Plates 27(1–6), 28(1–3).

General Key to *Maxillaria*

1. Plants without pseudobulbs; stems long and slender; leaves < 6 cm long... Key 5
1. Plants with definite pseudobulbs, stems short, *or* leaves > 10 cm long... 2
2(1). Plants fan-shaped with many 2-ranked leaves; pseudobulbs, if present, concealed .... Key 5
2. Plants not fan-shaped... 3
3(2). Pseudobulbs densely clustered; plants not ascending or pendent.... .... Key 1
3. Pseudobulbs separate on stem or rhizome, not densely clustered (may be densely clustered on young plants) 4
4(3). Pseudobulbs overlapping, space between pseudobulbs generally shorter than pseudobulbs .... Key 2
4. Pseudobulbs distant, not overlapping........................ 5
5(4). Pseudobulbs borne only at bases of stems .... Key 4
5. Pseudobulbs borne at intervals on stem .... Key 3

Key 1
Pseudobulbs densely clustered.

1. Pseudobulbs largely hidden by leaf sheaths.................... 2
1. Pseudobulbs at least partly exposed 4
2(1). Leaves fleshy, 15–20 cm long; C: white to yellow, lip spotted with purple; D: No, CR, Pma, SA (Cb, Mt, Pc); M: pl to 30 cm, S 13–15 mm .... *Maxillaria crassifolia*
2. Leaves membranous to leathery, < 15 cm long................ 3
3(2). Leaves leathery, obtuse; C: S, P pale yellow or greenish, lip basally lined with purple, midlobe yellow; D: No, CR, Pma, SA (Cb, Mt, Pc); M: pl to 8 cm, S 1.4–2 cm......... *Maxillaria brachybulbon*

3. Leaves membranous, acute; C: white, lip often reddish; D: No, CR, Pma (Cb, Mt); M: pl to 25 cm, S 9–16 mm.................... ....*Maxillaria angustissima*
   Easy to recognize by its hidden pseudobulbs, grassy leaves, and long chin.

4(1). Leaves with white spots; C: pale yellow becoming brown with age; D: CR, Pma, SA (Mt); M: pl to 12 cm, S 3–4 cm............... ....*Maxillaria reichenheimiana*
   Plate 27(3)
4. Leaves green, not spotted 5

5(4). Flowers "spidery," sepals and petals long and narrow, tapering to long, narrow tails... 6
5. Sepals and petals shorter, not markedly spidery............... 14

6(5). Flowers spotted... 7
6. Flowers not spotted 8

7(6). Sepals 2–3 cm long; C: yellow-orange with red spots; D: CR, SA (Mt); M: pl to 6 cm, S 1.8–2.1 cm..........*Maxillaria attenuata*
7. Sepals 12–16 cm long; C: S, P creamy white with few red spots, S greenish at tips, apex of lip yellow; D: CR, Pma (Wt); M: pl to 35 cm, S 12–16 cm ....*Maxillaria rodrigueziana*
   Plate 27(2)
   A spectacular species of wet areas, with a rather overpowering perfume.

8(6). Pseudobulbs slender, at least twice as long as wide............ 9
8. Pseudobulbs short and wide, more than half as wide as long ... 10

9(8). Column foot longer than column; chin very prominent; C: S creamy yellow, P white, lip yellow; D: CR, Pma (Mt, Pc); M: pl to 25 cm, S 5–5.7 cm ....*Maxillaria arachnitiflora*
9. Column foot shorter than column; chin only moderately developed; C: S, P cream white, lip orange; D: Pma, SA (Pc); M: pl to 45 cm, S 4–7 cm..*Maxillaria splendens*
   Common in some cloud forests of central Panama, but also found on the relatively dry Perlas Islands; our plant may be distinct from the South American *M. splendens*.

10(8). Plant small; leaf blade 5–20 cm long; flowers often borne above leaves; C: basally white, then tan, aging brown; D: No, CR (Mt); M: pl to 20 cm, S 3.5–5 cm.............. *Maxillaria pachyacron*

10. Plant large; leaves 15–25 cm long; flowers not borne above leaves 11

11(10). Midlobe of lip at least half as long as lip; apex of column with fringed hood over anther; C: S, P yellow within, paler externally, lip yellow with purple lines; D: No, CR, Pma (Cb, Pc); M: pl to 40 cm, S 4–7 cm .... *Maxillaria endresii*
This attractive species has been confused with the South American *M. luteoalba*.

11. Midlobe of lip about one-fourth length of lip; apex of column with a few teeth behind anther 12

12(11). Leaves narrowed below but without distinct petioles; C: S yellow, yellowish white, or tan, P yellow or white, lip white with purple lines on lateral lobes; D: No, CR, Pma (Cb, Mt, Pc); M: pl to 40 cm, S 2.5–4 mm .... *Maxillaria ringens*
Names have been published for several plants that are very closely related to *M. ringens*, including *M. brenesii*, *M. pubilabia*, and *M. rousseauae*; some of these may prove to be distinct species.

12. Leaves usually long-petiolate 13

13(12). Sepals 2–2.5 cm long; C: S white and yellow, P white, lip white with orange tip; D: Pma, SA (Mt); M: pl to 35 cm ....
.... *Maxillaria hennissiana*

13. Sepals 3–5.5 cm long; C: S white, P and lip yellow, lip with purple lines; D: No, CR, Pma (Mt); M: pl to 30 cm ....
.... *Maxillaria angustisegmenta*
This species has been confused with the larger *M. luteoalba* of South America.

14(5). Youngest pseudobulb with conspicuous sheathing leaves 15
14. Sheathing leaves few or none 19

15(14). Plants large to very large, leaves > 20 cm long 16
15. Plants small, < 10 cm tall 18

16(15). Leaves about 20 cm long; C: ocher-yellow with red spot on lip; D: No, Pma, SA (Cb); M: pl to 30 cm, S 2 cm .... *Maxillaria discolor*
16. Plants very large, leaves about 50 cm long 17

17(16). Lateral sepals ovate, abruptly acute; C: greenish yellow, lip pale yellow marked with red; D: No, CR, Pma (Mt); M: pl to 50 cm, S 2.3–2.6 cm .... *Maxillaria maleolens*

17. Lateral sepals narrowly lanceolate, narrowly acute; **C:** yellow, lip mostly red; **D:** No, CR, Pma, SA (Mt, Pc); **M:** pl to 45 cm, S 3.5–4.5 cm.................................... ***Maxillaria nasuta***

18(15). Leaves 4–8 mm wide; flowers 1.4–1.6 cm long; sepals and petals narrow and acute; **C:** S, P pale yellow or greenish, lip basally lined with purple, midlobe yellow; **D:** No, CR, Pma, SA (Cb, Mt, Pc); **M:** pl to 8 cm, S 1.4–2 cm .... ***Maxillaria brachybulbon***
18. Leaves 1.5–2 mm wide, flower about 5 mm long; sepals and petals rounded; **C:** white, apex of lip orange; **D:** CR (Pc, Mt); **M:** pl 2–5 cm, S 5 mm .... ***Maxillaria vittariifolia***

19(15). Petals ovate to obovate, distinctly widest above base........... 20
19. Petals lanceolate to narrowly oblong, widest near base or sides parallel 27

20(19). Dwarf plant to 5 cm tall; pseudobulbs narrowly ellipsoid; **C:** pale rose with red stripes apically; **D:** No, CR; **M:** pl to 10 cm, S 11–14 mm.................................... ***Maxillaria cobanensis***
20. Plants > 10 cm tall; pseudobulbs ovoid or oblong 21

21(20). Lateral lobes of lip acute or tapering........................ 22
21. Lateral lobes of lip rounded 24

22(21). Midlobe oblong, widest basally; **C:** white, lip orange with red-orange callus; **D:** No, CR, Pma (Cb, Mt); **M:** pl to 40 cm, S 1.7–1.8 cm .... ***Maxillaria hedwigae***
22. Midlobe subquadrate, sides parallel 23

23(22). Lateral sepals oblong, rounded; leaves broad and abruptly acute; **C:** yellowish green with red-brown spots on lip; **D:** No, CR, Pma, SA (Cb, Mt, Wt); **M:** pl to 20 cm, S 12–13 mm ....
 .... ***Maxillaria rufescens***
 This name has been applied to many superficially similar species, including *M. acutifolia* and *M. hedwigae*. It may be that none of the Central American plants is really *M. rufescens*.
23. Lateral sepals tapering, acute; leaves tapering, narrowly acute; **C:** sepals greenish yellow, petals orange-yellow, lip pale orange marked with red; **D:** No, CR, SA (Cb, Mt); **M:** pl to 25 cm, S 12 mm .... ***Maxillaria acutifolia***

24(21). Petals widest apically, obtuse; **D:** CR (Wt); **M:** pl 10 cm, S 2 cm .... ***Maxillaria piestopus***
24. Petals widest near middle, acute 25

25(24). Width of lip subequal across lateral lobes and midlobe; C: S, P yellow, lip red; D: CR (Wt); M: pl 15 cm, S 2.5 cm .... *Maxillaria rubrilabia*
25. Lip widest across lateral lobes 27

26(25). Sepals about 2–2.5 mm long; C: S, P yellowish to greenish yellow, spotted and striped with red, lip dark purple; D: No, CR, Pma, SA (Mt); M: pl to 50 cm .... *Maxillaria cucullata*
26. Sepals about 3 cm long; C: S, P ochraceous yellow with red-violet dots and stripes, midlobe reddish black; D: No, CR (Mt); M: pl to 40 cm, S 3 cm .... *Maxillaria hematoglossa*

27(19). Leaves with prominent, narrow petioles 4–10 cm long 28
27. Petioles short and indistinct 31

28(27). Leaf blades < 8 mm wide, narrowly elliptic; C: S reddish brown, P pale yellow flushed with red, lip yellow with 2 maroon spots; D: CR (Wt); M: pl 15–20 cm, S 14 mm .... *Maxillaria acostaei*
28. Leaf blades > 10 mm wide 29

29(28). Flowers red or red-brown 30
29. Flowers white 31

30(29). Flower stalk shorter than pseudobulb; C: S, P red or red-brown, lip dark red within; D: CR, Pma (Wt, Pc); M: pl 20 cm, S 1.8 cm .... *Maxillaria longipetiolata*
30. Flower stalk 2–4 times longer than pseudobulb; C: S red-brown to maroon-red, P pink or yellow-orange, lip yellow-cream with yellow callus, midlobe dark red-purple beneath; D: CR, SA (Mt); M: pl 30–50 cm, S 2.6–2.8 cm .... *Maxillaria cryptobulbon*
    To be expected in Panama; this has been confused with *M. brunnea*, but *M. cryptobulbon* has distinct petioles and sometimes sheathing leaves.

31(29). Dwarf plant, < 10 cm tall; sepals < 15 mm long; C: cream or brownish white, midlobe and callus yellow; D: CR, Pma (Pc, Mt); M: pl 4–10 cm, S 10–13 mm .... *Maxillaria brevipes*
31. Plants > 20 cm tall 32

32(31). Lateral sepals 3–5 cm long, narrowly obtuse or abruptly acute; C: S white, P and lip yellow, lip with purple lines; D: No, CR, Pma (Mt); M: pl to 30 cm, S 3–5 cm .... *Maxillaria angustisegmenta*
    This species has been confused with the larger *M. luteoalba* of South America.

32. Sepals 5–6 cm long, tapering to long, narrow tails; C: S, P creamy white, lip orange; D: Pma, SA (Pc); M: pl to 45 cm, S 4–7 cm .... *Maxillaria splendens*
   Common in some cloud forests of central Panama, but also found on the relatively dry Perlas Islands; this may be distinct from the South American *M. splendens*.

33(29). Petals oblong, obtuse or abruptly acute 34
33. Petals lanceolate, tapering, and acute or acuminate 37

34(33). Petals obtuse.. 35
34. Petals abruptly acute 36

35(34). Sepals 2.5–3.2 cm long; C: sepals and petals brownish yellow, tips brown, lip yellow with red-brown blotch; D: No, CR, Pma, SA (Cb, Mt, Pc); M: pl to 40 cm, S 3 cm .... *Maxillaria brunnea*
35. Sepals 1.6–2 cm long; C: S yellow flushed with red-brown on reverse, P yellow, lip yellow with red-brown lines on lateral lobes; D: Pma, SA (Cb, Mt); M: pl to 45 cm, S 1.6–2 cm .... *Maxillaria powellii*
   Similar to *M. ringens*.

36(34). Column foot subequal to column, chin prominent; C: white, S, P with red tips, lip with red midlobe and callus; D: No, CR (Mt); M: pl 10–20 cm, S 10–15 mm .... *Maxillaria ramonensis*
36. Column foot much shorter than column, chin small; C: S reddish brown, P pale yellow flushed with red, lip yellow with 2 maroon spots; D: CR (Wt); M: pl 15–20 cm, S 14 mm .... *Maxillaria acostaei*

37(33). Leaf blade elliptic with definite petiole; C: white, lip marked with red-violet; D: No, CR, Pma (Cb, Mt, Wt); M: pl to 20 cm, S 1.7–2 cm .... *Maxillaria confusa*
37. Leaf blade straplike, without definite petiole................. 38

38(37). Flower stalk less than twice as long as pseudobulb; petals narrowly triangular; C: sepals mostly yellow, petals white, lip white with red markings; D: No, CR; M: pl to 35 cm, S 2.5–2.7 cm............ .... *Maxillaria amparoana*
   Plate 27(1)
38. Flower stalk more than twice as long as pseudobulb; petals long-attenuate; C: S yellow, yellowish white, or tan, P yellow or white, lip white with purple lines on lateral lobes; D: No, CR, Pma (Cb, Mt, Pc); M: pl to 40 cm, S 2.5–4 cm.......... *Maxillaria ringens*
   Names have been published for several plants that are very

closely related to *M. ringens*, including *M. brenesii*, *M. pubilabia*, and *M. rousseauae*; some of these may prove to be distinct species.

Key 2

Pseudobulbs separate but overlapping, spaces between pseudobulbs shorter than pseudobulbs.

1. Leaves linear to cylindric; pseudobulbs inconspicuous or hidden 2
1. Leaves broader; pseudobulbs conspicuous, usually wider than leaves 4
2(1). Leaves flattened; C: pinkish white; D: CR, Pma, SA (Cb, Mt); M: pl to 60 cm, S 1.8–2.5 cm........................ *Maxillaria lueri*
2. Leaves fleshy, cylindric or nearly so 3
3(2). Leaves subcylindric, flattened above; C: whitish with red-brown stripes; D: No, CR, Pma, SA (Cb, Mt, Pc); M: pl to 15 cm, S 9–15 mm .... *Maxillaria uncata*
    The leaves vary according to exposure and moisture, but the species is easily recognized; there may be more than 1 species under this name in South America.
3. Leaves cylindric with narrow groove above; C: S, P pinkish green to reddish brown, lip purple basally, midlobe cream; D: ePma, SA (Cb); M: pl to 20 cm, S 10 mm .... *Maxillaria subulifolia*
4(1). Pseudobulbs oblong, grooved, each with 2–3 leaves; each flower stalk largely covered by bracts; lip fleshy, tapering, not or only weakly 3-lobed 5
4. Pseudobulbs various, usually with 1 leaf; flower stalks often exposed; lip weakly to clearly 3-lobed 7
5(4). Sepals about 2 cm long; column 9–11 mm long; C: white to yellow-green, lip yellow flushed with lavender; D: No, CR, Pma (Cb, Mt, Pc); M: pl to 30 cm, S 1.5–2 cm....... *Maxillaria friedrichsthalii*
5. Sepals 2–3 cm long; column > 12 mm long.................... 6
6(5). Column < 15 mm long; C: greenish, yellowish, or buff, lip with maroon spots on margins; D: No, CR, Pma, SA (Pc); M: pl 20–25 cm, S 1.6–2.5 cm...................... *Maxillaria scorpioidea*
6. Column > 15 mm long; C: brick red to yellowish green with red stains within, lip dark red; D: No, CR, Pma (Cb); M: pl to 30 cm, S 2.5–3 cm .... *Maxillaria aciantha*

7(4). Leaves narrowly straplike, < 10 mm wide (30–50 cm long) 8
7. Leaves > 10 mm wide, or < 30 cm long 9
8(7). Pseudobulbs fusiform; leaves up to 3 mm wide; C: S, P green with strong overlay of red, lip white with few small spots and maroon callus; D: CR, wPma (Cb, Pc); M: pl to 30 cm, S 1.8–2.7 mm .... *Maxillaria sanguinea*
Plate 28(2)
8. Pseudobulbs cylindric; leaves 3–6 mm wide; C: red or yellow spotted with red, lip yellow or white spotted with red; D: No, CR; M: pl to 60 cm, S 1.7–2.8 cm .... *Maxillaria tenuifolia*
9(7). Two leaves on each pseudobulb, pseudobulbs ellipsoid; lip subcircular, distinctly 3-lobed; callus hairy; C: white, lip yellow within, with reddish brown or reddish purple transverse lines; D: No, CR, Pma (Pc); M: pl to 60 cm, S 2.5–3.5 cm .... *Maxillaria camaridii*
9. One leaf on each pseudobulb 10
10(9). Midlobe of lip clawed, claw less than a fourth width of midlobe, petals and lip wavy or crisped; C: S pale yellowish green with rose-pink spots, P pale yellow with rose-pink spots, lip rose-violet with white margin; D: Pma (Mt); M: pl to 50 cm, S 5–5.2 cm .... *Maxillaria insolita*
At present this unusual species is known from a single plant, but without flowers the plant might easily be overlooked.
10. Midlobe of lip not distinctly clawed, petals and lip not wavy or crisped 11
11(10). Lip deeply saccate, continuous with column foot; lateral sepals partially united and deeply saccate at base; C: S, P white, lip yellow, anther red; D: CR, wPma (Cb, Mt); M: pl 6 cm, S 5 mm ....
.... *Maxillaria strumata*
(Syn. *Sepalosaccus strumatus*)
This was described as a distinct genus, but its flower structure is basically like that of *M. neglecta* and several other maxillarias.
11. Lip not deeply saccate, usually jointed at base; lateral sepals not deeply saccate at base 12
12(11). Lateral lobe of lip basal; flowers densely clustered, stalks less than twice as long as pseudobulb; flowers about 5 mm long; C: white, yellowish brown, rose, or pink; D: No, CR, SA; M: pl to 40 cm, S 11–12 mm .... *Maxillaria densa*
12. Lip lobed above middle or very shallowly lobed 13

13(12). Leaves much more than 10 mm wide 14
13. Leaves much less than 10 mm wide 18

14(13). Leaves broadly elliptic; width about one-fourth length; C: greenish yellow, base of lip marked with red; D: No, CR, Pma (Cb, Mt, Pc); M: pl to 50 cm, S 14–15 mm............... *Maxillaria diuturna*
14. Leaves straplike, width much less than one-fourth length....... 15

15(14). Lip oblong, without distinct lateral lobes; C: S, P muddy yellow externally, red-purple within, margins yellow, lS yellow basally, lip yellow with red-brown spots distally; D: No, rCR; M: pl to 60 cm, S 2.5 cm.................................. *Maxillaria houtteana*
15. Lip with distinct lateral lobes................................ 16

16(15). Midlobe of lip oblong, rounded; C: greenish white to yellow; D: No, CR (Mt); M: pl to 90 cm, S 1.8–2.1 cm .... *Maxillaria anceps*
16. Midlobe of lip triangular-oblong to lanceolate, tapering........ 17

17(16). Petals widest near middle; C: S, P purple-red or green with purple-red spots, lip white spotted with red; D: No, rCR; M: pl to 50 cm, S 2.1–2.5 cm................................ *Maxillaria curtipes*
17. Petals widest at base; C: white, lip cream with yellow apex; D: No, CR, Pma, SA (Cb, Mt, Pc); M: pl to 40 cm, S 2.2–2.5 cm ....*Maxillaria alba*

18(13). Leaves 15–30 cm long, narrowly acute; C: S red externally, pink within, P white to yellow, lip red, apical third yellow; D: No, CR, Pma (Cb, Mt, Pc); M: pl to 40 cm, S 1.5–1.8 cm ....*Maxillaria oreocharis*
18. Leaves shorter, < 20 cm long, obtuse....................... 19

19(18). Flowers about 6 mm long; C: greenish yellow with red callus; D: No, CR, Pma, SA (Mt); M: pl to 30 cm, S 6–7 mm ....*Maxillaria caespitifica*
Similar to *M. variabilis*, possibly not a distinct species.
19. Flower about 10 mm long................................ 20

20(19). Flowers orange-yellow with red blotch on lip; D: No, CR, Pma, SA (Mt); M: pl to 15 cm, S 10 mm............. *Maxillaria variabilis*
20. Flowers greenish or purplish cream with dark red lip; D: CR (Mt); M: pl to 25 cm, S 12 mm............... *Maxillaria costaricensis*
Though similar to *M. variabilis*, this species occurs in wetter habitats; dried specimens are easily recognized by the narrow, curved pseudobulbs (which are similar to those of *M. variabilis* in life).

Key 3

Pseudobulbs remote, not overlapping, spaces between pseudobulbs usually at least twice length of pseudobulb; plants often sprawling.

1. Stem creeping or pendent; flowers usually clustered near base of pseudobulb 2
1. Stem stiff, usually erect, plants thus becoming bushy; flowers often from stem between pseudobulbs or on young shoot above youngest mature pseudobulb. 8

2(1). Plant creeping, base of lip not saccate or bent; **C:** S, P green with reddish markings, lip yellow-orange with red spot; **D:** CR, Pma (Cb, Mt); **M:** pl to 15 cm, S 7–10 mm. .... *Maxillaria repens*
2. Base of lip saccate or bent 3

3(2). Leaves elliptic, obtuse; flowers usually borne along last mature segment of rhizome; **C:** S, P white or flushed with pink or maroon basally, lip orange-yellow; **D:** CR, Pma (Mt, Pk); **M:** S 6.5–8 mm .... *Maxillaria brevilabia*
3. Leaves lanceolate or straplike, always acute; flowers clustered near base of pseudobulb. 4

4(3). Lateral lobes of lip subcircular in side view, larger than midlobe; **C:** S, P white, midlobe may be yellow; **D:** CR (Mt, Pk); **M:** S 6–7 mm .... *Maxillaria concavilabia*
4. Lateral lobes of lip squarish, elliptic, or nearly lacking. 5

5(4). Lateral lobes of lip rounded, much smaller than midlobe. 6
5. Lateral lobes of lip conspicuous, angular, elongate 7

6(5). Flowers with prominent retrorse chins longer than column; **C:** orange-red or red; **D:** CR, Pma (Pc, Mt); **M:** S 7–10 mm .... *Maxillaria horichii*
6. Flowers without prominent chins; **C:** S, P white to light yellow, with orange-yellow lip; **D:** No, CR, SA; **M:** S 4.5–7 mm .... *Maxillaria parviflora*
   In our area known only from Isla del Coco.

7(5). Lip, when spread, 3–4 mm across; lowland species; **C:** S, P white, midlobe yellow; **D:** No, CR, Pma, SA (Cb, Mt); **M:** S 5 mm .... *Maxillaria neglecta*
7. Lip, when spread, 5–7 mm across; mostly found above 900 m elevation; **C:** S, P white, yellow, or orange, midlobe orange; **D:** CR, Pma (Cb, Mt, Pc); **M:** S 5.5–9 mm .... *Maxillaria pseudoneglecta*
   Plate 27(5)

8(1). Lip saccate or bent at base; flowers red, bright pink, or lilac 9
8. Lip straight, not strongly bent at base; flowers not brightly colored 12
9(8). Flowers 7–9 mm long, numerous 10
9. Flowers 1.2–2 cm long, usually few at a time 11
10(9). Flowers red, subglobose; buds obtuse; sepals about 7 mm long; C: coral red with red lip; D: No, CR, Pma (Mt); M: pl to 80 cm, S 7 mm .... *Maxillaria fulgens*
10. Flowers pink; buds acute; sepals 8–9 mm long; C: pink; D: CR, Pma (Mt); M: pl to 1.5 m................... *Maxillaria pittieri*
11(9). Sepals about 13 mm long; C: wine red; D: No, CR (Mt); M: pl to 60 cm, S 10–13 mm..................... *Maxillaria sigmoidea*
11. Sepals about 1.7 cm long; C: wine red; D: No, CR, Pma (Mt); M: pl to 40 cm, bushy .... *Maxillaria wrightii*
Plate 28(1)
12(8). Stems between pseudobulbs leaf-bearing throughout........... 13
12. Stems between pseudobulbs not leaf-bearing throughout but may have leaf blades near pseudobulbs........................ 20
13(12). Plants small, 20–30 cm tall............................... 14
13. Plants large, > 50 cm tall 15
14(13). Pseudobulbs unifoliate; C: dark purple; D: CR, SA (Cb, Pc); M: pl to 20 cm, S 8 mm..................... *Maxillaria ponerantha*
14. Pseudobulbs bifoliate; C: pale yellow-green, lip stained red on lower two-thirds; D: No, CR, SA (Cb, Mt); M: pl to 30 cm, S 10 mm ....*Maxillaria foliosa*
15(13). Lip shallowly 3-lobed, subquadrate-oblong, apex broadly rounded; C: orange-yellow; D: CR; M: pl to 80 cm, S 1.8–1.9 cm.......... ....*Maxillaria alfaroi*
15. Lip deeply 3-lobed 16
16(15). Lateral lobes of lip below middle of lip; C: white, lip edged with pink; D: CR, wPma (Mt); M: pl to 60 cm, S 1.7 cm............. .... *Maxillaria bracteata* (Syn. *M. vagans*)
16. Lateral lobes of lip above middle.......................... 17
17(16). Midlobe triangular-ovate, acute or subacute; C: white with yellow lip; D: CR (Pc); M: pl to 40 cm, S 4–5 cm..................... .... *Maxillaria suaveolens*
17. Midlobe retuse or rounded................................. 18
18(17). Sepals 3.5–4.5 cm long; C: S, P white or with pink to purple blotch on each, usually darker on outside, lip yellow to orange-yellow; D:

wPma (Mt, Pk); M: pl to 1.3 m, S 3.5–4.5 cm................
.... *Maxillaria ampliflora*
Plate 28(3)
18. Sepals 2–2.5 cm long..................................... 19
19(18). Petals widest below middle; blades of stem sheaths shorter than sheaths, curved; D: CR (Mt?); M: pl to 60 cm, S 2.2–2.4 cm .... *Maxillaria campanulata*
19. Petals widest above middle; blades of stem sheaths leaflike, mostly subequal to sheaths, spreading; C: yellow basally, above marked with red-brown; D: CR, Pma (Mt, Wt); M: pl to 70 cm, S 2.5 cm .... *Maxillaria bradeorum*
20(12). Sheaths of stem between pseudobulbs flattened and prominent but without full-size leaf blades............................... 21
20. Sheaths between pseudobulbs closely clasping stem, not flattened or prominent.. 27
21(20). Lower sheaths clasping stem, but upper sheaths flat and spreading, without leaf blades below pseudobulb; C: white, yellow within; D: Pma (Cb, Pc); M: pl to 40 cm, S 3–3.5 cm ....
.... *Maxillaria planicola*
Bracts are progressively larger on the upper part of each rhizome segment, with the largest bracts next to the pseudobulb.
21. Sheaths similar, or upper sheaths bearing leaf blades 22
22(21). Tips of sheaths curving away from stem..................... 23
22. Sheaths straight, not curving away from stem; lobes of lip rounded 24
23(22). Sheaths not jointed, sepals narrowly lanceolate; lobes of lip triangular; C: greenish white to orange suffused with red, or red, lip often with purple spots; D: CR, Pma (Mt); M: pl to 1 m, S 1.9–2.1 cm .... *Maxillaria umbratilis*
23. Sheaths jointed, recurved tips falling away; sepals elliptic; lobes of lip rounded; D: CR (Mt?); M: pl to 60 cm, S 2.2–2.4 cm ....
.... *Maxillaria campanulata*
24(23). Leaves elliptic, half as wide as long; lip deeply 3-lobed; C: straw; D: No, CR, SA (Mt); M: pl to 45 cm, S 2 cm ..... *Maxillaria paleata*
24. Leaves narrowly elliptic, about one-fifth as wide as long 25
25(24). Lip lance-ovate, not clearly lobed; C: greenish yellow, base of lip marked with red; D: No, CR, Pma (Cb, Mt, Pc); M: pl to 50 cm, S 14–15 mm .... *Maxillaria diuturna*
25. Lip 3-lobed.. 26

26(25). Sepals narrowly oblong, subacute; lip 3-lobed in upper third, base oblong; C: white with yellow lip; D: CR (Mt); M: pl to > 1 m, S 2.8 cm .... *Maxillaria vaginalis*
26. Sepals long-acuminate; lip 3-lobed near middle, base subcircular; C: white or pale yellow; D: CR, Pma (Mt); M: pl to 60 cm, S 2.5–4 cm .... *Maxillaria ctenostachya*

27(20). Pseudobulbs fusiform, largely hidden by sheaths 28
27. Pseudobulbs ovate, exposed 30

28(27). Leaves linear, < 5 mm wide; C: pinkish white; D: CR, Pma, SA (Cb, Mt); M: pl to 60 cm, S 1.8–2.5 cm .... *Maxillaria lueri*
28. Leaves distinctly flattened and elliptic, mostly > 5 mm wide ... 29

29(28). Lip narrow, about twice as long as wide; C: light yellow, lip with 2 reddish spots; D: CR, SA (Mt); M: pl (sprawling) to > 30 cm, S 1.9–2 cm .... *Maxillaria meridensis*
29. Lip little longer than wide, triangular-ovate; C: white or cream, lip reddish; D: CR, Pma, SA (Cb, Mt); M: pl (sprawling) to > 50 cm, S 2–2.8 cm .... *Maxillaria exaltata*

30(27). Stems thick, little branched; plants pendent or erect 31
30. Stems thin and wiry, much branched; plants distinctly bushy ... 36

31(30). Lip weakly constricted near or above middle................... 32
31. Lip distinctly and deeply 3-lobed........................... 33

32(31). Plant about 1 m or more tall, leaves 4–5 cm wide; C: ocher-yellow mottled with reddish brown; D: No, CR (Cb, Pc); M: S 2 cm .... *Maxillaria elatior*
32. Plant < 50 cm, leaves about 3 cm wide; C: greenish yellow, base of lip marked with red; D: No, CR, Pma (Cb, Mt, Pc); M: pl to 50 cm, S 14–15 mm .... *Maxillaria diuturna*

33(31). Lobes of lip basal or below middle of lip..................... 34
33. Lobes of lip above middle of lip 35

34(33). Lip wider across midlobe than across lateral lobes (when flattened), midlobe semicircular, toothed; C: S, P red-brown within, basally yellow, outside mostly yellow, margins flushed red-brown, lip red with yellow margins; D: CR (Mt); M: pl > 40 cm, S 2.2 cm .... *Maxillaria serrulata*
34. Lip wider across lateral lobes than across midlobe; midlobe obovate, not toothed; C: white, lip edged with pink; D: CR, wPma (Mt); M: pl to 60 cm, S 1.7 cm .... *Maxillaria bracteata* (Syn. *M. vagans*)

35(33). Lobes of lip triangular, subequal; **D:** CR; **M:** pl to > 1 m; S 2.2–2.5 cm .... *Maxillaria semiorbicularis*
35. Lateral lobes of lip rounded, smaller than midlobe; **C:** white with yellow lip; **D:** CR (Pc); **M:** pl to 40 cm, S 4–5 cm .... *Maxillaria suaveolens*
36(30). Lobes of lip subequal; sepals about 7 mm long; **C:** white to salmon pink or whitish with reddish stripes, lip yellow; **D:** CR, Pma (Mt); **M:** pl bushy, S 5–7 mm...................... *Maxillaria minor*
36. Lateral lobes of lip much smaller than midlobe 37
37(36). Midlobe of lip lanceolate or narrowly ovate, with prominent callus between lateral lobes; sepals 2.4–2.6 mm long; **C:** translucent flesh with dark red stripes; **D:** CR, Pma (Mt); **M:** pl to 30 cm.......... ....*Maxillaria tigrina*
(Syn. *Chrysocycnis tigrina*)
Though this species has been placed in *Chrysocycnis* and certainly resembles that genus in some features, I suspect that the resemblances to *Crysocycnis* are parallelisms and retain the species in *Maxillaria* for the present.
37. Midlobe of lip obovate, callus low or inconspicuous; sepals < 15 mm long; **C:** yellow; **D:** CR (Mt); **M:** S 10 mm................. .... *Maxillaria flava*
C: yellow; **D:** CR, Pma (Mt); **M:** S 9 mm .... *Maxillaria microphyton*
C: yellow to tan, maroon-striped; **D:** CR, Pma, SA (Mt); **M:** S 7 mm ....*Maxillaria wercklei*
Plate 27(4)
There may well be several species of *Maxillaria* with a branched, bushlike habit and yellowish or purple-striped flowers, but the whole complex needs study. The published descriptions of these 3 species seem very similar.

Key 4

Large plants with leafy stems, pseudobulbs borne only at bases or lacking.

1. Leaf sheaths flattened and spreading, projecting from stem (when leaves fallen); leaves with definite petioles or folded bases....... 2
1. Leaf sheaths not flattened or spreading, clasping stem; leaves without definite petioles or folded bases 3
2(1). Leaf blades with definite narrow petioles; **C:** white with red and

yellow on lip; **D:** CR, Pma (Mt); **M:** pl to 2 m, S 1.2–1.8 cm .... *Maxillaria trilobata*
2. Leaf blades folded basally but only slightly narrowed; **C:** white to light yellow; **D:** CR, Pma (Pc, Mt); **M:** pl 1 m, S 3– 6.5 cm .... *Maxillaria inaudita*
   Very distinctive in its large flattened canes, basally narrowed leaves, and long white sepals and petals.

3(1). Stems with swellings at intervals (apparently pseudobulbs hidden by sheaths).. 4
3. Stems of the same diameter throughout...................... 5

4(3). Lip narrow, about twice as long as wide; **C:** light yellow, lip with 2 reddish spots; **D:** CR, SA (Mt); **M:** pl (sprawling) to > 30 cm, S 1.9–2 cm .... *Maxillaria meridensis*
4. Lip triangular-ovate, little longer than wide; **C:** white or cream, lip reddish; **D:** CR, Pma, SA (Cb, Mt); **M:** pl (sprawling) to > 50 cm, S. 2–2.8 cm .... *Maxillaria exaltata*

5(3). Sepals 6–9 mm long, blunt or abruptly acute.................. 6
5. Sepals > 10 mm long, ovate to lanceolate, acute............... 7

6(5). Lip narrow basally; widest across lateral lobes when flattened; sepals ovate, blunt, about 9 mm long; **C:** yellow; **D:** ePma, SA (Mt); **M:** pl to 1.4 m, S 9 mm .... *Maxillaria aurea*
6. Lip wide near base, width subequal across lateral lobes and midlobe; sepals acute, about 6 mm long; **D:** wPma (Mt); **M:** pl to 80 cm, S 6 mm .... *Maxillaria conduplicata*

7(5). Lip subquadrate-oblong, not distinctly 3-lobed; **C:** orange-yellow; **D:** CR; **M:** pl to 80 cm, S 1.8–1.9 cm.......... *Maxillaria alfaroi*
7. Lip distinctly 3-lobed.. 8

8(7). Column long and slender, about 10 mm long; lip deeply 3-lobed, lateral lobes surpassing midlobe; **C:** white; **D:** CR, Pma (Mt); **M:** pl to 40 cm, S 1.7 cm......................... *Maxillaria falcata*
8. Column very short, < 4 mm long; lip shallowly 3-lobed, midlobe surpassing lateral lobes 9

9(8). Lateral lobes of lip basal; **C:** white with salmon pink lip; **D:** CR, wPma (Mt); **M:** pl to 1 m, S 11–12 mm........ *Maxillaria biolleyi*
Plate 27(6)
9. Lip 3-lobed well above middle............................. 10

10(9). Sepals < 2 cm long; **C:** canary yellow, lip dark red; **D:** CR, Pma (Mt); **M:** pl to 1.3 m, S 15 mm.............. *Maxillaria tonduzii*
10. Sepals > 2 cm long; **C:** white, lip yellow or yellow with purple lines;

D: CR, Pma (Mt); M: pl to > 1 m, S 2.8 cm..................
....*Maxillaria vaginalis*

Key 5
Plants without pseudobulbs; slender or fan-shaped.

1. Stems elongate, leaves scattered............................. 2
1. Stems short, plants fan-shaped............................... 8
2(1). Leaves tapering, acute 3
2. Leaves straplike or oblong, obtuse or notched 4
3(2). Leaves thin and leathery; column more than twice as long as wide; C: cream or yellow marked with purple spots; D: CR, Pma (Mt); M: pl to 1 m, S 13–14 mm .... *Maxillaria linearifolia*
3. Leaves thick and fleshy; column less than twice as long as wide; C: yellowish green; D: CR, Pma, SA (Cb, Mt, Pc); M: pl to 1 m, S 5.5 mm .... *Maxillaria adendrobium*
(Syn. *Neourbania adendrobium*)
4(2). Leaves fleshy, sheath dark and rough; C: white; D: No, CR, Pma, SA (Cb); M: pl to 20 cm, S 8–9 mm .... *Maxillaria nicaraguensis*
Similar to *M. adendrobium* but shorter and with blunt leaves.
4. Leaves thin and leathery, sheaths smooth 5
5(4). Leaves 1–2 cm wide; C: S, P pale yellow, lip reddish tan to reddish orange; D: Pma (Mt); M: pl to 50 cm, S 7–10 mm ....
....*Maxillaria allenii*
5. Leaves < 1 cm wide....................................... 6
6(5). Leaves about a third as wide as long, ovate-oblong; lip about two-thirds as wide as long; C: greenish yellow with purple callus; D: CR (Wt); M: pl to 50 cm, S 7–8 mm ..... *Maxillaria appendiculoides*
6. Leaves less than half as wide as long, oblong or lance-oblong ... 7
7(6). Lip narrowly fiddle-shaped, constricted above lateral lobes, midlobe much wider above base; C: white; D: CR, Pma (Cb, Mt); M: pl to 50 cm, S 9–10 mm.........................*Maxillaria valerioi*
7. Lip not fiddle-shaped, constricted in front of lateral lobes, but midlobe oblong, not wider above base; C: greenish yellow with reddish stains to pinkish, lip purplish; D: CR, Pma (Wt); M: pl to 50 cm, S 9 mm .... *Maxillaria dendrobioides*
Two or more species may well be currently treated under this name.

8(1). Leaves fleshy; plant pendent; C: greenish yellow, lip yellow with red or purple spots; D: No, CR, Pma, SA (Mt); M: pl to 60 cm, S 1.2– 1.8 cm .... *Maxillaria valenzuelana*
8. Leaves papery, plants usually erect 9
9(8). Sheaths spotted; leaves 2.4–5 cm wide; C: yellow becoming brownish red with age; D: CR, Pma, SA (Mt); M: pl to 60 cm, S 1.8–2.2 cm .... *Maxillaria chartacifolia*
An unusual *Maxillaria*; it often forms large mats.
9. Sheaths not spotted; leaves usually < 2.4 cm wide; D: CR, SA (Mt) .... *Maxillaria bicallosa*

Mormolyca. Epiphytic; pseudobulbs clustered, ovoid, smooth, slightly flattened, each with 1 terminal leaf; inflorescence basal, each flower on a separate stalk, stalks subequal to leaves; sepals and petals similar, spreading; lip 3-lobed with fleshy callus; column arched, without column foot; 1 species in our area; C: greenish yellow with maroon stripes, midlobe of lip red-brown; D: No, rCR; M: pl 15–30 cm, S 1.5–1.9 cm........... .... *Mormolyca ringens*
Plate 29(3)

Neomoorea. **Large** epiphyte; pseudobulbs clustered, ovoid, **strongly grooved**, each with 2 large pleated leaves; inflorescence basal, erect, with many flowers; sepals and petals similar, spreading; lip deeply 3-lobed; column without wings; 1 species known; C: sepals and petals white basally, then brownish red, lip yellow with maroon markings; D: Pma, SA (Cb); M: pl to 1 m, S 2– 2.8 cm .... *Neomoorea wallisii*
Plate 25(4)

The name *N. irrorata* is better known, but the older name, *wallisii*, has priority.

Ornithocephalus. Small, fan-shaped epiphytes without pseudobulbs; leaves laterally flattened, more or less fleshy; inflorescences lateral, with several to many flowers; sepals and petals similar, cupped or spreading; lip narrow, often 3-lobed, with open oil gland near base; column long-beaked, with the appearance of a bird's head or an elephant's trunk.

Key to *Ornithocephalus*

1. Rachis of inflorescence glandular-hairy or nearly spiny 2
1. Rachis of inflorescence smooth or microscopically hairy 3
2(1). Lip narrowly oblong with hornlike calli much wider than lip at base; **C:** S, P greenish or dull yellow, lip white; **D:** No, CR, Pma (Cb, Mt, PC); **M:** pl 4–12 cm, S 2 mm .... *Ornithocephalus bicornis*
2. Lip subquadrate, concave, nearly as wide as long; calli not wider than lip; **C:** S, P white, lip green; **D:** Pma (Mt); **M:** pl 4–12 cm, S 3 mm .... *Ornithocephalus cochleariformis*
3(1). Lip widest near apex, with subcircular callus at base; **C:** white, callus green; **D:** CR (Mt); **M:** pl to 7 cm, S 3 mm .... *Ornithocephalus lankesteri*
3. Lip widest basally 4
4(3). Lip narrowed in midportion and wider apically; basal lobes thin, spreading; **C:** lvs whitish green, fls dark green with white borders, callus white; **D:** CR, Pma (Cb, Mt); **M:** pl 5–8 cm, S 4 mm .... *Ornithocephalus powellii*
4. Lip narrow apically; basal lobes fleshy and erect; **C:** white with green callus; **D:** No, CR, Pma (Cb, Mt); **M:** pl to 7 cm, S 3 mm .... *Ornithocephalus inflexus*
Plate 30(3)

Pescatorea. Plants similar to *Chondrorhyncha* but larger (to 50 cm tall); flowers large; callus at base of lip convex with many rounded keels.

Key to *Pescatorea*

1. Blade of lip with rugose veins; **C:** S, P white, lip yellow; **D:** CR, Pma (Mt); **M:** pl to 50 cm, S 2.5–3.5 cm .... *Pescatorea cerina*
Plate 26(2)
1. Blade of lip smooth; **C:** S, P white with apical purple blotches on outside, lip purple; **D:** ePma, SA (Mt) .... *Pescatorea dayana*

Sphyrastylis. Small, fan-shaped epiphyte; without pseudobulbs, leaves laterally flattened, more or less fleshy; inflorescences lateral, with several to many flowers; sepals and petals similar, cupped or spreading; lip nar-

row, often 3-lobed, with open oil gland near base; column long-beaked, with the appearance of a bird's head or an elephant's trunk, but the "trunk" clearly wider above base; only 1 species known in Central America; C: white or greenish white; D: Pma (Mt); M: pl 4 cm, S 2.7–2.9 mm
.... *Sphyrastylis cryptantha*

I have collected plants in central Panama that may represent other species of *Sphyrastylis*, but the material is very sparse.

Stellilabium. Epiphytic; plants tiny; leaves conduplicate, very small, sometimes lacking at flowering; inflorescence flattened, becoming much longer than leaves, sometimes branching, each branch producing 1 flower at a time; flowers tiny; lip may have fleshy or hairy callus that clasps column; column short, with beaklike rostellum and hooked viscidium.

These tiny plants are probably frequent in wetter forests, but they are rarely noticed and are therefore poorly studied. They are often leafless at flowering, but the main photosynthetic organ appears to be the tapeworm-like inflorescence rather than the roots. The following key may not work well, and any plant you find may be something new.

Key to *Stellilabium*

1. Flowers without prominent bristles.......................... 2
1. Flowers with prominent bristles on column 3
2(1). Petals spreading, ovate, widest near base; C: yellowish green, maroon in center; D: CR (Mt); M: pl 3 cm, S 2 mm................
.... *Stellilabium monteverdense*
2. Petals curving downward, widest above middle; C: greenish with purplish stripes, appearing brownish; D: CR (Mt); M: pl to 15 cm, S 3.5 mm .... *Stellilabium campbellorum*
3(1). Column markedly 3-lobed, each lobe with tuft of branched (stellate) bristles 4
3. Column shallowly or not 3-lobed, with 3 tufts of unbranched bristles.. 6
4(3). Lobes of column short and rounded, subequal to bristles; C: green with black lip and bristles; D: CR (Mt); M: pl to 10 cm, S 2.8 mm
.... *Stellilabium bullpenense*
Plate 30(4)
4. Column deeply 3-lobed, lobes much longer than bristles 5

5(4). Blade of lip with sides parallel in middle, lip flat or nearly so; C: green tinged with purple, lip purple-black; D: CR (Mt); M: pl to 15 cm, S 3–3.5 mm .... *Stellilabium boylei*
5. Blade of lip with sides rounded in middle, markedly convex; C: dull oily green?; D: CR (Mt); M: pl to 50 cm, S 3 mm .... *Stellilabium distantiflorum*
6(3). Petals 1-veined; C: S, P green, lip cinnamon brown; D: CR, Pma (Mt); M: S 3 mm .... *Stellilabium minutiflorum*
6. Petals 3-veined 7
7(6). Bristles of column numerous, long, and spreading; D: CR (Mt); M: pl to 20 cm, S 2.5 mm................... *Stellilabium lankesteri*
7. Bristles short, in 3 distinct tufts; C: apparently yellow or green with purple center; D: CR (Mt); M: pl 3–6 cm, S 2.5 mm............. .... *Stellilabium standleyi*

Telipogon. Epiphytic or growing in moss; stems slender, new stems arising either from bases or higher on older stems; inflorescence terminal or appearing so, slender, flowers produced 1 at a time; sepals small, somewhat hidden, petals similar to lip, making flower appear triangular; lip or column often with bristles; viscidium forming hook, 4 pollinia.

Key to *Telipogon*

1. Lip without obvious callus at base, may have a smooth, hairy swelling not elevated at its apex, swelling may surround column as a collar, projecting forward from under stigma 2
1. Lip with obvious callus raised above lip at its outer margins ... 11
2(1). Swelling at base of lip with short horn on each side next to column; D: CR (Mt); M: pl 5–7 cm, P 10 mm .... *Telipogon setosus*
2. Swelling, if present, without short horn on each side at base 3
3(2). Swelling occupying basal half of surface, covered with long hairs; petals long-haired toward bases; C: yellow suffused with pale red-purple on P and lip with red-purple veins, base of P red-purple, basal half of lip, column, and bristles dark red-purple; D: CR (Mt); M: pl 5–6 cm, P to 2.5 cm.................. *Telipogon vampirus*
3. Swelling, if present, occupying one-fourth or less of lip surface; petals without long hairs at base 4
4(3). Stem elongate, new shoots arising well above base 5
4. Stems clustered, new shoots arising from base 6

5(4). Petals ovate, apices acute; column smooth or with 3 sparse clusters of spines; C: yellow, P with green or yellow vein lines and reticulations and a dark wine red blotch at base, column wine red; D: CR (Mt); M: pl 10–15 cm, P to 4 cm............*Telipogon gracilipes*
5. Petals broadly elliptic, obtuse; column with 3 dense clusters of spines; stems long; C: yellow to yellow-brown, usually with large brown spot at base of each petal and lip, vein lines green or brown with or without reticulations, column and bristles dark brown; D: CR (Mt); M: pl to 75 cm, P to 2.8 cm .... *Telipogon costaricensis*
6(4). Peduncle slender, < 5 cm tall; C: white with red-maroon vein lines and cross-veins, petals and lip with wine red blotch at bases, column wine red; D: CR (Mt); M: pl 3–4 cm, P to 11 mm................
.... *Telipogon elcimeyae*
Plate 29(6)
6. Peduncle thicker, > 8 cm tall; flowers yellow with brown veins and network 7
7(6). Flowers large, to 7 cm diameter; C: P dark yellow marginally, lighter yellow toward base with dark red-brown veins and a flush of wine red, lip yellow with red-brown veins, basal half wine red; D: CR (Mt); M: pl 8 cm, P to 3.2 cm .... *Telipogon ampliflorus*
7. Flowers small, not more than 4 cm diameter 8
8(7). Lines along veins very broad 9
8. Lines along veins narrow 10
9(8). Underside of column lobed, protruding much beyond stigma; C: yellow to yellow-brown with broad red-brown veins and many cross-veins, column and base of lip red-brown; D: CR (Mt); M: pl 4 cm, P to 2.3 cm .... *Telipogon ardeltianus*
9. Underside of column not lobed, not protruding noticeably beyond stigma; C: yellow-brown striped with red-brown, column and bases of lip and petals purple-red; D: CR (Mt); M: pl 5–6 cm, P to 13 mm .... *Telipogon caroliae*
10(8). Flowers > 3 cm diameter; base of lip forming red or brown collar around column, with slight longitudinal keel down front; C: yellow suffused with rose at bases and red-brown or green vein lines in petals and lip, with few to many cross-veins, column and bristles red to red-brown; D: CR (Mt); M: pl 5–7 cm, P to 2.1 cm...........
.... *Telipogon leila-alexandrae*
10. Flowers < 2 cm diameter; base of lip forming 3-angled light pink collar around column; C: yellow with green or red-brown veins and cross-veins, callus pink, column and bristles wine red; D: CR (Mt); M: pl 3–4 cm, P to 10 mm ....*Telipogon portilloi*

11(1). Column developed on each side to form obvious wings, wings fimbriate on upper margins, lower margin seated in callus of lip; C: lemon yellow, petals with green veins, lip with light brown veins, some branched, callus, column, and bristles wine red; D: CR (Mt); M: pl 3–4 cm, P to 9 mm.................. *Telipogon parvulus*

11. Column essentially cylindrical, without lateral lobes........... 12

12(11). Callus hollow, open on back, occupying a third of lip; C: yellow, petals flushed red-brown basally with red-brown vein lines, lip with large red-brown spot covering all but margins, callus dark red-brown, column and bristles red; D: CR, Pma (Mt); M: pl 5–6 cm, P to 2.2 cm .... *Telipogon biolleyi*
(Syn. *T. endresianum*)
Plate 29(5)

The commonest and most widespread *Telipogon* in our area.

12. Callus solid, occupying one-fourth or less of lip surface........ 13

13(12). Flowers maroon.. 14
13. Flowers yellow, cream, or tan 15

14(13). Petals obtuse with 11 veins, bases thickened and pubescent; C: dark maroon-red, callus wine red; D: CR (Wt); M: pl 3–4 cm, P to 12 mm....................................*Telipogon retanarum*
14. Petals acute with 5 veins, bases not thickened or pubescent; C: yellow covered by red–chocolate brown blotches with red veins and cross-veins; D: CR (Mt); M: pl 5–8 cm, P to 12 mm............
....*Telipogon guila*

15(13). Callus small, narrow, moon-shaped......................... 16
15. Callus ample, oblong or heart-shaped...................... 17

16(15). Flower large, to 3.5 cm diameter; lip with 11–17 veins; C: yellow flushed with pink toward bases, veins and reticulations red-brown, callus and column wine red, bristles red-brown; D: CR (Mt); M: pl 10–12 cm, P to 2.8 cm .... *Telipogon christobalensis*
16. Flower small, to 2 cm diameter; lip with 19 veins; C: creamy yellow with purple-brown veins and cross-connections, callus dark red-purple, column blackish red-purple, bristles red; D: CR (Mt); M: pl to 15 cm, P to 11 mm .... *Telipogon glicensteinii*

17(15). Shoots clustered or somewhat sprawling, stems < 20 cm long; bristle tufts unequal in length, laterals longer..................... 18
17. Plant sprawling, stems rarely < 20 cm long when flowering for first time; bristles long, in 3 dense tufts of equal length 20

18(17). Lateral lobes of callus not surrounding column; dorsal bristle tuft with very short bristles or nearly smooth; C: yellow, petals with green veins changing to red-brown toward base, lip with red-brown veins covered by cross-veins, callus and column wine red, bristles light red; D: CR (Mt); M: pl to 10 cm, P to 15 mm ....
.... *Telipogon pfavii*
18. Lateral lobes of callus surrounding column; dorsal bristles half the length of lateral bristles, dense 19

19(18). Underside of column much swollen, pink; C: lemon yellow with faint transverse, concentric reticulations at bases of petals and lip, callus green-brown; D: CR (Mt); M: pl 4–5 cm, P to 2 cm ....
.... *Telipogon cascajalensis*
19. Underside of column slightly swollen, red; C: lemon yellow suffused with tan, callus and column bright red; D: CR (Mt); M: pl to 10 cm, P to 1.9 cm .... *Telipogon ballesteroi*

20(17). Petals and lip rhombic-elliptic, not overlapping at margins; petals 7-veined; lip 15-veined; C: yellow with or without dull red-brown veins and cross-veins, petals and lip with red-brown suffusion covering more than half the surface, callus, column, and bristles purple-black; D: CR (Mt); M: pl to 70 cm, P to 2.1 cm ....
.... *Telipogon monticola*
20. Petals and lip very broadly ovate, overlapping on lower margins; petals 11–19-veined; lip 19–27-veined; C: from light to dark yellow with broad tan, brown, or red-brown veins, callus, column, and bristles dark brown; D: CR (Mt); M: pl to 20 cm, P to 2.4 cm
.... *Telipogon storkii*
   T. storkii subsp. *magnificus* has fewer veins on petals and lip and numerous cross-veins.

Teuscheria. Epiphytic; pseudobulbs ovoid, covered by stiff, cardboard-like sheaths, usually spotted, clustered or on creeping rhizome, each with 1 narrow, pleated leaf; inflorescence lateral with 1 flower; flower more or less funnel-shaped with prominent chin at base.

Key to *Teuscheria*

1. Pseudobulbs separate on creeping rhizome; lip hairy, margins not toothed or fringed, midlobe notched, longer than lateral lobes; C:

flowers pale pink; **D**: No, CR, Pma, SA (Cb, Mt); **M**: pl 15–20 cm, dS 13 mm .... *Teuscheria pickiana*
Plate 25(5)
1. Pseudobulbs clustered; lip not hairy, margins toothed or fringed, midlobe acute, subequal to lateral lobes; **C**: S, P olive green, lip white, rose-purple spot on callus; **D**: CR (Cb) ....
.... *Teuscheria horichiana*

Trigonidium. Epiphytic; pseudobulbs conic-ovoid, somewhat flattened, grooved; leaves straplike; inflorescence axillary, 1-flowered, with long, slender peduncles; sepals yellowish brown, large, connivent for half to two-thirds their length and then abruptly spreading or bent back, lip hidden within sepals, petals with blue apices that reach flower opening.

Key to *Trigonidium*

1. Pseudobulbs spread out on stout rhizome, each with 3–5 leaves; **C**: light greenish brown to cinnamon, veined brown or purplish, each P with purple spot; **D**: CR, Pma, SA (Mt); **M**: pl 30 cm, S 4.5–5 cm .... *Trigonidium insigne*
(Syn. *T. lankesteri*)
1. Pseudobulbs densely clustered, or overlapping if rhizome creeps, each with 2 leaves 2
2(1). Pseudobulbs densely clustered; ovary with pedicel 4.7–7 cm, flowering from mature pseudobulbs; **C**: yellow-brown with blue-gray "eyes" on petal tips; **D**: No, CR, Pma, SA (Cb, Pc); **M**: pl 35–40 cm, S 2.5–4.5 c .... *Trigonidium egertonianum*
Plate 30(5)
2. Rhizome creeping; ovary with pedicel 4–4.5 cm long; flowering on immature growth; **C**: greenish to pinkish yellow to brown, petals with metallic blue eyes; **D**: CR, Pma, SA (Cb, Pc); **M**: pl 30–40 cm, S 2.8–3.4 cm .... *Trigonidium riopalenquense*
This species is consistently smaller and more slender in all parts than the common *T. egertonianum*, but there is some overlap in plant size.

Warrea. Large terrestrial plants; pseudobulbs clustered, of several internodes, with several pleated leaves; inflorescence lateral, erect, of several

flowers; sepals and petals similar, spreading, lip weakly 3-lobed with fleshy callus near base; column without wings; 1 species in Central America; **C:** pinkish flesh, or petals and lip marked with violet; **D:** No, CR, Pma (Mt); **M:** pl to 80 cm, S 4–4.5 cm . . . . *Warrea costaricensis*
Plate 29(4)

A large plant with showy flowers, but these all face the ground.

Warreopsis. Terrestrial; pseudobulbs clustered, narrowly conical, each with 2–3 terminal leaves and some sheathing leaves; inflorescence basal, erect, of several flowers; sepals and petals similar, spreading, lip stalked with fleshy basal callus; column without wings; 1 species in our area; **C:** purple, lip violet, column white; **D:** CR, wPma (Mt); **M:** pl to 80 cm, S 7–10 mm . . . . *Warreopsis parviflora*

Xylobium. Epiphytic; pseudobulbs from globose to long and slender, smooth when young; leaves pleated, 1–3 on each pseudobulb; inflorescence lateral; flowers usually greenish, lip fleshy.

Key to *Xylobium*

1. Pseudobulbs slender, about 10 times longer than wide, each with 2 leaves; **C:** whitish or pale yellow; **D:** No, CR, Pma, SA (Cb, Mt); **M:** pl to 50 cm, S 1.5–2.2 cm . . . . *Xylobium elongatum*
Plate 25(6)
1. Pseudobulbs much shorter, usually with 1 leaf 2
2(1). Pseudobulbs narrowly conic, gradually tapering; **C:** yellow; **D:** No, CR, Pma (Cb, Mt); **M:** pl 30 cm, S 1.5–2 cm . . . .
. . . . *Xylobium powellii*
2. Pseudobulbs globose to ovoid, abruptly tapering 3
3(2). Pseudobulbs globose; inflorescence scarcely longer than pseudobulb; **C:** S, P white to pale pink with flecks of red, lip purple-brown to pinkish; **D:** Pma, SA (Cb); **M:** pl 40 cm, S 10 mm . . . .
. . . . *Xylobium colleyi*
3. Pseudobulbs ovoid; inflorescences distinctly longer than pseudobulbs 4
4(2). Lobes of lip papillose; lip without keel on underside; **C:** pale flesh

with maroon overlay, midlobe of lip red; **D**: No, CR, SA; **M**: pl 50 cm, S 2.5–2.8 cm .... ***Xylobium squalens***
4. Lobes of lip smooth; midlobe with keel on underside; **C**: white, lip with reddish veining; **D**: No, CR, Pma, SA (Cb, Mt); **M**: pl 50 cm, S 10–14 mm .... ***Xylobium foveatum***

9 Miscellaneous Orchids with Corms or Pseudobulbs

This chapter includes a varied collection of orchid genera, both epiphytic and terrestrial, that do not fit in any of the previous chapters. The plants treated here have definitely thickened stems *or* are epiphytes with a single leaf on each stem. The thickened stems may be either aerial pseudobulbs or underground corms. A couple of odd species of *Malaxis* have slender stems and many leaves, but the key in Chapter 10 should direct you to the proper place.

Key to Genera of Miscellaneous Orchids with Corms or Pseudobulbs

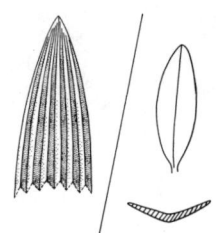

1. Leaves distinctly pleated 2
1. Leaves conduplicate, fleshy or subherbaceous 11

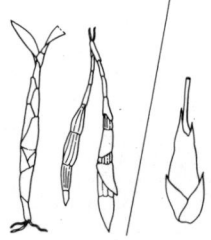

2(1). Pseudobulbs club-shaped or cigar-shaped from a narrow base; leaves scattered along upper pseudobulb 3
2. Pseudobulbs or corms never club-shaped 5

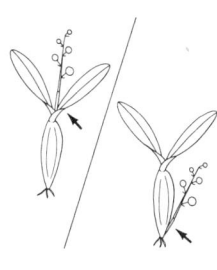

3(2). Inflorescence terminal; flower with spur or prominent, acute chin................................ ***Galeandra***
3. Inflorescence lateral; flower without spur, chin blunt if present 4

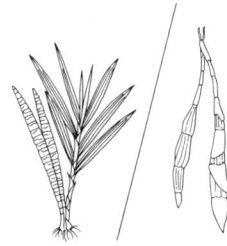

4(3). Pseudobulbs cigar- or horn-shaped, erect, tapering to tip, spiny when leaves have fallen; inflorescence branched, longer than pseudobulb; lip from narrow base.......... ....***Cyrtopodium punctatum***
4. Pseudobulbs club-shaped, more or less pendent, never spiny; inflorescence not branched, shorter than pseudobulb; lip wide at base; flowers large and fleshy .... ***Chysis***

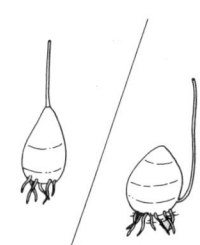

5(2). Inflorescence terminal; stem a conic pseudobulb or a corm 6
5. Inflorescence lateral; stem a gladiolus-like corm, sometimes elongate and rhizome-like.................... 7

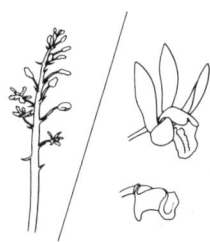

6(5). Flowers small (sepals < 10 mm) and green or bronze, without prominent chin................ ***Liparis nervosa***
6. Flowers much larger (sepals > 2 cm), with prominent chin .... ***Galeandra***

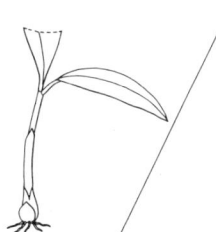

7(5). Petioles forming angular tube................. ***Govenia***
7. Petioles never forming tube....................... 8

8(7). Coarse plant 1–1.5 m tall; inflorescence many-flowered; flowers with prominent, rounded chin . . . . ***Eulophia alta***
8. Plants usually much less than 1 m tall; inflorescence with few flowers open at a time; flowers not as above 9

9(8). Leaves obovate or broadly elliptic, abruptly tapering, sparsely hairy; flowers white or pinkish with small spur at base . . . . ***Calanthe calanthoides***
9. Leaves elliptic or narrowly elliptic, tapering gradually at base and apex, not hairy; flowers usually brightly colored, without spur 10

10(9). Lip deeply 3-lobed, base of midlobe narrower than column, forming stalk at least half the length of midlobe, lateral lobes divergent, longer than wide . . . . ***Spathoglottis plicata***
10. Lip 3-lobed, but midlobe without long, narrow stalk, lateral lobes wider than long. . . . . ***Bletia***

11(1). Inflorescence terminal. 12
11. Inflorescence lateral 15

12(11). Leaves wide and subherbaceous or leathery; flowers small and green; often terrestrial 13
12. Leaves straplike, clearly conduplicate or fleshy; usually epiphytic. 14

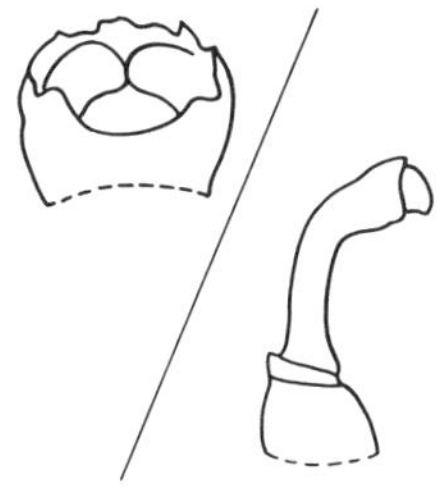

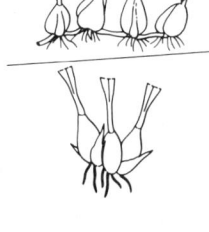

13(12). Column very short; inflorescence may appear umbellate; tiny viscidia associated with pollinia .... **Malaxis**

13. Column long and slender; inflorescence never umbellate; pollinia naked .... **Liparis**

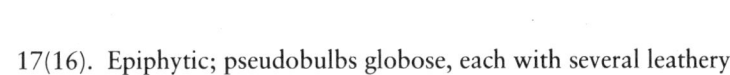

14(12). Each stem with 1 leaf; flowers rose-pink ... **Arpophyllum**

14. Each pseudobulb with 2 to several leaves; flowers yellow .... **Polystachya**

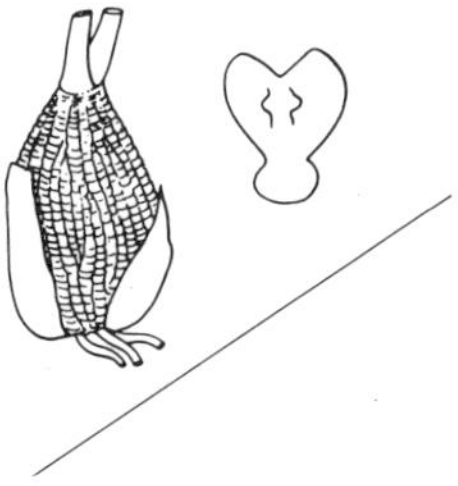

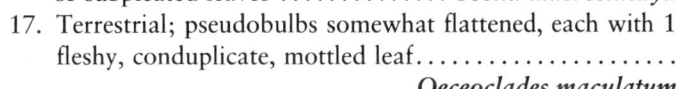

15(11). Pseudobulbs large and warty; leaves thick and leathery; column slender and arched; lip with 2 high calli near base; pollinia 2 .... **Eriopsis biloba**

15. Without above combination of features 16

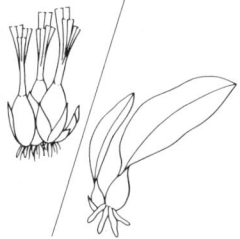

16(15). Pseudobulbs well separated on rhizome, usually 4-angled, each with 2 fleshy leaves .... **Bulbophyllum**

16. Pseudobulbs densely clustered 17

17(16). Epiphytic; pseudobulbs globose, each with several leathery or subpleated leaves .... **Coelia macrostachya**

17. Terrestrial; pseudobulbs somewhat flattened, each with 1 fleshy, conduplicate, mottled leaf..................... .... **Oeceoclades maculatum**

Arpophyllum. Epiphytic; with creeping rhizome, stems slightly thickened, each stem with a single terminal leaf; leaves leathery or fleshy, straplike; inflorescence a terminal raceme of pink or rose-purple flowers very like those of *Elleanthus*.

Key to *Arpophyllum*

1. Flowers > 8 mm long; flower cluster short and wide, about twice as long as wide; **C:** pink; **D:** No, CR (Mt); **M:** lvs to 30 cm long, 2.5 cm wide .... *Arpophyllum medium*
1. Flowers < 8 mm long; flower cluster 4–5 times as long as wide; **C:** rose-purple; **D:** No, CR, SA (Mt); **M:** lvs to 60 cm long, 4 cm wide .... *Arpophyllum giganteum*
Plate 31(1)

A. alpinum, of Mexico and northern Central America, probably does not occur in our area. It is a much smaller plant (about 15 cm tall) with a short, subglobose inflorescence.

Bletia. Terrestrial or lithophytic; with gladiolus-like corms, each with a few narrow, pleated, usually deciduous leaves; inflorescence lateral, of several to many flowers; pollinia 8.

Key to *Bletia*

1. Lip shallowly 3-lobed, about 4 cm long; flowers self-pollinating and not opening widely; **C:** purple, white in center; **D:** No, CR, SA (Mt); **M:** lvs to 30 cm, S 3.3 cm .... *Bletia purpurata*
(Syn. *Crybe rosea*)
Commonly self-pollinating; the flowers would be showy if they opened.
1. Lip distinctly 3-lobed, < 3.5 cm long........................ 2
2(1). Lip < 2 cm long, base cordate or squarish; corms rounded; sepals spreading; **C:** pink or rose-pink, rarely white; **D:** No, CR, Pma, SA (Cb, Mt, Pc); **M:** lvs to 50 cm, S 1.5–1.7 cm ...... *Bletia purpurea*
Plate 31(2)
Common, often seen on rocks and road cuts.

2. Lip > 2 cm long, base narrow or rounded; sepals not spreading widely... 3

3(2). Lip deeply lobed, base of midlobe narrower than apex; sepals and petals not connivent; corms about as wide as long; C: sepals purplish, petals white with purplish tips, lip white, midlobe dark purple; D: No, CR (Mt); M: lvs to 45 cm, S 3 cm ... *Bletia edwardsii*
3. Lip shallowly lobed, base of midlobe as wide as apex; sepals and petals connivent, forming hood over lip; corms long and narrow; C: S red-violet, P white with red-violet tips, midlobe of lip red-violet; D: No, CR, Pma, SA (Mt); M: lvs to 30 cm, S 3.3 cm............
.... *Bletia campanulata*
Widespread, often self-pollinating; apparently infrequent in our area.

Bulbophyllum. Epiphytic; pseudobulbs well separated on creeping rhizome, somewhat 4-angled; leaves 2, leathery or fleshy; inflorescence lateral, fleshy, with many small green flowers with red or purple markings. *Bulbophyllum* is primarily an Old World genus with 3 or 4 species in Central America.

Key to *Bulbophyllum*

1. Petals bristle-tipped; column not thicker apically; peduncle longer than leaves; C: greenish spotted with purple, lip yellow and dark purple; D: No, CR, Pma (Mt, Pc); M: pl 15–20 cm, S 6 mm
.... *Bulbophyllum aristatum*
1. Petals rounded; column wider above; peduncle shorter than leaves
.... 2

2(1). Rachis of inflorescence thick and fleshy (thicker than length of flower); dorsal sepal broadly elliptic, abruptly acute, petals less than half as long as sepals; C: yellowish green spotted with red-violet; D: No, CR, Pma, SA (Cb, Pc); M: pl 15–20 cm, S 5–6.5 mm........
.... *Bulbophyllum oerstedii*
The name *B. pachyrachis* has been used for this species, but that name applies to a distinct West Indian species. *B. wagneri*, described from central Panama, may be a self-pollinating form of *B. oerstedii*.

2. Rachis of inflorescence not markedly thick and fleshy; dorsal sepal elliptic-lanceolate, acuminate, petals more than half as long as sepals; C: wine red; D: No, CR (Cb); M: pl 12 cm, S 6 mm ......
.... *Bulbophyllum vinosum*

Calanthe. Terrestrial; corms short, on or in soil, each shoot with 2 wide, pleated leaves; inflorescence lateral, erect, with many flowers; sepals and petals similar, spreading; lip widest near apex, base with short finger-like spur; column short, thick, with 8 club-shaped pollinia; 1 species in our area; C: white, pinkish, or faintly flushed with purple; D: No, CR, Pma, SA (Mt); M: lvs to 45 cm, S 11–12 mm........ *Calanthe calanthoides*
Plate 31(3)

Chysis. Epiphytic; pseudobulbs club-shaped, usually curved and often hanging downward, youngest with several pleated leaves; inflorescence lateral, of several large fleshy flowers; column thick, with 8 pollinia. *Chysis aurea* var. *maculata* is reported from central Panama. I have seen plants north of Gamboa and on Barro Colorado Island, but I have seen no flowers and I cannot determine how or if this plant differs from *C. costaricensis*.

Key to *Chysis*

1. Callus of lip with 3 fleshy keels; C: orange-yellow with red spots, lateral lobes of lip white with red spots; D: No, CR (Mt); M: ps to 30 cm, S 2.2–2.4 cm .... *Chysis tricostata*
1. Callus of lip with 5 fleshy keels; C: golden yellow with reddish stripes; D: No, CR (Mt); M: ps to 25 cm, S 2.5–3.5 cm ....
.... *Chysis costaricensis*
Plate 31(4)

Coelia. Epiphytic; pseudobulbs clustered, globose, smooth when young, each with several narrow terminal leaves that are several-veined but not clearly pleated or conduplicate; inflorescence basal, erect, with several large, papery bracts, with many small flowers, each flower with long, narrow bract and prominent saclike nectary at base; sepals and petals

similar, lip simple; pollinia 8; 1 species in our area; C: flowers light rose-purple; D: No, CR, Pma (Mt); M: lvs to 80 cm, S 8–15 mm ....
.... *Coelia macrostachya*
(Syn. *Bothriochilus macrostachyus*)
Plate 31(5)

Cyrtopodium. Usually on rocks or steep slopes; pseudobulbs clustered, club- to cigar-shaped, of many internodes, tapering to a sharp point; leaves several, scattered on upper pseudobulb, pleated, deciduous, pseudobulb spiny after leaves have fallen; inflorescence basal, produced from young shoot, erect and branching, with conspicuous bracts; sepals and petals similar, spreading, lip stalked, deeply 3-lobed; column straight, wingless; pollinia 2, with stipe; 1 species in our area; C: flowers yellow with brown blotches; D: No, CR, Pma, SA (Pc); M: ps to 80 cm, S 2 cm
.... *Cyrtopodium punctatum*
Plate 32(1)

An inhabitant of dry, rocky areas, in Panama known only from the Perlas Islands.

Eriopsis. Large epiphyte; pseudobulbs clustered, dark green, rough-surfaced, each with 2–3 tough, leathery leaves; inflorescence basal, erect or ascending, with many flowers; sepals and petals similar, spreading, lip deeply 3-lobed; column arching, without wings; 1 species in Central America; C: S, P yellow suffused and bordered with red, lip yellow marked with red, midlobe white with purple spots; D: No, CR, Pma, SA (Mt); M: S 1.6–1.8 cm .... *Eriopsis biloba*
Plate 32(2)

Probably a member of the Maxillarieae but with only 2 pollinia.

Eulophia. Terrestrial; in our area with an underground corm; leaves several, pleated; inflorescence lateral, erect, of many flowers; sepals and petals similar, sepals standing up like rabbit ears, petals hooded over column; lip 3-lobed, saccate at base; 1 species in Central America; C: greenish suffused with brown or reddish brown; D: No, CR, Pma, SA (Cb, Pc); M: pl to 1.5 m, S 2–2.3 cm .... *Eulophia alta*
Plate 32(3)

Frequently found in grassy or weedy areas.

Galeandra. Epiphytic or terrestrial; epiphytes with ovoid or cigar-shaped pseudobulbs of several internodes, covered with sheaths and sheathing leaf bases; leaves on upper part of pseudobulb; terrestrials with globose or ovoid corms on or in substrate, with 1 or a few slender terminal leaves; leaves pleated, deciduous; inflorescence terminal, of 1 to several flowers; sepals and petals similar, spreading or cupped; lip with spur or prominent chin at base, more or less 3-lobed, enfolding column; anther beaked; pollinia 2, with stipe.

Key to *Galeandra*

1. Epiphytic; pseudobulbs narrowly ovoid to cigar-shaped; C: S, P greenish maroon, lip violet; D: No, CR, Pma (Cb); M: ps to 20 cm, S 2.5–3 cm .... *Galeandra batemanii*
Plate 32(4)

    The name G. *baueri* has been used for Central American plants but that name is based on a plant from French Guiana, apparently quite distinct from ours.

1. Terrestrial; corms globose or broadly ovoid; usually leafless at flowering 2

2. Flower with prominent chin but no spur; column foot at about a right angle to column axis; C: green, lip marked with red or purple; D: CR, Pma, SA (Pc); M: lvs to 75 cm, S 2.5–3 cm ....
.... *Galeandra beyrichii*

2. Flower with slender spur; column foot curving back from base of column; C: rose; D: Pma, SA (Pc); M: lvs to 20 cm, S 11–12 mm .... *Galeandra styllomisantha*
(Syn. *G. juncea*)

Govenia. Terrestrial; with flattened or ovoid corms of several internodes, each with 2 leaves (in our area); leaves thin, pleated, stalked, stalks tubular, enclosed by sheaths; inflorescence lateral, emerging from sheaths around leaf stalks, erect, of few to many flowers; sepals and petals similar, spreading, or petals and dorsal sepal hooded over column; lip simple, curved; column arched, more or less winged above, with prominent foot at base; pollinia 4 with small stipe. The entire *Govenia* plant is tender and easily bruised. There may be several species in our area, but they have not been well studied. Several names based on plants of Mexico and the West Indies have been used in our area. *G. liliacea* does occur in Costa Rica, but it is unlikely that *G. capitata*, *G. superba*, or *G. utriculata* are to be found there. There are probably several unnamed species. Plate 32(5).

Key to *Govenia*

1. Lip with fringe of small hairs; **C**: sepals and petals yellow with pink spots, lip pink; **D**: No, CR, wPma (Mt); **M**: pl 30–36 cm, S 10–13 mm .... *Govenia ciliilabia*
1. Lip without fringe; **C**: white, petals barred with pink or magenta, lip with 3 to several red-purple spots on **underside** at apex of lip; **D**: No, CR, rPma (Mt); **M**: pl to 60 cm, S 2–3 cm .... *Govenia liliacea*

Liparis. Either terrestrial or epiphytic; if epiphytic, pseudobulb conic-ovoid, of several internodes, with 2–3 broad, pleated leaves; if terrestrial, corm short, ovoid, in soil, with 1 terminal leaf that is leathery, not pleated; inflorescence terminal, of several to many green flowers; column arched; pollinia 4, without appendages of any sort.

Key to *Liparis*

1. Leaves deeply cordate, about as wide as long; sepals, petals, and lip acute or acuminate; **C**: deep maroon; **D**: No, rCR; **M**: pl to 26 cm, S 8.5–10.8 mm .... *Liparis fantastica*
1. Leaves tapering basally, longer than wide; sepals, petals and lip obtuse or broadly acute 2

2(1). Leaves pleated, 2–4; old pseudobulb exposed; **C**: green flushed with red-purple; **D**: No, CR, Pma, SA (Cb, Mt); **M**: pl 20–30 cm, S 5–7 mm .... *Liparis nervosa*
(Syn. *L. elata*)
Plate 33(2)

Frequently found on stumps and logs.

2. One leaf, leathery or herbaceous; corm underground; C: S, P yellowish, lip greenish purple or greenish brown; D: No, CR, SA (Mt); M: lvs 5–15 cm, S 11–15 mm.................*Liparis vexillifera*

From its description, *L. fratrum* appears to be a member of the *Malaxis blephariglottis* complex and at least as much a misfit in *Liparis* as in *Malaxis*. It may be a third member of that complex in Costa Rica, or it may be a synonym of *M. blephariglottis* or *M. tipuloides*.

Malaxis. Terrestrial or epiphytic; with or without corms; leaves soft-herbaceous, often wavy or crisped; inflorescence terminal, in terrestrial species sheathed by a tubular petiole; flowers small, petals usually very narrow, lip often with basal cavity or cavities, column very short; pollinia 2 or 4, ovoid or club-shaped, with microscopic viscidia. The flowers of *Malaxis* are small and usually green, so hobbyists do not often cultivate the genus. The species of Central America have not been carefully studied, and there are probably more species than this key indicates.

Malaxis ranges nearly throughout the world, being absent only from New Zealand. It is a large and bewildering group that clearly needs study. In our area one may easily recognize 3 distinct groups and 1 anomalous species:

1. *fastigiata* group: Terrestrial, with a narrow, ellipsoid or conic pseudobulb; leaves 2; inflorescence dense and flat-topped (subumbellate). In these plants the narrow pseudobulb is usually above the soil surface, and the dense, flat-topped inflorescence may continue elongating until the rachis is quite long, but it always has a flat-topped group of flowers at the tip. Most of our species belong here.
2. *unifolia* group: Terrestrial; with globose corm covered by fibrous sheaths; leaf 1; inflorescence usually long and slender. In these the corm is usually buried. In some species with 1 leaf the inflorescence may be dense or subumbellate; includes *M. carnosa*, *M. majanthemifolia*, *M. pandurata*, *M. pittieri*, and *M. soulei*. The tiny, epiphytic *M. wendlandii* is odd but may be an offshoot of the *unifolia* group.
3. *blephariglottis* group: Epiphytic; without corm or pseudobulb; leaves several to many, 2-ranked; inflorescence slender. In details and column structure this group seems completely distinct from the rest of *Malaxis*. It may deserve a genus of its own; includes *M. blephariglottis*, *M. tipuloides*, *Liparis fratrum*?

Key to *Malaxis*

1. Plants without pseudobulbs; leaves 2-ranked, several to many ... 2
1. Plants with pseudobulbs or narrow corms, each with 1 or 2 leaves 3

2(1). Lip broad at base and then constricted below widest part of blade; plants mostly < 20 cm tall; lip 3.5–4 mm long; **C:** pale translucent green or lip flushed with dull orange; **D:** CR, Pma (Mt, Pk); **M:** pl to 15 cm, S 3.5–4 mm .... *Malaxis blephariglottis*
2. Lip gradually expanding from narrow base; plants 25–60 cm tall; lip 7–8 mm long; **C:** translucent green; **D:** No, CR, Pma, SA (Mt); **M:** pl to 60 cm, S 6 mm .... *Malaxis tipuloides*
Plate 33(4)

3(1). Each corm or pseudobulb normally with 1 leaf 4
3. Each corm or pseudobulb normally with 2 leaves 9

4(3). Lip quadrangular-oblong, longer than wide and widest at apex; dwarf plant 2–3 cm tall; epiphytic; bulbs covered with network of fibers; **C:** pale green; **D:** No, CR, Pma, SA (Mt); **M:** pl to 5 cm, S 2.7–3.5 mm. .... *Malaxis wendlandii*
This curious little plant has been classified as a *Malaxis* and as a *Liparis*, but it fits neither very well.
4. Lip widest basally; plant larger, terrestrial; bulb covered by thin sheaths 5

5(4). Rachis of inflorescence becoming elongate, but flowers densely clustered, cluster more or less flat-topped; **C:** green or yellowish green; **D:** No, CR (Mt); **M:** pl 8–15 cm, S 3 mm .... *Malaxis aurea*
5. Inflorescence elongate, flowers never forming flat-topped cluster 6

6(5). Apex of lip notched or 3-toothed 7
6. Apex of lip a single blunt lobe or tooth..................... 8

7(6). Lip tridentate at apex; inflorescence a dense spike; **C:** yellowish green; **D:** No, CR, Pma, SA (Mt); **M:** pl to 10 cm, S 1.5–1.8 mm ....*Malaxis soulei*
7. Lip trilobate in front; inflorescence a loose raceme; **D:** CR, Pma (Mt); **M:** lvs 2–3 cm, S 2.2–2.7 mm .... *Malaxis pittieri*

8(6). Leaf broad and deeply notched at base; **C:** pale green, margins of lip brownish; **D:** No, CR, Pma (Mt); **M:** pl to 15 cm, S 3 mm....... .... *Malaxis majanthemifolia*

8. Leaf oblong or elliptic from a tapering base; C: greenish yellow; D: No, CR, SA (Mt); M: pl to 10 cm, S 1.7 mm ....*Malaxis carnosa*

9(3). Apex of lip notched or 3-toothed 10

9. Apex of lip rounded or tapering, not at all notched or toothed 13

10(9). Lip subquadrate, blade widest near apex, with 2 linear basal lobes; C: yellowish green; D: wPma (Mt); M: pl to 15 cm, S 2.5–4 mm .... *Malaxis woodsonii*

10. Lip widest near base, with short or triangular basal lobules 11

11(10). Lip with erect or recurved basal lobules; C: greenish; D: No, CR, Pma, SA (Mt); M: pl to 15 cm, S 4.5 mm .... *Malaxis excavata*

11. Lip without basal lobules, or these rounded and indistinct 12

12(11). Apex of lip acute; D: CR, Pma (Mt); M: lvs to 14 cm, S 3–4 mm .... *Malaxis simillima*

12. Apex of lip subtruncate, slightly concave; D: CR (Mt); M: pl to 18 cm .... *Malaxis adolphi*

13(9). Lip subcircular to oblong or fiddle-shaped, widest above base 14

13. Lip ovate to triangular, widest near base.................... 16

14(13). Lip longer than wide, widest near apex; D: No, CR (Cb, Mt); M: lvs to 7.5 cm, S 5 mm .... *Malaxis pandurata*

14. Lip as wide as long... 15

15(14). Lip subcircular, apex rounded or with very slight apical lobule; C: greenish brown; D: No, CR, Pma, SA (Mt); M: pl to 15 cm, S 3.5 mm................................... *Malaxis histionantha*
Plate 33(3)

The name *M. parthonii* has been used for this species in Central America, but that name belongs to a quite different species of Brazil.

15. Lip with triangular apical lobe about a third the length of lip; D: CR (Mt); M: pl 9 cm, S 2 mm .... *Malaxis tonduzii*

16(13). Lip ovate-lanceolate, abruptly long-acuminate; dwarf epiphyte, < 7 cm tall; D: CR (Wt); M: pl to 6.5 cm, S 7–8 mm... *Malaxis nana*

16. Lip not long-acuminate; > 10 cm tall, terrestrial.............. 17

17(16). Lip with narrow basal lobes longer than wide; C: green; D: CR, SA (Mt); M: pl to 15 cm, S 4 mm................*Malaxis crispifolia*

17. Basal angles of lip rounded to triangular, not longer than wide 18

18(17). Basal angles of lip rounded; apex of lip narrowed; C: green; D: No,

CR, Pma, SA (Mt); **M:** pl to 15 cm, S 3.5 mm ....
.... *Malaxis fastigiata*
18. Basal angles of lip triangular; apex of lip rounded; **C:** S, P light green, lip dark green; **D:** No, CR (Mt); **M:** pl to 30 cm, S 4 mm
.... *Malaxis brachyrrhynchos*

Oeceoclades. Terrestrial; pseudobulbs clustered, ovoid, each with 1 conduplicate terminal leaf; leaves fleshy, mottled with darker green; inflorescence lateral, erect, with many small flowers; sepals and petals similar, cupped; lip 3-lobed with prominent basal spur thicker at tip; column arched; pollinia 2, with stipe; 1 species in Western Hemisphere; **C:** S, P green or reddish ocher, lip green with pink markings; **D:** Pma, SA (Pc); **M:** pl 20 cm, S 9 mm....................... *Oeceoclades maculata*
Plate 32(6)

The only American representative of an African genus, this species is spreading and has appeared in Panama, the West Indies, and Florida within the last few years and will doubtless spread to Costa Rica and other Central American countries. The flowers are self-pollinating, and the seedlings grow rapidly in the soil and are relatively large when they appear.

Polystachya. Small epiphytes; pseudobulbs clustered, of several internodes, hidden by sheaths, each with 1 or 2 conduplicate terminal leaves; inflorescence terminal, often branched, flowers closely spaced and opening several at a time; flowers with lip uppermost, may have prominent chin at base; sepals and petals spreading, petals narrower than sepals, lip 3-lobed, usually covered by mealy hairs (pseudopollen); column short, thick; with 4 pollinia and tiny stipe. Basically an African genus with some American representatives. The Central American polystachyas are monotonously similar. Botanists have found several older names for the 2 commonest American species in recent decades. One hopes that yet older names will not appear.

Key to *Polystachya*

1. Peduncle and rachis of inflorescence finely fuzzy; inflorescence simple or with small basal branches; **C:** yellowish green, callus white;

276 Miscellaneous Orchids with Corms or Pseudobulbs

        D: No, CR, Pma, SA (Cb, Mt, Pc); **M**: pl to 10 cm, S 2.5–3 mm
.... ***Polystachya masayensis***
        Leaves deciduous in drier areas.
  1. Peduncle and rachis smooth; inflorescence normally with several branches.. 2
2(1). Chin very prominent, column foot more than half length of lip; lateral sepals wider than long; **C**: greenish yellow to yellow; **D**: No, CR, SA (Mt); **M**: pl to 20 cm, S 4–5 mm.... ***Polystachya concreta***
        (Syns. *P. flavescens*, *P. luteola*, *P. minuta*)
        Recognized by the long chin that gives the flowers a cone-like form.
  2. Chin short, column foot less than half length of lip; lateral sepals longer than wide; **C**: yellowish green to whitish green; **D**: No, CR, Pma, SA (Cb, Pc); **M**: pl to 20 cm, S 3–3.5 mm ....
.... ***Polystachya foliosa***
        (Syn. *P. cerea*)
        Plate 33(1)
    The commonest *Polystachya* in our area.

Spathoglottis. Terrestrial; with gladiolus-like corms in soil and narrow, pleated leaves; inflorescence lateral, of many flowers opening a few at a time; sepals and petals similar, spreading; lip deeply 3-lobed, base of midlobe narrow, lateral lobes divergent, longer than wide; column arching, with 8 pollinia; **C**: rose-purple or pink; **D**: Pma (Cb) ....
.... ***Spathoglottis plicata***
        Plate 31(6)
An Asiatic genus with a self-pollinating form naturalized here. The flowers shown in Plate 31 are paler than the form naturalized in Panama, and they also open more widely than self-pollinating flowers.

10 Miscellaneous Orchids without Pseudobulbs, Mostly Terrestrial

Some orchids with slender stems and several to many leaves are members of the subtribes Laeliinae, Oncidiinae, or (other) Maxillarieae and are treated in Chapters 4, 5, and 8. All other orchids with slender stems are treated in this chapter. Many of them have either distinctly pleated leaves or soft-herbaceous leaves, but there are a few with conduplicate leaves, and some have fleshy or leathery leaves that cannot be classed in any of the major leaf types. There are advantages in grouping genera with their relatives (especially for those who know more about the groups) and advantages in an alphabetic arrangement. I arrange the genera alphabetically, but I briefly describe the major groups treated here and list their genera. The interested reader may learn the groups as such and skip part or all of the key.

Cranichidinae. Terrestrial or sometimes epiphytic; usually with rosettes of basal leaves, the leaves soft-herbaceous or weakly pleated, usually stalked; column acute; pollinia clublike, brittle. *Baskervilla, Cranichis, Ponthieva, Pseudocentrum, Pterichis, Solenocentrum.*

Cypripedioideae (lady's slippers). This group is so distinctive that it forms a separate subfamily, and some authors treat it as a distinct family. The shape of the flowers is such that they can scarcely be confused with any other group of plants. *Phragmipedium, Selenipedium.*

Gastrodieae. These are saprophytic orchids; that is, they depend on organic material in the soil that is available to them through their partnership with fungi. These plants never have leaves or green coloring. Any leafless orchid without corms but with fleshy roots that does not match one of these is probably one of the Spiranthinae or Prescottiinae that has leaves during the growing season but is leafless at flowering. *Uleiorchis, Wullschlaegelia.*

Goodyerinae. Rhizome (base of stem) prostrate but similar to aerial stem; roots scattered on rhizome; leaves soft-herbaceous, spiral, often variegated; pollinia sectile. *Aspidogyne, Erythrodes, Goodyera, Kreodanthus, Ligeophila, Platythelys.*

Prescottiinae. Terrestrial plants usually of high elevations; the soft-herbaceous leaves forming basal rosette; sepals and petals similar; lip often concave or saccate; column short, pollinia soft and mealy. *Aa, Gomphichis, Prescottia.*

Sobraliinae. Stems elongate, with ovate or elliptic, pleated leaves, or with narrow or conduplicate leaves in a few species with smaller plants. The inflorescence is terminal in our species and may have 1 large flower at a time (*Sobralia*) or several to many small flowers together (*Elleanthus*). Except for the smaller species of *Elleanthus*, the plants of these 2 genera are almost identical, and one must look for an old inflorescence to distinguish plants without flowers. *Elleanthus, Sobralia.*

Spiranthinae. Terrestrial or epiphytic; roots fleshy; leaves basal, soft-herbaceous, may be lacking at flowering time; inflorescence terminal; flowers with lip lowermost; anther dorsal; pollinia 4, soft. *Beloglottis, Brachystele, Coccineorchis, Cyclopogon, Deiregyne, Discyphus, Eltroplectris, Eurystyles, Funkiella, Galeottiella, Hapalorchis, Lankesterella, Lyroglossa, Mesadenella, Mesadenus, Pelexia, Sarcoglottis, Schiedeella, Spiranthes, Stenorrhynchos.*

Tropidieae. This relatively small group occurs throughout the tropics; 3 species are known in Central America. The flower structure is much like that of the Cranichideae, but the plants have long stems with pleated leaves. The inflorescences are often both terminal and lateral. *Corymborkis, Tropidia.*

Key to Genera of Miscellaneous Orchids without Pseudobulbs

1. Lip deeply cup-shaped or saclike; column with shield-like staminode at apex and 2 fertile anthers, 1 at each side of column (lady's slippers)........................... 2
1. Lip various but never deeply cup-shaped or saclike with opening constricted; column normally with a single fertile anther at apex.................................... 3

Cypripedioideae

2(1). Leaves thin, somewhat pleated; leafy stem very long and sometimes branched; lip yellow..... ***Selenipedium chica***
2. Leaves conduplicate, leathery; leafy stem short.......... .... ***Phragmipedium***

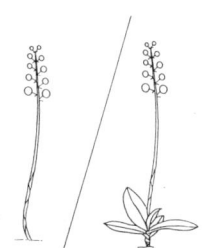

3(1). Leaves totally lacking............................ 4
3. Leaves present, if absent at flowering (in Cranichideae), there should be remains of old leaves at base of inflorescence.. 6

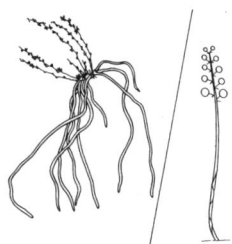

4(3). Epiphytic; plant body primarily a mass of gray-green roots; stem may be very short or quite evident.......... .... ***Campylocentrum***
4. Terrestrial plants with roots buried in soil........... 5

Gastrodieae

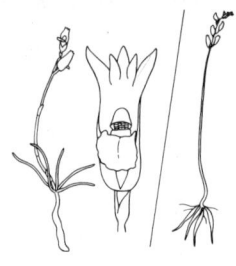

5(4). Flowers few (1–5), each > 1 cm long; sepals and petals strongly united; plant with bumpy underground tuber and slender roots ....***Uleiorchis ulaei***
5. Flowers many, each < 5 mm long; sepals and petals not markedly united; plant with spindle-shaped roots ....***Wullschlaegelia***

6(3). Climbing or sprawling vines; leaves fleshy, convolute in development; pollen masses very soft .... ***Vanilla***
6. Plant habit various but never a vine 7

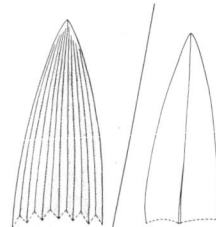

7(6). Leaves distinctly pleated 8
7. Leaves not distinctly pleated, but soft-herbaceous, leathery, or conduplicate 14

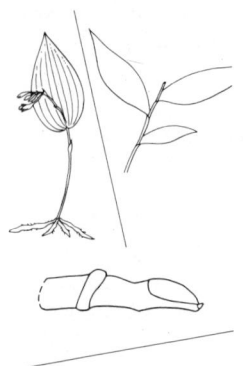

8(7). Leaf, solitary, fanlike, nearly as wide as long, tapering abruptly, with silver streaks along veins................ ....***Monophyllorchis***
8. Leaves much longer than wide, usually several to many 9

9(8). Column and anther pointed, with distinct viscidium at apex; pollinia sectile............................ 10
9. Column blunt; anther caplike on column apex; pollinia soft 11

Tropidieae

10(9). Column < 10 mm long; inflorescence terminal.........
.... *Tropidia polystachya*
10. Column > 10 mm long; inflorescences mostly lateral
.... ***Corymborkis***

11(9). Flowers relatively small; lip strongly united with column along midline; pollinia 4, without caudicles or viscidia
.... ***Palmorchis***
11. Flowers small to quite large; lip not strongly united with column along midline; pollinia 8, may have caudicles or viscidia 12

Sobraliinae

12(11). Leaves pleated 13
12. Leaves conduplicate, with single median fold ....
.... ***Elleanthus***

13(12). Flowers large or very large, > 1 cm long; usually borne 1 at a time from a cluster of bracts; resembling a *Cattleya* flower .... ***Sobralia***
13. Flowers small, usually < 1 cm long; several to many borne at once in a raceme or headlike cluster ....
.... ***Elleanthus***

14(7). Leaves scattered along stem, may be leathery 15
14. Leaves generally basal or whorled, soft-herbaceous; flowers often numerous........................... 21

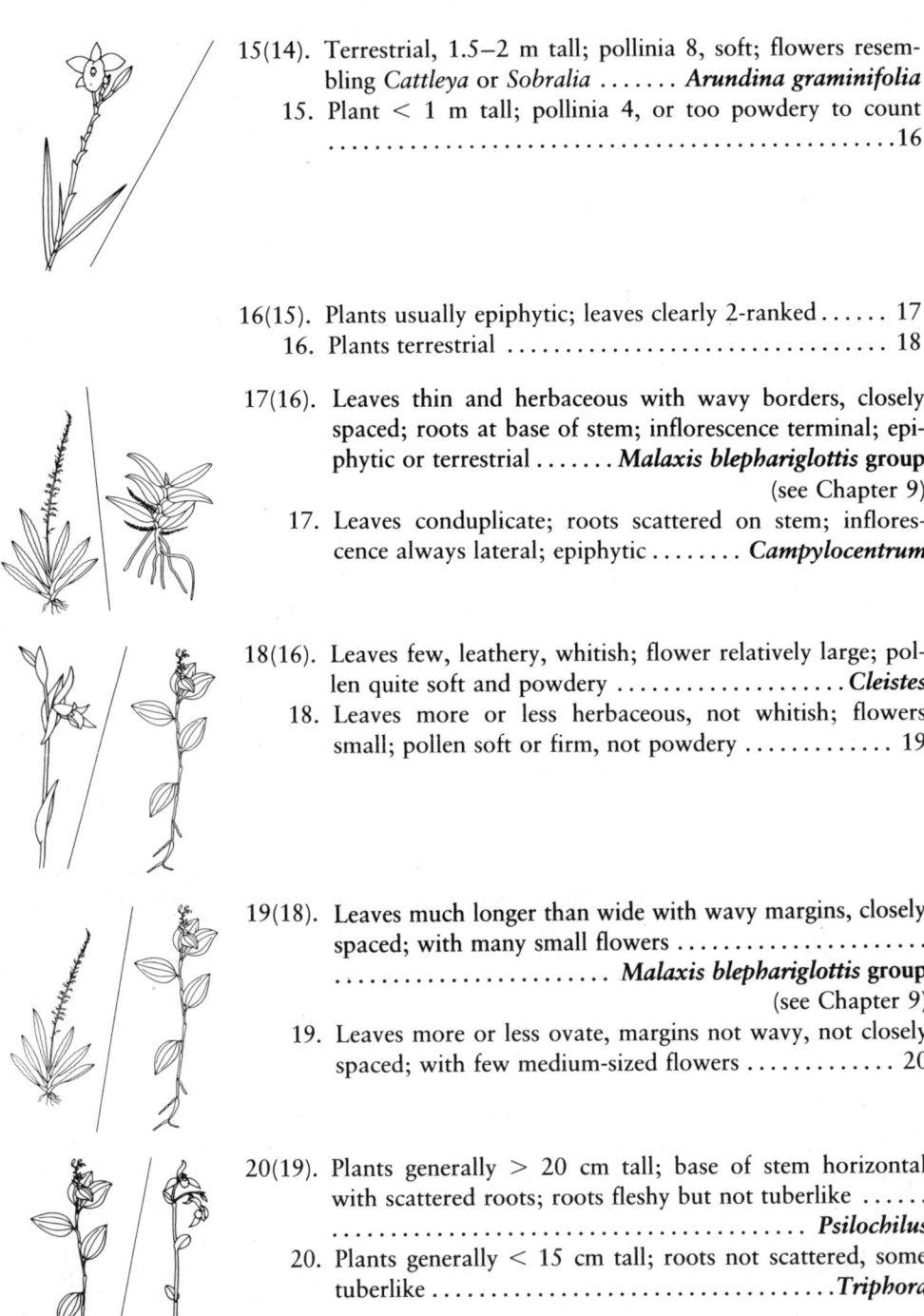

15(14). Terrestrial, 1.5–2 m tall; pollinia 8, soft; flowers resembling *Cattleya* or *Sobralia* .... ***Arundina graminifolia***
15. Plant < 1 m tall; pollinia 4, or too powdery to count 16

16(15). Plants usually epiphytic; leaves clearly 2-ranked...... 17
16. Plants terrestrial 18

17(16). Leaves thin and herbaceous with wavy borders, closely spaced; roots at base of stem; inflorescence terminal; epiphytic or terrestrial .... ***Malaxis blephariglottis* group**
(see Chapter 9)
17. Leaves conduplicate; roots scattered on stem; inflorescence always lateral; epiphytic .... ***Campylocentrum***

18(16). Leaves few, leathery, whitish; flower relatively large; pollen quite soft and powdery .... ***Cleistes***
18. Leaves more or less herbaceous, not whitish; flowers small; pollen soft or firm, not powdery 19

19(18). Leaves much longer than wide with wavy margins, closely spaced; with many small flowers ....
.... ***Malaxis blephariglottis* group**
(see Chapter 9)
19. Leaves more or less ovate, margins not wavy, not closely spaced; with few medium-sized flowers 20

20(19). Plants generally > 20 cm tall; base of stem horizontal with scattered roots; roots fleshy but not tuberlike ......
.... ***Psilochilus***
20. Plants generally < 15 cm tall; roots not scattered, some tuberlike .... ***Triphora***

21(14). The 2 halves of anther widely separated, each pollinium attached to a separate viscidium, pollinia sectile; petals often deeply bilobed; lip usually with prominent spur...... .... ***Habenaria***
21. Anther acute, the halves never widely separated; pollinia sectile or mealy with a single apical viscidium; not with the above combination of features (Cranichideae) 22

22(21). Base of stem prostrate, soft, and herbaceous, with roots scattered along stem; leaves often spotted or otherwise marked 23
22. Base of stem not prostrate, usually short, with roots clustered 28

Goodyerinae

23(22). Lip without spur .... *Goodyera*
23. Lip with definite saccate spur at base 24

24(23). Rostellum split or notched by removal of viscidium... 25
24. Viscidium surrounding rostellar remnant like a thimble 26

25(24). Column elongate, axis much longer than anther; anther about as wide as long.................... ***Kreodanthus***
25. Column short, axis about as long as anther; anther much longer than wide........................ ***Erythrodes***

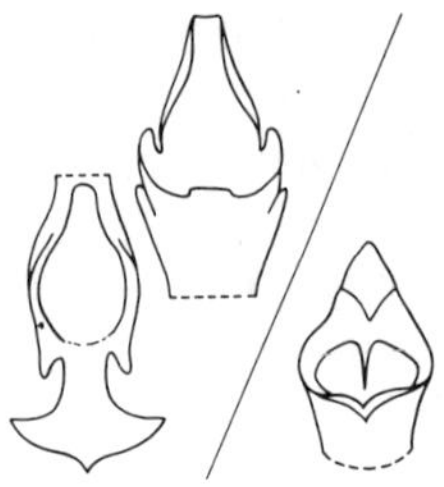

26(24). Stigma not divided into 2 lobes; rostellum jointed; sepals about 8 mm long, 4 mm wide; spur usually 1.5–1.8 cm long .... *Ligeophila*
26. Stigma divided into 2 lobes; rostellum not jointed; sepals to 6 mm long, < 3 mm wide; spur usually much shorter than sepals 27

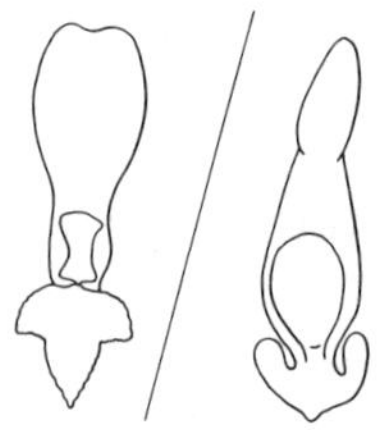

27(26). Lip constricted between base and blade; flowers white, fleshy; sometimes epiphytic .... *Platythelys*
27. Lip not markedly constricted between base and blade; flowers usually green or greenish, not fleshy .... *Aspidogyne*

28(22). Lip on lower side of flower 29
28. Lip on upper side of flower 48

Spiranthinae

29(28). Flowers densely clustered (capitate or subcapitate); plants epiphytic, up to 5 cm tall .... *Eurystyles*
29. Inflorescence dense or loose but always elongate; mostly terrestrials 30

30(29). Bracts on peduncle and below flowers papery, semitransparent .... *Deiregyne*
30. Bracts on peduncle and below flowers green, or only margins translucent 31

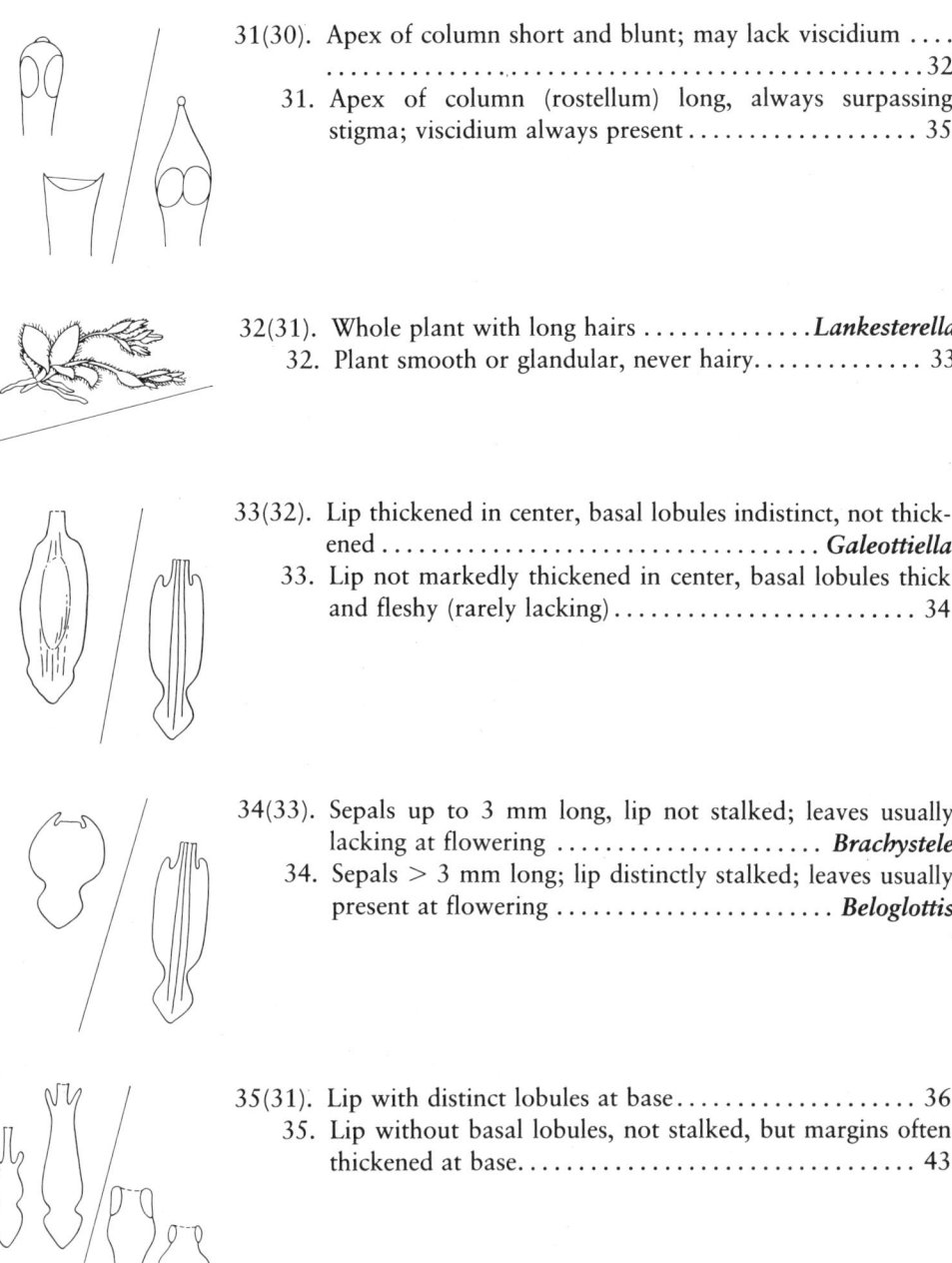

31(30). Apex of column short and blunt; may lack viscidium 32
31. Apex of column (rostellum) long, always surpassing stigma; viscidium always present 35

32(31). Whole plant with long hairs .... *Lankesterella*
32. Plant smooth or glandular, never hairy.............. 33

33(32). Lip thickened in center, basal lobules indistinct, not thickened .... *Galeottiella*
33. Lip not markedly thickened in center, basal lobules thick and fleshy (rarely lacking)........................ 34

34(33). Sepals up to 3 mm long, lip not stalked; leaves usually lacking at flowering .... **Brachystele**
34. Sepals > 3 mm long; lip distinctly stalked; leaves usually present at flowering .... **Beloglottis**

35(31). Lip with distinct lobules at base.................... 36
35. Lip without basal lobules, not stalked, but margins often thickened at base................................ 43

286   Miscellaneous Orchids without Pseudobulbs

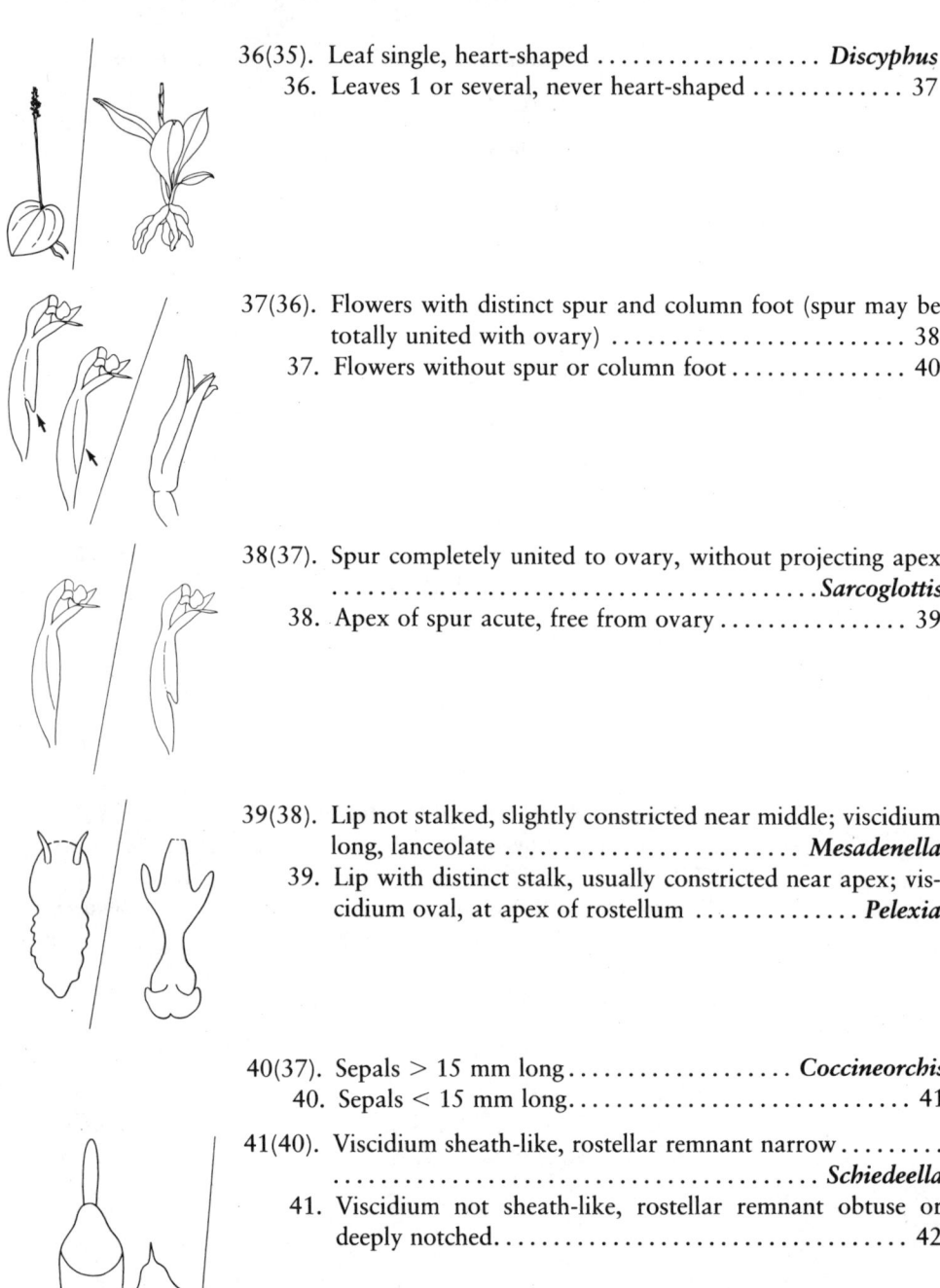

36(35). Leaf single, heart-shaped .... ***Discyphus***
36. Leaves 1 or several, never heart-shaped 37

37(36). Flowers with distinct spur and column foot (spur may be totally united with ovary) 38
37. Flowers without spur or column foot 40

38(37). Spur completely united to ovary, without projecting apex .... ***Sarcoglottis***
38. Apex of spur acute, free from ovary 39

39(38). Lip not stalked, slightly constricted near middle; viscidium long, lanceolate .... ***Mesadenella***
39. Lip with distinct stalk, usually constricted near apex; viscidium oval, at apex of rostellum .... ***Pelexia***

40(37). Sepals > 15 mm long .... ***Coccineorchis***
40. Sepals < 15 mm long 41
41(40). Viscidium sheath-like, rostellar remnant narrow .... ***Schiedeella***
41. Viscidium not sheath-like, rostellar remnant obtuse or deeply notched 42

42(41). Viscidium clasped by rostellar arms, rostellar remnant deeply notched.............................. *Spiranthes*
42. Viscidium apical, rostellar remnant obtuse.............
   .... *Cyclopogon*

43(35). Margins of lip thickened basally.................... 44
43. Margins of lip not thickened basally............... 46

44(43). Lip with red thickening in center............. *Funkiella*
44. Lip without red thickening in center................ 45

45(44). Lip constricted near middle, lyrelike in outline ....
   .... *Lyroglossa*
45. Lip undivided or indistinctly constricted, more or less lanceolate............................... *Stenorrhynchos*

46(43). Lip without constriction, elliptic or lanceolate...........
   .... *Mesadenus*
46. Lip constricted near apex 47

47(46). Flowers large, > 15 mm long, with spur that is free at apex .... *Eltroplectris*
47. Flowers small, up to 15 mm long, without spur .... **Hapalorchis**

48(28). Pollinia club-shaped, brittle; column pointed 49
48. Pollinia flattened, very soft; column generally blunt (Prescottiinae) 54

Cranichidinae

49(48). Lip with spur or saccate nectary at base 50
49. Lip without spur or saccate nectary 52

50(49). Lip with 2 elongate basal lobes, base saccate .... **Baskervilla**
50. Lip without basal lobes or these indistinct, lip and/or lateral sepals forming spur 51

51(50). Spur formed by lateral sepals, with an extension of lip within; petals entire .... ***Pseudocentrum hoffmannii***
51. Spur formed by lip; sepals 2-lobed but free; petals 2-lobed .... ***Solenocentrum costaricense***

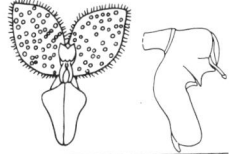

52(49). Lip and petals clearly united to column for more than half length of column; petals together forming a "pseudolip" that is larger and more obvious than true lip............
.... *Ponthieva*

52. Lip and petals free from column or only slightly united; petals not close together, not forming "pseudolip" ... 53

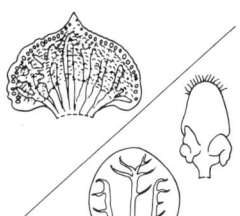

53(52). Lip more or less cordate-ovate, attenuate, warty; plant of high elevations............................*Pterichis*

53. Lip rather obovate or orbicular, with fleshy veins but not warty.....................................*Cranichis*

Prescottiinae

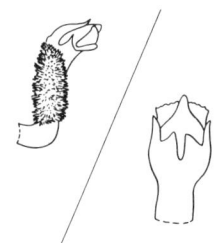

54(48). Column hairy, bent near base and again near apex (nearly S-shaped); petals hairy on margins; lip not deeply concave
........................*Gomphichis costaricensis*

54. Column smooth, straight; petal margins smooth; lip deeply concave.................................. 55

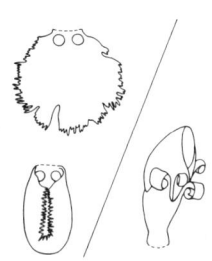

55(54). Sepals free to base; lip fringed; brownish, papery bracts much longer than flowers .... *Aa paleacea*

55. Sepals basally united; margin of lip smooth; bracts green, usually subequal to flowers.................*Prescottia*

Aa. Terrestrial; roots fleshy; leaves several, basal, soft-herbaceous, basally narrowed but not clearly stalked, lacking at flowering; inflorescence lateral, erect, with papery bracts below, flowers many, each with narrow, papery bract longer than flower; petals narrower than sepals; lip saccate, margins fringed; column very short, blunt; anther dorsal; pollinia 4, soft; 1 species in Central America; **C:** white; **D:** CR, SA (Pk); **M:** pl 20–50 cm, S 3–4.5 mm .... *Aa paleacea*

Arundina. Tall terrestrial; stems clustered, canelike, up to 2 m tall, with numerous 2-ranked conduplicate, straplike leaves; inflorescence terminal, producing 1 flower at a time; floral bracts small, triangular, conduplicate; flowers similar to *Cattleya* or *Sobralia* with 8 soft, ovoid pollinia; a native of Asia but commonly cultivated in the American tropics and occasionally naturalized, especially along roadsides; C: rose-purple, lip darker; M: pl to 2 m, S 4.5 cm .... *Arundina graminifolia*
Plate 34(6)

Aspidogyne. Terrestrial; stem herbaceous, basally sprawling, then erect; leaves several, spiral, soft-herbaceous or weakly pleated, somewhat stalked; inflorescence terminal, of several to many flowers; sepals and petals similar; lip lowermost, with saclike nectary at base, blade often 2-lobed; anther dorsal, pollinia sectile, viscidium forming sheath around rostellum. Two named species are reported from our area, but the specimens I have seen that were so labeled are all referable to other genera.

Key to *Aspidogyne*

1. Spur slender, 7–8 mm long; midlobe bilobed, the lobules subparallel; C: white; D: CR (Cb); M: pl 35 cm, S 4.5 mm ....*Aspidogyne* species
Plate 39(1)
1. Spur saclike, < 3 mm long; midlobe acute or rounded.......... 2
2(1). Lip < 6 mm long, gradually constricted to about half width of apical lobe; C: brownish marked with white; D: No, rCR; M: pl to 25–40 cm, S 2.5 mm.................*Aspidogyne tuerckheimii*
2. Lip > 7 mm long, abruptly constricted to less than one-fourth width of apical lobe; C: lvs dark red-green with white, fls greenish white; D: No, rCR; M: pl 20–25 cm, S 5–6 mm ....*Aspidogyne stictophylla*

Baskervilla. Terrestrial; leaves stalked, soft-herbaceous (weakly pleated); inflorescence erect; flowers green, whitish green, or brownish green; petals asymmetrical, basally united with column, then abruptly bent back; lip uppermost, markedly united with column, base forming blunt spur or chin, side lobes between column and rest of lip, midlobe scooplike; column pointed, anther on dorsal side; pollinia 4, brittle, club-shaped.

Beloglottis

Key to *Baskervilla*

1. Petals widest where bent, but not distinctly lobed; lateral sepals < 3 mm wide; saclike base of lip attached about 2 mm from column base; **C**: S, P green, lip white; **D**: CR (Mt); **M**: pl 35–45 cm, S 5.5–6 mm .... ***Baskervilla* species**
Plate 40(1)
1. Petals with a distinct lateral lobe; lateral sepals 4–5 mm wide; saclike base of lip attached < 1 mm from base of column 2
2(1). Petals > 2 mm wide; **C**: S green, P and lip cream or white, or lip brownish with white center; **D**: Pma (Mt); **M**: pl 40–50 cm, S 7.5–9.5 mm .... ***Baskervilla colombiana***
2. Petals usually < 2 mm wide; **C**: S green, P and lip white; **D**: No, CR (Mt); **M**: pl 28–35 cm, S 6–9 mm ...... ***Baskervilla nicaraguensis***

The unnamed species with narrow flower parts is clearly distinct. The Panamanian plants are distinctly more robust than those of Costa Rica and have slightly larger flowers with wider parts, but the differences are slight and the measurements overlap. I hesitate to suggest name changes, as I have not seen plants from Nicaragua or Colombia.

Beloglottis. Terrestrial; roots thick and fleshy; leaves several, somewhat stalked, usually present at flowering; with many small tubular flowers; sepals and petals spreading only near apex; lip stalked; column short, rostellar remnant deeply notched.

Key to *Beloglottis*

1. Lip without basal lobules; **C**: green; **D**: No, CR (Mt); **M**: pl to 35 cm, S 5–6 mm .... ***Beloglottis ecallosa***
1. Lip with definite lobules at base 2
2(1). Basal lobules of lip much shorter than one-fifth of lip length, lip constricted near apex, apical lobe wider than constriction; **C**: white or whitish; **D**: CR, Pma (Cb, Mt); **M**: pl to 20 cm, S 4.5–5 mm .... ***Beloglottis subpandurata***
2. Basal lobules of lip longer than one-fifth of lip length, lip without distinct constriction near apex 3
3(2). Lip rhombic, widest near middle, narrowed toward base and apex; **C**: white or white with green midveins; **D**: No, CR, Pma (Cb); **M**:

pl to 25 cm, S 4.3–6 mm ....*Beloglottis costaricensis*
Plate 38(2)
3. Lip arrowhead-shaped, widest near base, distinctly narrowed to apex 4
4(3). Lip up to twice as long as wide; dorsal sepal up to 3 times as long as wide; viscidium oblong; inflorescence lax; C: S white, basally green; D: No, CR; M: pl to 30 cm, S 5 mm .... *Beloglottis hameri*
4. Lip about 4 times as long as wide; dorsal sepal narrow, about 4 times as long as wide; viscidium ovate; inflorescence dense; C: white; D: CR, SA; M: pl to 30 cm, S 3 mm ....
....*Beloglottis bicaudata*

Brachystele. Small terrestrial; leaves several, narrow, somewhat stalked, present or absent at flowering; with many tiny subglobose flowers; sepals and petals similar, not spreading; lip squarish with conspicuous basal lobes; column short, viscidium oval, rostellar remnant obtuse; C: green with white lip; D: No, CR, Pma, SA; M: pl to 20 cm, S 2.5 mm ....
.... *Brachystele guayanensis*

Campylocentrum. Epiphytic; with or without leaves, reed-stem if leafy, terminal growth continued but may branch; if present, leaves conduplicate, 2-ranked; inflorescence racemose, flowers small with definite spurs.

These are the American representatives of the *Angraecum* complex that is numerous in Africa and Madagascar. The leafy species look like miniature vandas, with lateral inflorescences and indefinite terminal growth. The leafless species are simply tangles of gray-green roots and scarcely look like plants even when in flower. Many species have characteristic "toothbrush" inflorescences, with the flowers densely clustered and all facing in the same direction. Both leafy and leafless species of *Campylocentrum* are often found in citrus trees, hedges, and other cultivated trees and shrubs.

Key to *Campylocentrum*

1. Plants with obvious leaves, these flat and much wider than thick...
....2
1. Plants without leaves, or these rudimentary, may be small and cylindrical... 8

Campylocentrum

2(1). Leaves > 3 cm long.................................... 3
2. Leaves < 2.5 cm long................................... 4

3(2). Inflorescences longer than leaves, loosely flowered (fewer than 7 flowers per cm); sepals and petals < 3 mm long; C: white; D: No, CR, Pma (Mt); M: pl to 30 cm, S 1.7–2 mm............................
....*Campylocentrum schiedei*
3. Inflorescences shorter than leaves, densely flowered; sepals and petals at least 4 mm long; C: yellowish white; D: No, CR, Pma, SA (Cb, Mt, Pc); M: pl to 25 cm, S 4–5 mm......................
....*Campylocentrum micranthum*
Plate 39(5)
Commonly found on the twigs of citrus and guava.

4(2). Inflorescences less than half as long as leaves; C: cream or greenish cream; D: No, CR, wPma (Mt, Wt); M: pl 2–5 cm, raceme < or sub=lvs, S 2 mm......................*Campylocentrum brenesii*
4. Inflorescence as long as or longer than leaves................. 5

5(4). Spur 2 mm or more long, longer than ovary; lip 3-lobed; C: white; D: CR, Pma (Mt); M: pl 5–10 cm, raceme > lvs, S 1.5 mm
.... *Campylocentrum longicalcaratum*
5. Spur 1.5 mm or less long, subequal to ovary.................. 6

6(5). Leaves > 13.5 mm long; C: white, spur green; D: Pma, SA (Cb); M: pl 5 cm, raceme > lvs, S 1.5 mm ...*Campylocentrum schneeanum*
6. Leaves < 12.5 mm long..................................... 7

7(6). Margins of leaf sheaths fringed; C: white or cream; D: Pma (Pc, Mt); M: pl 1–6 cm, raceme < lvs, S 1 mm.....................
....*Campylocentrum tenellum*
7. Margins of leaf sheaths entire; C: white; D: CR (Mt, Wt); M: pl 6–10 cm, raceme > lvs, S 1 mm ....*Campylocentrum parvulum*

8(1). Stems elongate with obvious spaces between nodes or leaves; roots scattered along stem; leaves tiny, narrowly conical, soon lost; C: greenish white; D: CR, Pma SA (Cb); M: stems to 25 cm, raceme 10–15 mm, S 2 mm................. *Campylocentrum poeppigii*
Like other leafless species, this may be fairly common but is rarely noticed.
8. Stems very short, roots clustered........................... 9

9(8). Flowers few (< 5), on a hairlike inflorescence; C: greenish yellow;

D: No; M: raceme 2–5 cm, S 2.5 mm ....
.... *Campylocentrum porrectum*
(Syn. *Harrisella porrecta*)
Not recorded from our area, but easily overlooked.
9. Flowers many; inflorescence not hairlike 10
10(9). Roots thick and flattened, > 4 mm wide; C: yellowish or pinkish white; D: CR, Pma, SA (Cb); M: raceme 2–3.5 cm, S 4–5 mm
.... *Campylocentrum pachyrrhizum*
This inconspicuous plant may appear in hedges where least expected and is probably much more widespread than published records indicate.
10. Roots thin, cylindrical, < 2 mm wide 11
11(10). Spur short, < 1.3 mm, saclike; C: cream; D: No, CR, Pma, SA (Cb, Pc); M: raceme 5–7 cm, S 1 mm .... *Campylocentrum fasciola*
Plate 39(6)
11. Spur long, > 2 mm, tapering to narrow apex 12
12(11). Lip triangular, widest near base, with glandular hairs; C: white, spurs green; D: Pma, SA (Cb); M: raceme 3 cm, S 1.5 mm ....
.... *Campylocentrum tyrridion*
12. Lip widest near middle, concave, without glandular hairs; C: greenish yellow; D: ePma (Pc); M: raceme 2–2.5 cm, S 1 mm ....
.... **Campylocentrum dressleri**

Cleistes. Terrestrial; roots slender but with some fleshy root-tubers; stem slender, erect; leaves few to several, convolute in development but not pleated, leathery and whitish; inflorescence terminal, with few flowers, bracts leaflike; sepals and petals similar, usually somewhat spreading; lip simple, with fleshy callus along midline; column large, fleshy, with terminal anther and very soft pollen mass.

Key to *Cleistes*

1. Dorsal sepal < 3 cm long; lip distinctly 3-lobed, the midlobe toothed or fringed; C: pink; D: CR (Mt/Wt); M: pl 35–55 cm
.... *Cleistes costaricensis*
1. Dorsal sepal 6–7 cm long; lip not lobed, the margins not toothed or

fringed; C: pink or rose-purple with yellow in throat; D: Pma, SA (Mt); M: pl to 40 cm............................*Cleistes rosea*
Plate 33(6)

Occasional in open grassy areas, where it may be frequent one year and scarce the next.

Coccineorchis. Terrestrial or epiphytic; leaves several, somewhat stalked, present at flowering; several flowers crowded at end of inflorescence; flowers relatively large, tubular; lip stalked, with basal lobules, lanceolate to oblong, more or less constricted; column long, viscidium sheath-like, long and narrow, rostellar remnant linear-lanceolate.

Key to *Coccineorchis*

1. Lowermost floral bract distinctly longer than flowers........... 2
1. Lowermost floral bract shorter than flowers or subequal........ 3

2(1). Flowers smooth; C: bracts and fls yellow; D: No, CR (Mt); M: pl to 40 cm, S 2.5 cm ....*Coccineorchis standleyi*
2. Flowers glandular without; C: red, lip yellow; D: CR, Pma (Mt); M: pl to 35 cm, S 1.9 cm............... *Coccineorchis bracteosa*
Plate 37(4)

3(1). Pedicel about as long as ovary; bracts shorter than flowers; C: fls dark red, lip yellow; D: CR, Pma, SA (Mt); M: pl to 45 cm, S 1.8–2 cm ....*Coccineorchis cernua*
3. Pedicel much shorter than ovary; bracts about as long as flowers; C: bracts and fls red; D: CR, Pma (Mt); M: pl to 50 cm, S 2.1–2.2 cm .... *Coccineorchis navarrensis*

Corymborkis. Terrestrial; stems slender, reedlike, with many 2-ranked or spiral pleated leaves; inflorescence axillary or terminal and axillary, short, branched, of several flowers; sepals and petals similar, subparallel; lip simple; column arching, anther parallel with column axis; pollinia sectile.

Key to *Corymborkis*

1. Flowers about 15 mm long; anther pointed; C: yellow; D: CR, Pma (Mt); M: pl 0.4–2 m .... *Corymborkis flava*
Plate 36(3)
1. Flowers about 2.5 cm long; anther blunt; C: green and white; D: No, CR (Mt); M: pl 0.5–2 m .... *Corymborkis forcipigera*

Cranichis. Terrestrial; with somewhat fleshy roots; leaves few to several, soft-herbaceous or weakly pleated, stalked; inflorescence terminal, of several to many flowers; petals narrower than sepals, often fringed; lip uppermost, concave, often with raised veins; column pointed, with anther on dorsal side; pollinia 4, club-shaped, brittle.

Key to *Cranichis*

1. Petals distinctly fringed 2
1. Petals not fringed 4
2(1). Lip subcircular, nearly as wide as long; C: S green, P white with green tip, lip light green with dark green stripes; D: No, CR, Pma, SA (Mt); M: pl to 35 cm, S 4 mm .... *Cranichis ciliata*
2. Lip much longer than wide, or 3-lobed 3
3(2). Petals narrow; lip narrowly ovate, without fringe; C: S green or green with purple veins, P purple-brown, lip white with brownish stripes; D: No, CR, wPma, SA (Mt); M: pl to 20 cm, S 4.2–4.5 mm
.... *Cranichis wageneri*
3. Petals ovate, obtuse, fringed, lip 3-lobed, tip of midlobe fringed; C: white, lip with red bar or spot near apex, basal lobes orange or yellow; D: CR (Mt); M: pl to 30 cm, S 6 mm ....
.... *Cranichis lankesteri*
Plate 40(2)
4(1). Lip at least basally saccate 5
4. Lip shallowly concave but not deeply saccate 7
5(4). Lip helmet-shaped, not at all lobed; usually with 1 or 2 leaves; C: lvs dark green with pale veins, fls crystalline white, green spot in lip; D: CR (Mt, Pk); M: pl to 15 cm, S 6–6.5 mm ....
.... *Cranichis reticulata*
Grows in moss on vertical rock faces and tree trunks.

5. Lip basally saccate but not helmet-shaped; several leaves scattered along stem 6
6(5). Blade of lip 3-lobed, toothed; C: white with green veins in base of lip; D: CR (Mt); M: pl to 20 cm, S 3.5–6 mm .... *Cranichis saccata*
   6. Blade of lip not lobed, without teeth; C: S green, P and lip white; D: CR (Mt); M: pl 10–27 cm, S 3.5–5 mm...................... .... *Cranichis acuminatissima*
7(4). Lip narrowed basally, margin thickened apically; C: white, green markings in lip; D: No, CR; M: pl to 10 cm, S 3.5–4 mm .... *Cranichis sylvatica*
   7. Lip not markedly narrowed basally, margins not thickened....... 8
8(7). Petals widest basally; margins of lip fleshy and folded in basally; inflorescence bracts leaflike; C: white with green spots on lip; D: No, CR, Pma, SA (Mt); M: pl to 20 cm, S 3.3–3.5 mm .... *Cranichis muscosa*
   8. Petals widest apically; margins of lip not fleshy or folded in basally; inflorescence bracts not leaflike; C: greenish; D: No, CR, Pma, SA; M: pl to 25 cm, S 2.5–2.6 mm .... *Cranichis diphylla*

Cyclopogon. Terrestrial or epiphytic; leaves several, stalked, present at flowering; inflorescence of several to many small flowers, dense or lax; flowers tubular, lateral sepals sometimes spreading; lip stalked, with basal lobules; column slender; viscidium oval, rostellum and rostellar remnant straplike.

Key to *Cyclopogon*

   1. Basal margins of lip forming curved, acute lobules 2
   1. Basal lobules of lip rounded, with small, finger-like calli near centers 3
2(1). Leaves shorter than inflorescence; midlobe of lip narrower and shorter than base of lip, with cushion-like callus; C: greenish white with brownish suffusion; D: Pma, SA; M: pl to 60 cm, S 13 mm .... *Cyclopogon millei*
   2. Leaves subequal to inflorescence; midlobe of lip as wide as base or wider, without callus; often epiphytic; C: pale green with white lip; D: No, CR, Pma; M: pl to 10 cm, S 5–7 mm .... *Cyclopogon prasophyllum*

3(1). Midlobe of lip about half as wide as base of lip, small, more or less oval; **C:** whitish to greenish, with purple veins and margins on dS and P; **D:** No, CR, SA; **M:** pl to 40 cm, S 5–6 mm .... ***Cyclopogon cranichoides***
3. Midlobe of lip about as wide as base or wider................. 4
4(3). Petals with small thickenings at apex; **C:** lvs dark or reddish green with pale pink specks, S, P green, lip white; **D:** CR, Pma, SA (Mt); **M:** pl to 40 cm, S 5–5.5 mm.............. ***Cyclopogon olivaceus***
4. Petals without apical thickenings............................. 5
5(4). Apex of lip blunt or notched, without callus; **C:** greenish with white lip; **D:** CR, Pma, SA (Mt); **M:** pl to 40 cm, S 6.5–8 mm.......... .... ***Cyclopogon miradorensis***
5. Apex of lip 3-lobed, with thick callus on midnerve; **C:** greenish with whitish lip; **D:** No, Pma (Mt); **M:** pl to 35 cm, S 4.5–5 mm .... ***Cyclopogon elatus***
Plate 38(3)

Deiregyne. Terrestrial or lithophytic; leaves several, narrow; inflorescence with conspicuous bracts; sepals and petals connivent, spreading apically, dorsal sepal much wider than laterals; column long and slender, hairy beneath; 1 species known in Nicaragua, may occur in Costa Rica; **C:** S green basally, white above, P whitish, lip white, green in center; **D:** No; **M:** pl to 70 cm, S 8–12 mm................ *Deiregyne hemichrea*
Plate 38(1)

Discyphus. Terrestrial; with 1 basal, heart-shaped leaf flat on ground, leaf about as wide as long; inflorescence erect, with several small tubular flowers; lip long-stalked with thick basal lobules, slightly constricted near middle, with long, united column foot; viscidium oblong, rostellum straplike, rostellar remnant deeply notched; stigma 2-lobed, cupped; 1 species known; **C:** greenish white; **D:** Pma, SA (Mt); **M:** pl 10–18 cm, S 4–4.5 mm................................... ***Discyphus scopulariae***

Elleanthus. Terrestrial or epiphytic; stems slender, usually long and reedlike, with many 2-ranked, pleated or narrow conduplicate leaves; inflorescence terminal, of several to many flowers produced at once or in succession; flowers small, usually brightly colored; sepals and petals similar, cupped, lip saccate basally, usually with 2 large, fleshy glands; column short, without wings; pollinia 8, ovoid. The flowers of *Elleanthus* may be brightly colored, and most appear to be pollinated by hummingbirds.

Key to *Elleanthus*

1. Leaves distinctly pleated, wide or narrow but never straplike 2
1. Leaves conduplicate, usually narrow; bracts always 2-ranked, flowers always white 15
2(1). Inflorescence dense and headlike, about as wide as long 3
2. Inflorescence much longer than wide (very young inflorescences may be nearly as wide as long)................................... 5
3(2). Head 5 cm wide or more; bracts not long and narrow; commonly has "jelly" among flowers; **C:** rose-purple; **D:** No, CR, Pma, SA (Cb, Mt, Pc); **M:** pl to 80 cm, S 9 mm........................ .... *Elleanthus cynarocephalus*
3. Head < 4 cm wide; bracts may be long and narrow, without "jelly" 4
4(3). Head > 2 cm wide; **C:** bracts red-purple, fls white; **D:** CR, Pma (Cb, Mt); **M:** pl to 50 cm, S 7–10 mm .... *Elleanthus lentii* Plate 35(5)
4. Head < 2 cm wide; **C:** white; **D:** No, CR, Pma (Mt); **M:** pl to 30 cm, S 8 mm............................... *Elleanthus caricoides*
5(2). Stems much branched; leaves 1.5–2.5 cm wide 6
5. Stems not or rarely branched................................. 7
6(5). Flowers densely clustered; **C:** orange or orange-red; **D:** No, CR, Pma (Mt); **M:** pl to 30 cm, S 6 mm .... *Elleanthus aurantiacus*
6. Flowers loosely clustered; **C:** red, coral-red, or rose-purple; **D:** CR, Pma (Mt); **M:** pl to 40 cm, S 8–9 mm .... *Elleanthus tonduzii* Plate 35(6)
7(5). Leaves lanceolate or narrowly elliptic, < 3 cm wide............ 8
7. Leaves ovate or broadly elliptic, > 3 cm wide 12
8(7). Flowers yellow, orange, or rose-purple...................... 9
8. Flowers white 10
9(8). Bracts long-acuminate, much longer than flowers; **C:** yellow; **D:** CR, Pma, SA (Cb, Mt, Wt); **M:** pl to 60 cm, S 8 mm............. .... *Elleanthus xanthocomos*
9. Bracts short-acuminate, slightly longer than flowers; **D:** No, CR; **M:** pl to 30 cm, S 5 mm................... *Elleanthus wercklei*

10(8). Bracts spiral; C: white with yellow in throat; D: No, CR, Pma (Mt); M: pl to 30 cm, S 8 mm .... *Elleanthus caricoides*
10. Bracts 2-ranked and conspicuous, inflorescence flattened 11
11(10). Lip gradually expanding from base; inflorescences suberect; C: white; D: CR, Pma, SA (Mt); M: pl to 40 cm, S 6–7 mm .... *Elleanthus lancifolius*
11. Lip abruptly expanding from a narrow claw; inflorescences nodding; C: white; D: CR, Pma (Mt, Wt, Pc); M: pl to 35 cm, S 7–9 mm .... *Elleanthus laxus*
12(7). Leaves whitish or grayish; C: rose-purple; D: CR (Mt); M: pl to > 1 m, S 8 mm .... *Elleanthus glaucophyllus*
12. Leaves green 13
13. Bracts purple or magenta; flowers white or purple; large plant with dense inflorescence > 3 cm wide; C: purple or white with orange or brown in throat; D: CR, Pma, SA (Cb, Mt); M: pl to > 1 m tall, S 10 mm .... *Elleanthus robustus*
13. Bracts green; flowers yellow or orange 14
14(13). Lip subcircular, as wide as long; sepals not mealy without; C: red-orange; D: No, CR, Pma (Mt); M: pl to 60 cm, S 4–6 mm .... *Elleanthus hymenophorus* (Syn. *E. curtii?*)
14. Lip oblong, longer than wide; sepals mealy without; C: red-orange; D: No, CR; M: pl to 25 cm, S 4–5 mm .... *Elleanthus albertii*
15(1). Leaves narrowly ovate or oblong, obtuse or retuse 16
15. Leaves linear to narrowly straplike 17
16(15). Floral bracts prominent, longer than flowers; stems clustered; sepals without dark scales; C: white; D: CR, Pma (Mt); M: pl to 10 cm, S 3 mm .... *Elleanthus muscicola*
16. Floral bracts much shorter than flowers; rhizome creeping; dark scales on flowers; C: white with dark scales outside, lip apically violet; D: CR, Pma (Mt, Pk); M: pl to 10 cm, S 3.5 mm .... *Elleanthus jimenezii* (Syn. *Epilyna jimenezii*)
This easily recognized species seems clearly a member of *Elleanthus*.
17(15). Floral bracts well separated, not overlapping 18
17. Floral bracts densely clustered, overlapping 19
18(17). Leaves 1–2 mm wide; inflorescence surpassing upper leaves; bracts

10–15 mm long; **C:** white; **D:** Pma, SA (Cb, Mt); **M:** pl to 45 cm, S 3.5–4 mm.................................*Elleanthus fractiflexus*
18. Leaves < 1 mm wide; upper leaves subequal to inflorescence; bracts 5–8 mm long; **C:** white; **D:** No, CR, Pma (Wt); **M:** pl 12–20 cm, S 2.5–2.7 mm............................*Elleanthus poiformis*
Plate 36(1)
19(17). Leaves mostly < 1 mm wide................................ 20
19. Leaves mostly > 1.5 mm wide............................. 21
20(19). Leaves < 2 cm long; delicate, sprawling plant, upper stems rooting and thus spreading; **C:** white; **D:** CR, Pma (Cb, Mt); **M:** pl 5–7 cm, S 2–2.5 mm............................ *Elleanthus stolonifer*
20. Leaves > 3 cm long; **C:** white; **D:** CR, Pma (Pc, Mt); **M:** pl to 15 cm, S 3–3.5 mm......................*Elleanthus tillandsioides*
21(19). Inflorescence 1–1.5 cm wide; plant < than 20 cm tall; **C:** white; **D:** No, CR, Pma (Cb); **M:** pl 15–20 cm, S 3–3.5 mm ....
...................................*Elleanthus graminifolius*
The name *E. linifolius* has been used for this plant in Central America, but it refers to a species of Peru and Ecuador.
21. Inflorescence 2 cm across; plant 20–25 cm tall; **C:** white; **D:** Pma (Cb)...................... *Elleanthus* species near *isochiloides*

Eltroplectris. Terrestrial; leaves several, basal, wide, slightly narrowed basally; inflorescence with several large flowers; flowers with slender spur free at apex, lateral sepals spreading; column with long partially free foot, viscidium sheath-like, rostellum linear with narrow base; lip without basal lobules; 1 species in our area; **C:** white or segments with rose margins; **D:** No, CR, SA (Pc/Mt); **M:** pl 20–40 cm, S 1.8–2.3 cm......
.... *Eltroplectris roseoalba*

Erythrodes. Usually terrestrial; stem herbaceous, basally sprawling, then erect; leaves several, spiral, soft-herbaceous or weakly pleated, somewhat stalked; inflorescence terminal, of several to many flowers; sepals and petals similar; lip lowermost, with slender or saclike nectary at base, blade often 2-lobed; anther dorsal, pollinia sectile, stigma (rostellum) split by removal of viscidium.

Key to *Erythrodes*

1. Plants epiphytic; stem thick and succulent, petiole flat, to 12 mm wide in larger leaves; inflorescence densely flowered; C: white; D: Pma, CR (Cb, Mt); M: pl 20–45 cm, S 4–4.5 mm............... .... ***Erythrodes* species**
1. Plants terrrestrial; stem slender; petioles narrow (< 5 mm wide); infloresence may be loosely flowered 2
2(1). Plants 40–80 cm tall; ovary and inflorescence axis densely hairy; C: fls white, S brownish basally; D: CR, Pma, SA (Cb, Mt); M: S 6–7 mm .... ***Erythrodes killipii***
    Plate 39(2)
    These plants often form large colonies; the plants of central Panama usually have larger leaves that are more clustered on the upper stem.
2. Plants 20–35 cm tall, leafy stem usually < 15 cm.............. 3
3(2). Apex of lip transversely oblong or rhombic, about 2 mm wide; spur saclike, 2–3 mm long; leaves mottled dark and pale green; C: fls white; D: CR, Pma (Mt).................. ***Erythrodes vesicifera***
3. Apex of lip transversely ribbon-like, 3–5 mm wide, twisted; spur slender 4
4(3). Spur about 5 mm long; leaves mottled dark and pale green; C: white, S reddish or purplish tinged outside; D: CR (Cb, Pc); M: pl 30–35 cm, S 4.5–5 mm .... ***Erythrodes calophylla***
4. Spur 2.5–3 mm long 5
5(4). Leaves basally clustered (a loose rosette), dark velvety green, may have pale specks; peduncle of inflorescence longer than leafy stem; C: white, basally brown; M: pl about 15 cm, S 4 mm .... ***Erythrodes purpurea***
5. Leaves scattered on stem about 10 cm long, mottled dark and pale green; peduncle shorter than leafy stem; D: No, CR (Pc); M: pl 20–30 cm, S 4 mm........................... ***Erythrodes lunifera***
    E. lunifera is based on a Guatemalan specimen; the Costa Rican plant identified as this species agrees fairly well with the original description but has not been compared with Guatemalan plants.

Eurystyles. Small epiphytes; several leaves in rosette, usually glossy, often whitish, somewhat narrowed basally, 3–4.5 cm long; inflorescence nodding or pendent, flowers densely clustered at apex, bracts conspicuous, usually fringed; flowers very small, tubular, not twisted; column short, viscidium oval, rostellar remnant notched or obtuse.

Key to *Eurystyles*

1. Bracts not fringed; flowers smooth externally; C: S, P pale green flushed with pink, lip pale green; D: CR, Pma; M: S 5.5 mm
   .... *Eurystyles* species
1. Bracts fringed; flowers more or less glandular externally 2
2(1). Lip straplike, without constriction; D: CR; M: S 7 mm ....
   .... *Eurystyles standleyi*
2. Lip constricted near apex 3
3(2). Column without rostellum or viscidium (self-pollinating); D: Pma
   .... *Eurystyles* species
3. Column with distinct rostellum and viscidium 4
4(3). Basal part of lip distinctly wider than midlobe; rostellum long, triangular; C: white; D: CR; M: S 5–6 mm ...... *Eurystyles auriculata*
4. Basal part of lip about as wide as midlobe; rostellum very short; C: greenish white; D: CR, Pma (Wt); M: S 4.5–5 mm ....
   .... *Eurystyles cotyledon*
   Plate 38(4)

Funkiella. Terrestrial; roots fleshy and tuberlike, scattered on long rhizome; without leaves at flowering; inflorescence with 1 to 3 medium-sized flowers, lateral sepals spreading; lip constricted near apex, with red thickening in center, margins thickened basally; column slender, viscidium sheath-like, rostellar remnant lanceolate; C: whitish, center of lip red; D: CR; M: pl 25–35 cm, lS 1.7–2 cm .... ***Funkiella stolonifera***

Galeottiella. Terrestrial; roots fleshy; leaves narrow, withered at flowering; inflorescence slender, with several tiny tubular flowers; lip with distinct stalk, thickened in center, with basal angles, without constriction; column short, rostellum with viscidium obtuse, viscidium oval, rostellar remnant notched.

Key to *Galeottiella*

1. Apex of lip 3-lobed; sepals up to 3 mm long; C: greenish white becoming reddish brown with age, callus with red stripes; D: No, CR; M: pl 25–30 cm, S 0.7–1.2 mm..... ***Galeottiella minutiflora***
1. Apex of lip tongue-like, not lobed; sepals 4–5 mm long; D: CR; M: pl 10–30 cm, S 4–5 mm................***Galeottiella nutantiflora***

Gomphichis. Terrestrial; roots fleshy; leaves several, basal, soft-herbaceous, basally narrowed but not clearly stalked; inflorescence terminal, erect, flowers many; sepals and petals similar; lip shallowly 3-lobed, lateral lobes enfolding column; column bent basally and near apex (almost S-shaped), blunt; anther dorsal; pollinia 4, soft; 1 named species in Central America; C: greenish white; D: CR (Pk); M: pl to 30 cm, S 4.5–5 mm....................................***Gomphichis costaricensis***

Goodyera. Terrestrial; stem herbaceous, basally sprawling, then erect; leaves several, spiral, soft-herbaceous or weakly pleated, somewhat stalked; inflorescence terminal, of several to many flowers; sepals and petals similar; lip lowermost, without saclike nectary; column short; anther dorsal; pollinia sectile, stigma (rostellum) split by removal of viscidium.

Six names are based on Costa Rican material, yet I have seen only 3 different species, and 1 of these is clearly not one of the 6 named from Costa Rica.

Key to *Goodyera*

1. Margin of lip smooth or nearly so; leaves ovate or elliptic-ovate, 2.5–4 cm long, dark green with pale veins; C: fls greenish or pinkish white; D: CR, Pma (Mt); M: pl 18–25 cm, S 3–4 mm, lip 3.2–3.5 mm...................................***Goodyera ovatilabia***
1. Margin of lip distinctly erose-crisped; leaves elliptic, markedly asymmetric, mostly 4–8 cm long........................... 2
   2(1). Sepals 6–7 mm long; lip basally globose and then constricted, midlobe oblong, about 3 mm long; C: white; D: No, CR (Mt); M: pl 30–40 cm, infl to 40 cm, fls on upper 4–6 cm ... ***Goodyera major***
   This species has been called *G. major* in El Salvador and

Nicaragua, but the lip is rather unlike that illustrated in Paul C. Standley and Julian A. Steyermark's *Flora of Guatemala* (Fieldiana: Botany [Chicago: Field Museum of Natural History]); it is larger in both plant and flower than the other Costa Rican species.

2. Sepals about 4 mm long; lip triangular-ovate, acute; C: white or greenish white; D: CR, Pma (Mt); M: pl 25–35 cm, infl to 30 cm, fls on upper 6–10 cm .... **Goodyera erosa**

Habenaria. Terrestrial; with slender roots and 2 globose root-stem tubers; leaves few to several, spiral, basal or scattered on stem, upper leaves usually smaller; inflorescence terminal, of few to many flowers; sepals similar, usually spreading, petals usually bilobed or toothed on 1 side; lip usually deeply 3-lobed with a long, slender spur at base; column short, stigma 2-parted, each part on separate stalk; halves of anther widely separated; pollinia 2, sectile, each with long caudicle and viscidium.

Key to *Habenaria*

1. Petals entire, toothed, or lower lobe less than half as long as upper (wider) lobe 2
1. Petals deeply 2-lobed, lower lobe at least half as long as upper lobe 6
2(1). Lip entire or with low, blunt teeth near base 3
2. Lip with distinct acute lobes at base 4
3(2). Sepals obtuse; petals square-oblong, more or less 3-toothed apically; lip with low, blunt teeth at base; C: green; D: No, CR, Pma, SA (Mt, Pc); M: pl to 80 cm, S 6.5–7 mm, spur 10 mm ....
.... *Habenaria floribunda*
(Syn. *H. odontopetala*)
3. Sepals and petals acute; petals lanceolate or lance-ovate; C: greenish or greenish yellow; D: No, CR, Pma, SA (Mt, Pc); M: pl to 60 cm, S 6–6.5 mm, spur 12–16 mm .... *Habenaria alata*
4(2). Lateral lobes of lip subequal to midlobe; D: CR (Mt); M: pl to 30 cm, S 4 mm, spur 10 mm .... *Habenaria lankesteri*
4. Lateral lobes of lip much less than half length of midlobe 5

5(4). Lip bent near base; petals 2-lobed; C: green; D: CR (Mt, Pc); M: pl to 42 cm, S 5–7 mm, spur 1.9–2.9 cm .... *Habenaria aviculoides*
5. Lip not bent; petals entire; C: dark green; D: Pma (Cb, Pc); M: pl to 45 cm, S 4–10 mm, spur 1.5–3 cm .... *Habenaria avicula*
6(1). Lateral lobes of lip much longer than midlobe; lower petal lobe much longer than upper petal lobe 7
6. Lateral lobes of lip subequal to or slightly longer than midlobe; lower petal lobes subequal to upper lobes 10
7(6). Spur > 4 cm long; C: S green, P and lip white; D: No, CR, Pma, SA; M: pl to 90 cm, S 11–14 mm, spur 13 cm .... *Habenaria quinqueseta*
7. Spur < 2.5 cm long; flowers green 8
8(7). Leaves narrow and ascending, inconspicuous; flower < 10 mm diameter; C: dark green; D: No, Pma, SA; M: pl 30–45 cm, S 3–4 mm, spur 7–15 mm .... *Habenaria mesodactyla*
8. Leaves wide and spreading, prominent; flower > 10 mm diameter 9
9(8). Spur < 15 mm long; C: pale green; D: No, CR, Pma, SA (Mt); M: pl 20–35 cm, S 6–8 mm, spur 10–14 mm .... *Habenaria entomantha*
Plate 36(6)
9. Spur > 15 mm long; C: yellowish green; D: No, CR (Mt); M: pl to 40 cm, S 8–9 mm, spur 2.5 cm .... *Habenaria novemfida*
10(6). Lateral lobes of lip much shorter than midlobe; D: CR; M: pl 15–20 cm, S 3–4 mm, spur 3 mm .... *Habenaria verecunda*
10. Lateral lobes of lip subequal to midlobe 11
11(10). Leaves largely basal; C: light green; D: No, CR, Pma, SA (Mt); M: pl to 20 cm, S 6–8 mm, spur 16 mm .... *Habenaria distans*
11. Leaves scattered along stem 12
12(11). Leaves narrow, ascending, inconspicuous; C: green or yellow; D: Pma, SA (Pc, Mt); M: pl to 30 cm, S 3.5–4 mm, spur 13 mm .... *Habenaria leprieuri*
12. Leaves wide, spreading, prominent 13
13(12). Spur > 4 cm long 14
13. Spur < 4 cm long 15
14(13). Sepals about 13–15 mm long; C: white or greenish; D: No, CR, Pma, SA (Pc); M: pl to 1 m, S 13–15 mm, spur 13–15 mm .... *Habenaria bractescens*
14. Sepals 6–10 mm long; D: Pma (Pc); M: pl to 65 cm, S 6–10 mm, spur 4–5 cm .... *Habenaria bicornis*

15(13). Spur > 2 cm long 16
15. Spur < 2 cm long 17
16(15). Flowers normally 1–3; C: greenish cream or lip yellow; D: No, CR, Pma, SA (Mt, Pc); M: pl to 35 cm, S 9–10 mm, spur 4–6 cm .....
.... *Habenaria trifida*
(Syn. *H. pauciflora*)
16. Flowers normally 5 to many; C: white; D: No, CR, Pma (Mt); M: pl to 20 cm, S 5–8 mm, spur 2 cm .... *Habenaria clypeata*
17(15). Leaves narrowly lanceolate, ascending; growing in or near water; C: green or yellowish green; D: No, CR, Pma, SA (Cb, Mt); M: pl to 90 cm, S 5–5.5 mm, spur 10 mm .... *Habenaria repens*
17. Leaves broadly lanceolate, spreading; not aquatic; C: white; D: No, CR, Pma, SA (Mt, Pc); M: pl to 1.2 m, S 8–9 mm, spur 2 cm
.... *Habenaria monorrhiza*

Hapalorchis. Small, delicate terrestrials; leaves stalked, basal; inflorescence slender, few-flowered, all flowers facing same direction; flowers small, tubular, lip not stalked, without basal lobules, more or less constricted near apex; column slender, rostellum straplike with apical viscidium, rostellar remnant notched.

Key to *Hapalorchis*

1. Lip distinctly constricted, widest across midlobe, obtuse, base of lip about twice as long as wide; D: SA ... *Hapalorchis cheirostyloides*
1. Lip shallowly or slightly constricted, midlobe subequal to base, base of lip more than half as wide as long 2
2(1). Lateral sepals 1-nerved; lip distinctly constricted with lobes overlapping, midlobe without ridge, not toothed on margins, base of lip slightly thickened in center; D: No, SA ...... *Hapalorchis lineatus*
2. Lateral sepals 3-nerved; lip shallowly constricted, midlobe toothed with marginal ridge, base of lip thin; D: CR, SA; M: pl 13 cm, S 5.7–6 mm .... *Hapalorchis pumilus*

Kreodanthus. Similar to *Erythrodes*, column axis much longer and arching; lateral sepals usually reflexed; anther relatively short, rostellum abruptly narrower than stigma.
Kreodanthus secundus has been reported from Costa Rica and Pan-

ama; the only plants I have seen, from high mountains in Costa Rica, have much smaller flowers than are described for *K. secundus* (based on a Mexican plant). There are also 3 new species to be described from western Panama, Costa Rica, and El Salvador; this last could well be the supposed *K. secundus* of Costa Rica. I cannot offer a useful key to *Kreodanthus*.

Lankesterella. Small epiphytes; with rosette of several basal leaves, leaves narrowed toward apex and base, hairy; inflorescence lax, with few loosely clustered flowers, bracts and flowers hairy, flowers tubular with prominent chin; lip constricted; column erect, rostellum lanceolate with narrow base, viscidium narrow, lanceolate, rostellar remnant linear; 1 named species in Central America; **C:** S pale green, lip and petals white; **D:** CR, SA (Mt, Wt) .... *Lankesterella orthantha*

The plant shown in Plate 38(5) appears to be an undescribed *Lankesterella* with 1-flowered inflorescences rather than *L. orthantha*.

Ligeophila. Terrestrial; stem herbaceous, basally sprawling, then erect; leaves several, spiral, soft-herbaceous or weakly pleated, somewhat stalked; inflorescence terminal, of several to many flowers; sepals and petals similar, petals connivent with dorsal sepal; lip lowermost, with long, narrow spur at base, blade 3-lobed, midlobe recurved; anther dorsal, pollinia sectile; rostellum said to be hinged and movable; 1 species in Central America; **C:** S, P green shading to red at tips, lip white; **D:** No, CR, Pma, SA (Mt); **M:** pl to 80 cm, S 8–10 mm ....
.... *Ligeophila clavigera*

Some plants of *Ligeophila* have lips that are not divided into 3 lobes, but they otherwise agree with *L. clavigera*. The constriction of the lip varies in width in the other plants seen; there may be a single, variable species of *Ligeophila* in Central America.

Lyroglossa. Terrestrial; leaves absent at flowering (sheath-like bracts present); inflorescence few- to several-flowered, lax; flowers small, tubular, lateral sepals slightly spreading; lip without stalk, margins thickened basally, strongly constricted near middle; column short, rostellum lanceolate, viscidium apical, rostellar remnant finger-like with narrow base; **D:** No; **M:** pl 20 cm, S 6–7 mm .... *Lyroglossa pubicaulis*

Mesadenella. Terrestrial; leaves several, elliptic to lanceolate, stalked; inflorescence long, lax, with many flowers; flowers small, erect to arched, tubular, lip narrowed basally but not stalked, slightly constricted near middle with long, fleshy, finger-like calli at base; column with long foot united with ovary; rostellum and rostellar remnant narrow with narrow base, viscidium narrow and long.

Key to *Mesadenella*

1. Flowers erect; lip lanceolate, acute, margin of midlobe wavy; **C:** white; **D:** No; **M:** pl 15–25 cm, S 4–6 mm .... *Mesadenella petenensis*
1. Flowers arched; lip widest near apex, more or less blunt, margins of midlobe not wavy; **C:** purplish brown, P and lip cream; **D:** No, CR (Cb); **M:** pl to 18 cm, S 4.2–5 mm .... **Mesadenella tonduzii**

Mesadenus. Terrestrial; usually leafless at flowering; inflorescence of many small tubular flowers, sepals and petals very narrow, spreading only apically, with small chin at base; lip slightly stalked, lanceolate or elliptic, without constriction; column slender, rostellum and viscidium small, rostellar remnant small, obtuse; **C:** green; **D:** No (CR?); **M:** pl to 30 cm, S 3.2–7 mm .... *Mesadenus polyanthus*

Monophyllorchis. Terrestrial; roots fleshy; stems erect, with single ovate or heart-shaped pleated leaf, often with white or silver streaks on ridges; inflorescence terminal, of few flowers, opening 1 at a time; sepals and petals similar, not spreading; lip 3-lobed; column slightly arched, anther terminal with fleshy beak; pollinia 4, soft; 1 species in Central America; **C:** pale green or purplish; **D:** CR, Pma (Cb, Mt); **M:** pl to 25 cm, S 1.7 cm .... **Monophyllorchis maculata**
Plate 36(5)

The plants I have seen in Panama agree with the description of *M. maculata*, but the name *M. microstyloides* has also been used in Central America. That species, described from Colombia, supposedly has smaller leaves without silver streaks.

Palmorchis. Terrestrial; stems slender and reedlike; leaves several or many, 2-ranked or spiral, pleated; inflorescence terminal or lateral, of several flowers, usually produced 1 at a time; sepals and petals similar, spreading; lip enfolding column and basally united with column along midline; column slender; pollinia 4, soft. This American genus is so distinctive that it merits a tribe by itself, and it has several quite primitive features. The fruit is fleshy and divided into 3 chambers, and the seeds are relatively large with a hard seed coat. The seeds are probably dispersed by birds or other animals. The plants in a population all flower on the same days, and the delicate flowers last for only a few hours. When not in flower, the plants look much like forest floor grasses or seedling palms, and these orchids have been poorly collected.

Key to *Palmorchis*

1. Lip with several keels or ridges 2
1. Midlobe of lip rounded and fleshy, without keels or ridges 3
2(1). Lip hairy, wide, enfolding column; plants generally about 1 m tall; inflorescences both lateral and terminal; leaves not shiny; C: S pale greenish cream, P and lip white; D: CR, Pma, SA (Cb, Pc); M: pl 0.5–1 m, S 2–2.2 cm .... *Palmorchis powellii*
   Found in both seasonally dry forests and very wet forests.
2. Lip hairless, not concealing column; plants 20–30 cm tall; inflorescence terminal; leaves shiny; C: white, lip marked with rose-purple; D: Pma, SA (Cb, Pc); M: pl 25–30 cm, S 1.7–1.9 cm .... *Palmorchis nitida*
3(1). Lateral lobes of lip triangular-trapezoidal, not fleshy; C: white, lip with 3 rose-purple lines; D: No, CR, Pma (Cb, Pc); M: pl 35–45 cm, S 12 mm .... *Palmorchis silvicola*
   Plate 36(2)
3. Lateral lobes of lip rounded, thick, and fleshy; C: white, may have rose-purple within lip; D: CR, Pma (Pc, Mt); M: pl to 30 cm, S 7–9 mm .... *Palmorchis trilobulata*

Pelexia. Terrestrial; leaves few to several, basal, usually stalked; inflorescence erect, usually of many flowers; flowers usually large, with prominent nectary partly or completely united with ovary, but with tip free as a visible chin; lateral sepals spreading; column slender, rostellum linear with oval, apical viscidium.

Key to *Pelexia*

1. Spur completely united with ovary, chin at base of flower....... 2
1. Spur free at apex 4
2(1). Lip up to 6 mm long; flowers < 10 mm; D: No ....
 .... *Pelexia obliqua*
2. Lip > 6 mm long; flowers > 10 mm 3
3(2). Lip 12–13 mm long; C: S greenish, P and lip white; D: No, CR, SA; M: pl to 65 cm, S 6–8.5 mm.................... *Pelexia adnata*
3. Lip > 18 mm long; flowers arched; C: waxy greenish yellow; D: CR (Mt); M: pl to 30 cm, S 2.1–2.3 cm.......... *Pelexia smithii*
 (Syn. *Sarcoglottis valida*)
 Plate 37(2)
 This species resembles *Sarcoglottis* so strongly that it casts some doubt on the separation of the 2 genera.
4(1). Spur obtuse at apex, rounded.............................. 5
4. Spur with acute apex....................................... 6
5(4). Lip shallowly constricted near apex; midlobe triangular, with thick callus; D: SA; M: pl to 40 cm, S 6–8.5 mm...... *Pelexia callifera*
5. Lip deeply constricted near apex; midlobe as wide as middle portion of lip, without callus; C: S green, P greenish white, lip yellowish cream, D: No, CR, Pma, SA (Cb, Mt); M: pl to 80 cm, S 12 mm .... *Pelexia laxa*
6(4). Spur to 2 mm long; C: S green, P greenish white, lip white; D: No, CR, Pma, SA (Cb, Mt); M: pl to 40 cm, S 1.6–1.9 cm ....
 .... *Pelexia funckiana*
 Plate 37(1)
6. Spur 4–10 mm long.. 7
7(6). Inflorescence dense; leaves reaching inflorescence; C: S, P greenish white, lip white with yellow terminal lobe; D: No, CR (Cb, Mt); M: pl 20 cm, S 1.5–3.3 cm...................... *Pelexia congesta*
7. Inflorescence lax; leaves shorter than inflorescence; C: S pinkish brown, P white, lip yellow with white terminal lobe; D: No, CR, SA .... *Pelexia olivacea*

Phragmipedium. Terrestrial or epiphytic; stems short with several 2-ranked, conduplicate leaves; leaves straplike, leathery; inflorescence terminal, of several flowers; lateral sepals united, petals narrow; lip deeply saccate with margins turned inward; column short, with 1 fertile anther on each side and a shield-like sterile anther above, stigma broad and domelike.

Key to *Phragmipedium*

1. Petals spreading, less than twice as long as lip; **C:** S pale cream-green, may have green veins, P cream or each with rose submarginal band, lip brownish green; **D:** CR, Pma, SA (Cb, Mt, Pc); **M:** lvs to 80 cm, S 2.5–5 cm .... ***Phragmipedium longifolium***
Plate 34(4)

    This variable species usually occurs near streams. Very narrow-leaved forms occur, especially in the Caribbean drainage in rocky streambeds. Much larger plants may be found on steep cuts on the Pacific slope.

1. Petals pendent, many times longer than lip; **C:** S and P bases greenish yellow with red-brown streaks and veins, P shading to purple-brown, lip greenish brown with yellow and red-brown markings; **D:** No, CR, Pma, SA (Mt); **M:** lvs to 25 cm, S 8–13 cm, P to 60 cm .... ***Phragmipedium caudatum***
Plate 34(3)

    This striking species, known from only a few localities, usually occurs as an epiphyte in wet montane forests. The Central American form is somewhat different from South American plants; the name *P. warscewiczianum* has been used for our plants, but that name refers to a South American plant, so Central American plants have no valid name as a distinct species.

Phymatidium. Diminutive epiphyte; without pseudobulbs; leaves laterally flattened; inflorescence lateral, racemose; column without prominent rostellar beak; lip fringed, with concave basal callus; 1 species in Panama; **C:** white with green callus; **D:** Pma (Pc/Mt); **M:** lvs 4–7 mm, S 1.7–1.8 mm .... ***Phymatidium panamense***

This tiny plant has been found only once. It should probably be assigned to the genus *Eloyella*, known from Colombia and Ecuador, rather than to *Phymatidium*, otherwise a Brazilian genus.

Platythelys. Terrestrial or epiphytic; stem herbaceous, basally sprawling, then erect; leaves several, spiral, soft-herbaceous or weakly pleated, somewhat stalked; inflorescence terminal, of several to many flowers; sepals and petals similar; lip lowermost with saclike nectary at base, blade distinctly constricted; anther dorsal, pollinia sectile, viscidium forming sheath around rostellum.

Key to *Platythelys*

1. Epiphytic, usually with thickened rhizome; lip apiculate but not clearly 3-lobed; **C:** S, P pale green, lip white; **D:** No, CR, Pma (Cb); **M:** pl 4–12 cm, S 3–3.5 mm.............. *Platythelys venustula*
Plate 39(3)
1. Normally terrestrial, with thin rhizome; lip more or less 3-lobed...
....2
2(1). Leaves narrowly ovate, dark green, usually with white pattern; **C:** white, petal tips red; **D:** No, CR, Pma, SA (Mt); **M:** pl 12–20 cm, S 4 mm.................................... *Platythelys maculata*
2. Leaves ovate, not variegated; **C:** S, P pale green, lip white; **D:** No, CR, Pma (Cb, Mt); **M:** pl to 20 cm .... *Platythelys vaginata*

Ponthieva. Terrestrial or epiphytic; roots somewhat fleshy; leaves few to several, soft-herbaceous or weakly pleated, narrow basally; inflorescence terminal, of several to many flowers; sepals spreading, petals united with column, together forming a pseudolip; lip uppermost, united with column, small and concave; column pointed, with anther on dorsal side; pollinia 4, club-shaped, brittle.

Key to *Ponthieva*

1. Leaves hairy; epiphytic; sepals and petals about 1 cm long; **C:** S olive green to yellow or reddish, P olive to reddish yellow, lip yel-

lowish green; **D:** No, CR, Pma (Mt); **M:** pl to 15 cm, S 12–15 mm
....*Ponthieva maculata*
(Syns. *P. brenesii, P. formosa*)
Plate 40(3)

Costa Rican plants are usually epiphytic; the variation in size and color suggests that there may be more than 1 species, but I have been unable to distinguish clear species in Central America or to distinguish our plants from the South American *P. maculata*.

1. Leaves smooth; terrestrial; sepals and petals < 7 mm long 2

2(1). Lip oblanceolate with narrow, recurved lateral lobes; petals asymmetrically T-shaped (blades expanded on each side), bases united to column to beyond lip; lower leaves long-petiolate; **C:** dS brownish green, lS white with brownish green margins, P creamy yellow, lip yellowish green; **D:** No, CR (Mt); **M:** pl to 20 cm, S 5–5.5 mm
......................................*Ponthieva tuerckheimii*
2. Lip nearly as wide as long, lateral lobes, if present, broad and rounded; petal margins strongly expanded on 1 side only; leaves with short or no petioles.. 3

3(2). Lip attached near column apex, distinctly 3-lobed; **C:** white; **D:** No, Pma (Mt); **M:** pl to 30 cm, S 6–7 mm...... *Ponthieva ephippium*
3. Lip attached below middle of column, apiculate but not or weakly 3-lobed; **C:** S greenish or brownish, P and lip white; **D:** No, CR, Pma, SA (Mt, Pc); **M:** pl to 20 cm, S 4.5–5.5 mm ....
....*Ponthieva racemosa*

Prescottia. Terrestrial; roots fleshy; leaves few to several, basal, soft-herbaceous, basally narrowed or clearly stalked; inflorescence terminal, erect, with many flowers; sepals and petals partially united, often reflexed, petals narrower than sepals; lip saccate; column very short, blunt; anther dorsal; pollinia 4, soft.

Key to *Prescottia*

1. Leaves several, ovate, petioles much shorter than blades, < 10 cm long; flowers white **D:** No, CR, Pma, SA (Mt); **M:** pl to 30 cm, S 2 mm ....*Prescottia oligantha*
1. Petioles subequal to blades or longer, leaves > 15 cm long; flowers green... 2
2(1). Leaves 1 or 2, often somewhat cordate at base, > 10 cm wide; **C:**

  pale green **D:** No, CR, Pma, SA (Cb, Pc); **M:** pl 60–95 cm, lip 4.5–5 mm .... *Prescottia cordifolia*
                   Plate 40(6)
 2. Leaves several, elliptic or narrowly ovate, tapering to stalk, < 10 cm wide 3
3(2). Leaves < 3 cm wide; bracts much longer than flowers; **C:** pale green; **D:** CR (Pc/Mt); **M:** pl 30–40 cm .... *Prescottia* species
 3. Leaves 3.5–6 cm wide; bracts subequal to flowers; **C:** pale green; **D:** No, CR, Pma, SA (Mt); **M:** pl to 45 cm, lip 2.5–3.5 mm ....*Prescottia stachyodes*

Pseudocentrum. Terrestrial; roots somewhat fleshy; leaves few to several, soft-herbaceous or weakly pleated, stalked; inflorescence terminal, of several to many flowers; lateral sepals united to form deep, saclike spur, petals small and inconspicuous; lip uppermost with long basal extension into spur; column pointed with anther on dorsal side; pollinia 4, club-shaped, brittle; **C:** S green, lip and column white; **D:** CR, wPma (Mt); **M:** pl to 70 cm, S 10 mm .... *Pseudocentrum hoffmannii*

Psilochilus. Herbaceous terrestrials; stems basally sprawling, leafy stems erect; leaves more or less ovate, soft-herbaceous or weakly pleated; plants with the aspect of *Commelina*; inflorescence terminal, producing 1 flower at a time; flowers short-lived, most plants flowering on the same days; pollen very soft.

Key to *Psilochilus*

 1. Lip with narrow basal stalk about one quarter of its length; leaves dark green or with silver stripes; **C:** S green, P greenish cream, lip white, midlobe rose-purple; **D:** Pma, SA (Mt, Wt); **M:** pl 15–30 cm, S 2–2.7 cm .... *Psilochilus physurifolius*
                   Plate 39(4)
 1. Lip tapering to base, without distinct stalk 2
2(1). Lip with prominent median thickening from base to midlobe; **C:** S, P green, lip white with green callus; **D:** No, SA; **M:** pl 15–20 cm, S 8.5 mm.. *Psilochilus modestus*
 2. Lip not markedly thickened, may have 3 faint median ridges; **C:** S

greenish, P greenish white, lip white; D: No, rPma; M: pl 10–35 cm, S 1.8–1.9 cm .... *Psilochilus macrophyllus*

Pterichis. Terrestrial; roots somewhat fleshy; leaves few, soft-herbaceous or weakly pleated, stalked; inflorescence terminal, of several to many flowers; petals narrower than sepals, connivent with dorsal sepals; lip uppermost, enfolding column, broad, with raised veins; column pointed, with anther on dorsal side; pollinia 4, club-shaped, brittle.

Key to *Pterichis*

1. Lip chartreuse or very pale sulfur yellow, apex apiculate-subulate and recurved, with 2–3 veins on each side, these 1- to 3-branched; petals usually adhering to dorsal sepal; C: S, P greenish yellow, lip chartreuse with green spots and veins or pinkish yellow with maroon spots and veins; D: CR, Pma, SA (Pk); M: pl to 50 cm, S 6–7 mm .... *Pterichis habenarioides*
(Syn. *P. costaricensis*)
Plate 40(4)
1. Lip red or greenish maroon with marginal yellow papillae; apex obtuse, not recurved, with 4–6 veins on each side, each forked once; petals free from dorsal sepal; C: S, P red-brown, lip pinkish red with darker veins; D: CR (Pk); M: pl to 30 cm, S 7–9.5 mm .... *Pterichis leo*

Sarcoglottis. Large terrestrials; leaves several, scattered or basal, narrowed basally, stalked or not, often with pale spots or stripes, often lacking at flowering; inflorescence stout, erect, of few to several flowers; flowers large, basally erect, curved, tubular; lateral sepals spreading or recurving; lip stalked with basal lobules, usually constricted near apex, large nectary completely united with ovary; column short, column foot long, united with ovary; rostellum straplike with apical oval viscidium; flowers fleshy, green with darker markings, fragrant.

Key to *Sarcoglottis*

1. Leaves scattered along stem; rhizome elongate with long internodes; D: Pma (Mt); M: pl to 55 cm, S 1.6–1.9 cm .... *Sarcoglottis woodsonii*

1. Leaves forming rosette at base of stem; rhizome short 2
2(1). Lateral sepals more or less erect, about 2 cm long, 2–3 mm wide; midlobe of lip 7 mm long, 8 mm wide, basal lobules 3 mm long; C: green; D: CR, Pma (Cb, Pc); M: pl 12–24 cm.................
.... *Sarcoglottis neglecta*
2. Lateral sepals curved, about 3 cm long, 5–6 mm wide; midlobe 8–10 mm long, 8 mm wide, basal lobules about 4 mm long; C: S, P waxy green, lip yellow-green with dark green veins, or whole fl bronze-flushed; D: Pma (Cb, Pc); M: pl 45–50 cm...............
.... *Sarcoglottis hunteriana*
Plate 37(3)

Schiedeella. Terrestrial; leaves 1 or few, wide, elliptic or ovate, stalked, usually withered at flowering; inflorescence of few to several small flowers; flowers horizontal, tubular; lip stalked, with basal lobules, more or less constricted near middle; column slender, rostellum with narrow base, lanceolate, viscidium sheath-like, rostellar remnant linear.

Key to *Schiedeella*

1. Lip with red thickening in center; C: white with violet-red or brown-red thickening on lip; D: No, CR; M: pl to 30 cm, S 5–7 mm .... *Schiedeella fauci-sanguinea*
1. Lip without central red thickening 2
2(1). Sepals longer than lip; basal lobules of lip fleshy, finger-like..... 3
2. Sepals subequal to lip; basal lobules rounded................. 4
3(2). Flowers erect to suberect; lateral sepals > 10 mm long; inflorescence short, dense; lip narrow basally and widest above middle; C: white with pink tinge; D: No, CR; M: pl to 15 cm, S 9–10 mm
.... *Schiedeella trilineata*
3. Flowers horizontal; lateral sepals up to 9 mm long; inflorescence long, lax; C: S purplish, lip white with brown nerves; D: No, CR (Mt); M: pl to 20 cm, S 4.5–5 mm .... *Schiedeella llaveana*
Plate 38(6)
4(2). Lip oblong-quadrate, straplike; C: pinkish white; D: No, CR (Mt); M: pl to 17 cm, S 5–6 mm .... *Schiedeella valerioi*
4. Lip nearly quadrate with 2 small wings near middle; C: yellowish green with white lip; D: No, CR; M: pl to 25 cm, S 5.5–6 mm
.... *Schiedeella wercklei*

Selenipedium. Terrestrial; stems long, slender, sometimes branched; leaves pleated, 2-ranked; inflorescence terminal, of many flowers, but only 1 developing at a time on each inflorescence, with conspicuous bracts; lateral sepals united, petals narrow; lip deeply saccate with margins turned inward; column short with 1 fertile anther on each side and shield-like sterile anther above, stigma broad and domelike; 1 species in Central America; **C:** S, P greenish brown, lip yellow flushed with brown apically, stained with red-purple within; **D:** Pma, SA (Cb, Pc); **M:** pl 2–5 m, S 2 cm .... ***Selenipedium chica***
Plate 34(5)

Habitat destruction has made this one of the rarest species in Panama. It is known on the Pacific slope of central Panama and in the upper Chagres basin.

Sobralia. Terrestrial or epiphytic; stems slender, usually long and reedlike, usually with many 2-ranked pleated leaves; inflorescence terminal, of several flowers usually produced 1 at a time from a cluster of sheathing bracts; flowers small to quite large, sepals and petals similar, usually spreading; lip enfolding column; column straight, may have armlike wings; pollinia 8, soft and irregular. The flowers of *Sobralia* are very delicate and membranous, and most species last for only a few hours. The buds of all plants of a given species open simultaneously several days after some environmental cue (probably a sudden drop in temperature due to a rainstorm). On a *Sobralia* day the plants are in flower everywhere. The next day there are none. The identification of *Sobralia* species from museum specimens is difficult because the delicate flowers are hard to preserve. A careful study of the genus is much needed. Most of the species reported in Costa Rica and Panama are treated in this key, but I have seen a number of plants that I cannot identify, and I am sure the key could be improved.

Key to *Sobralia*

1. Margin of lip clearly fringed; flowers small (lip < 4 cm long), white or yellow 2
1. Margins of lip wavy, ruffled, or toothed, but without fringe of narrow, hairlike lobes 5

2(1). Plant with 1 leaf on each stem; **C:** S, P white or greenish yellow, lip marked with yellow; **D:** No, CR, Pma, SA (Cb, Pc); **M:** pl 15–20 cm, S 4 cm .... ***Sobralia fragrans***

2. Plant with 3 or more leaves on each stem 3
3(2). Marginal fringe very short; lip without keels; C: white, lip pale yellow; D: Pma (Mt); M: pl to 50 cm, S 3.5 cm ...... *Sobralia allenii*
3. Marginal fringe long; lip with definite fringed keels 4
4(3). Leaves < 1 cm wide; C: yellow, orange-yellow in throat; D: CR; M: pl 18–26 cm, S 2.2 cm .... *Sobralia pfavii*
4. Leaves > 2 cm wide; C: cream, most of lip orange-yellow; D: No, CR, Pma, SA (Cb); M: pl to 40 cm, S 3.4–3.7 cm ....
.... *Sobralia suaveolens*
Plate 35(1)
5(1). Leaves < 15 mm wide; lip with ovate basal calli; C: rose-pink; D: Pma (Cb, Mt); M: pl to 40 cm, S 2–2.5 cm ...... *Sobralia callosa*
Plate 35(2)
5. Leaves > 15 mm wide, usually much more; lip without ovate basal calli 6
6(5). Sepals markedly united at base; flower not opening widely; central portion of lip fleshy and hard to spread; with calli in throat, but blade of lip without raised keels; C: S, P white, lip blood red above, beneath white with red vein; D: No, CR; M: pl to > 1 m, S 6.5 cm? .... *Sobralia helleri*
6. Sepals largely free, or without above combination of features.... 7
7(6). Flowers borne from loose leafy bracts about 8 cm long; base of lip wide, spreading; lip with 9 prominent keels, some raised and some reaching to near apex; C: white, lip striped with purple, may be suffused with purple; D: CR, Pma (Mt); M: pl to 1.5 m, S 4.5 cm
.... *Sobralia undatocarinata*
7. Bracts of inflorescence smaller or tightly in-rolled; without above combination of features................................... 8
8(7). Flowers largely white or yellow.............................. 9
8. Flowers largely lilac, rose-purple, or violet 19
9(8). Leaves < 3 cm wide.. 10
9. Leaves usually > 3 cm wide.................................. 11
10(9). Lip with 2 fringed lines or keels; leaves very narrow, < 1 cm wide; C: yellow, orange-yellow in throat; D: CR; M: pl 18–26 cm, S 2.2 cm .... *Sobralia pfavii*
10. Lip without fringed lines or keels, leaves > 1 cm wide; C: white, yellow-brown in center; D: CR (Mt); M: pl to 60 cm, S 4.5 cm
.... *Sobralia macra*
11(9). Flowers relatively small, lip < 5 cm long..................... 12
11. Flowers larger, lip > 5 cm long............................. 16

12(11). Leaf sheaths with black hairs, upper sheaths compressed; C: white, lip spotted with red-violet; D: No, CR, Pma (Mt); M: pl to 60 cm, S 4–5 cm .... *Sobralia lindleyana*

12. Leaf sheaths smooth or warty, not markedly compressed above.... 13

13(12). Lip with 5 prominent keels, may be hairy 14

13. Lip with only 2 or 3 keels or ridges 15

14(13). Lip distinctly hairy, keels low throughout; normally has 2–3 flowers and several buds at once; C: creamy yellow with orange beard in throat, lip with purple border; D: No, CR (Cb, Mt); M: pl to 80 cm, S 3–3.5 cm .... *Sobralia luteola* (Syn. *S. pleiantha*)

14. Lip not hairy, keels highest near tip; may have 2 flowers at once; C: greenish white with brown-red nerves; D: No, CR (Mt); M: pl to 50 cm, S 3 cm .... *Sobralia mucronata*

15(13). Lip with 2 low median ridges and warty basal callus; C: yellowish white; D: CR, SA (Mt); M: pl to 40 cm, S 2–2.2 cm .... *Sobralia corazoi*

15. Lip with 3 low keels; C: white or greenish yellow; D: Pma, SA (Mt); M: pl to 80 cm, S 5 cm........................ *Sobralia valida*

16(11). Margin of lip strongly crisped, distal margins of lip and petals erose, lip hairy; C: white with yellow-orange throat; D: rCR, SA; M: pl to 1 m, S 5.5–6 cm .... *Sobralia fimbriata*

16. Margin of lip only slightly wavy or crisped, lip weakly and petals not at all erose, lip not hairy 17

17(16). Petals much wider than sepals; leaf sheaths minutely hairy; C: white with yellow in throat; D: CR, Pma, SA (Cb, Pc); M: pl to > 1 m, S 7 cm .... *Sobralia powellii*

17. Petals subequal to sepals or narrower....................... 18

18(17). Stems compressed above; lip notched in front, without keels; C: creamy white with yellow in throat; D: CR, Pma, SA (Cb, Mt, Pc); M: pl to 50 cm, S 6.5 cm .... *Sobralia macrophylla*

18. Stems cylindrical throughout; lip rounded in front, with 3 keels; C: white or greenish yellow; D: Pma, SA (Mt); M: pl to 80 cm, S 5 cm .... *Sobralia valida*

19(8). Leaves narrow, < 4 cm wide; flowers bright magenta or pink, small (lip < 5 cm long), tube narrow and short, with lip expanding abruptly 20

19. Leaves usually wider; flower larger, or if small, not bright magenta or pink, tube wider and expanding gradually 21

20(19). Lip about 3 cm long; C: magenta; D: CR, Pma (Cb, Mt); M: pl 15–50 cm, S 3 cm .... *Sobralia amabilis*
(Syn. *Fregea amabilis*)
20. Lip about 4.5 cm long; C: bright pink or magenta, throat white with brown-orange spots; D: CR, Pma (Mt); M: pl to 60 cm, S 3.3 cm .... *Sobralia wercklei*

21(19). Leaf sheaths with small dark or red hairs; C: S, P bronzy brown, lip rose-purple with pale ruffled margin; D: CR, Pma, SA (Cb, Mt); M: pl to 1 m, S 4.5–5 cm .... *Sobralia atropubescens*
(Syn. *S. decora* var. *aerata*)
Plate 35(3)
21. Leaf sheaths smooth or warty, without conspicuous hairs 22

22(21). Lip very large, about 6 cm wide; leaves narrow, up to 3.5 cm wide; stem spotted; C: rose-purple, white blotch on lip, brownish in throat; D: No, CR (Mt); M: pl to > 1 m, S 8 cm ....
.... *Sobralia macrantha*
Plate 35(4)
22. Lip smaller; leaves usually > 3.5 cm wide.................... 23

23(22). Stem sheaths distinctly warty 24
23. Stem sheaths smooth or faintly warty...................... 26

24(23). Bracts much longer than flowers; sepals basally united to form definite tube; C: dark purple-violet; D: CR, SA; M: pl to 60 cm, S 2.3–2.5 cm... *Sobralia lepida*
24. Bracts much shorter than flowers; sepals not markedly united basally... 25

25(24). Leaves 5–6 cm wide, not markedly folded; lip about 5 cm long; C: S, P, and margin of lip magenta-red, lip yellow in throat, brown-ocher between throat and margin; D: No, CR, Pma (Cb, Mt); M: pl to 1.5 m, S 7.5 cm .... *Sobralia warscewiczii*
May not be distinct from *S. bradeorum*.
25. Leaves up to 8 cm wide, stiff and strongly veined, usually folded; with 3 raised nerves from base to apex of lip; C: rose-purple, lip darker with yellow throat; D: No, CR (Mt, Pc); M: pl to 2 m, S 6–6.5 cm.. *Sobralia bradeorum*

26(23). Lip strongly ruffled, with 7 crisped keels; C: S, P rose-purple, lip purple with yellow keels; D: Pma (Mt); M: pl to 75 cm, S 5.5–6.5 cm .... *Sobralia bouchei*

26. Lip wavy but not strongly ruffled, without crisped keels; C: rose-purple, lip darker;
D: No, rCR .... *Sobralia decora*
D: Pma, SA (Cb) .... *Sobralia fenzliana*
D: CR (Cb). .... *Sobralia neglecta*
These may be distinct species, but I do not have enough information to distinguish them.

Solenocentrum. Epiphytic in wet forest; roots somewhat fleshy; leaves several, soft-herbaceous or weakly pleated, stalked; inflorescence terminal, of many flowers, with loose sheathing bracts below; sepals 2-lobed, petals notched or 2-lobed; lip uppermost with long spur basally; column pointed, with anther on dorsal side; pollinia 4, club-shaped, brittle; 1 species known; C: pale green, white in center; D: CR, Pma (Mt); M: pl 35–40 cm, S 7–9 mm .... *Solenocentrum costaricense*
Plate 40(5)

Spiranthes. Terrestrial; usually leafless at flowering; stem delicate, erect; inflorescence of many flowers, spiral, usually dense; flowers small, tubular; lip short-stalked with 2 fleshy lobules at base, usually not or only slightly constricted near middle; column short, rostellum straplike, viscidium oblong, rostellar remnant deeply notched. Not definitely recorded southeast of Nicaragua, but 3 species occur there and any of them might be found in Costa Rica.

Key to *Spiranthes*

1. Lip slightly constricted near middle; midlobe minutely toothed, obtuse; basal lobules hairy; C: white; D: No; M: pl to 50 cm, S 8–9 mm .... *Spiranthes graminea*
1. Lip without constriction, minutely toothed along margins 2
2(1). Lip ovate or broadly lanceolate, acute; basal lobules hairy; D: No
.... *Spiranthes amesiana*
2. Lip oblong, obtuse; basal lobules without hairs; D: No ....
.... *Spiranthes torta*

Stenorrhynchos. Terrestrial or epiphytic; with or without leaves at flowering, leaves basal, short-stalked, lanceolate; inflorescence erect, of several to many flowers, usually dense, bracts colored; flowers relatively large, erect to suberect, with definite chin at base of sepals, tubular or spreading apically; lip not stalked, margins thickened basally, not or only weakly constricted in middle; column massive, with long foot united to ovary, rostellum and rostellar remnant with narrow base, long and narrow, viscidium sheath-like, narrow.

Key to *Stenorrhynchos*

1. Terrestrial; leaves withered at flowering; ovary and sepals glandular; C: red, pink, or greenish cream; D: No, CR, Pma, SA (Cb, Pc); M: pl to 50 cm, S 1.7–2 cm .... *Stenorrhynchos lanceolatum*
Plate 37(6)
1. Epiphytic; leaves present at flowering; ovary and sepals smooth; C: bracts red or dark red, S, P pinkish red or red, lip white; D: No, CR, Pma, SA (Mt); M: pl to 25 cm, S 11–15 mm ....
.... *Stenorrhynchos speciosum*
Plate 37(5)

Triphora. Terrestrial; with fleshy root tubers, stems soft and herbaceous, each with several herbaceous or weakly pleated leaves, or leafless; inflorescence terminal, usually of several flowers; sepals and petals similar, spreading or not; lip 3-lobed, enfolding column; column slender, anther with fleshy beak; pollinia 4, very soft. These inconspicuous little plants have delicate, short-lived flowers and are rarely collected.

Key to *Triphora*

1. Plants without foliage leaves, stem bracts scalelike, scarcely wider than the stem. 2
1. Plants with definite foliage leaves 3
2(1). Plants each with 1 or 2 flowers on short pedicels; D: Pma (Pc?); M: pl 6–7.5 cm, S 8 mm....................... *Triphora wagneri*
 2. Plants each with several flowers, lower flowers with long pedicels; D: No, Pma, SA (Cb); C: S greenish tinged with brown, P pale yellow-green, lip pale green or white; M: pl 8–20 cm, S 7–8 mm .... *Triphora gentianoides*

3(1). Lateral lobes of lip much longer than wide, length subequal to midlobe, curved; **C:** lvs and fls purple; **D:** CR, wPma (Pc); **M:** pl 3–7 cm, S 11 mm.............................*Triphora ravenii*
Sometimes found on mossy tree trunks.
3. Lateral lobes of lip about as wide as long, much shorter than midlobe... 4
4(3). Midlobe of lip triangular, margins toothed, lobes of lip not overlapping; **D:** CR (Cb?); **M:** pl 10 cm, S 10–12 mm....*Triphora nitida*
4. Midlobe of lip ovate, overlapping with lateral lobes when flattened, not toothed; **C:** white, lip rose; **D:** No, Pma (Cb?); **M:** pl to 15 cm, S 1.7 cm................................ *Triphora mexicana*
Very similar to the North American *T. trianthophora*.

Tropidia. Terrestrial; stems slender, reedlike, with many 2-ranked or spiral pleated leaves; inflorescence terminal, branched, of several flowers; sepals and petals similar, subparallel; lip simple; column short and thick, anther parallel with column axis; pollinia sectile; 1 species in Central America; **C:** white; **D:** No, CR, SA (Cb, Pc); **M:** pl to 35 cm, S 6–7 mm
.... *Tropidia polystachya*
Plate 36(4)

Tropidia is likely to occur in scrubby or dry forests that seem to be poor orchid habitat. I suspect that the South American *Tropidia* may be a distinct species.

Uleiorchis. Terrestrial saprophytes; fleshy underground rhizome or tuber, roots slender; without leaves; inflorescence erect, with few flowers; sepals and petals united, forming a tube; lip fleshy; column straight, anther terminal, pollinia 4, sectile; 1 species known; **C:** whitish with tinge of lilac; **D:** Pma, SA (Pc); **M:** pl to 10 cm, fl 1.6 cm ....
....*Uleiorchis ulaei*

In central Panama known from a single site in wet forest (now largely cut) but may be more widespread. Both flowers and fruits are inconspicuous.

Vanilla. Fleshy, climbing vines; leaves convolute in development but leathery to fleshy; inflorescences lateral or terminal, usually of several flowers but opening 1 at a time, flowers short-lived; lip with cluster of scales on blade; pollen soft and slightly sticky; fruit long and beanlike, fleshy, usually splitting open when fully ripe; seeds tiny, hard. Because the flowers are short-lived and are often produced in the treetops, *Vanilla* is poorly represented in museum collections and is not well understood. I have found a species with relatively small white flowers near the Caribbean coast of Panama, but I have no idea of its identity.

Key to *Vanilla*

1. Leaves 2–3 cm wide, 5–7 times longer than wide; **C:** S, P yellowish green, lip ocher with orange-yellow veins; **D:** No, CR, Pma (Cb, Pc); **M:** lvs 14 cm long, 2 cm wide, S 5.8–6 cm .... *Vanilla odorata*
   Such narrow-leaved plants found in western Panama are probably this species, but I have seen no flowers.
1. Leaves at least 4 cm wide, 3–4 times longer than wide 2
2(1). Leaves broadly elliptic, widest near middle; inflorescence with prominent or leaflike bracts....................................... 3
2. Leaves widest near base, sides parallel or narrowly elliptic; inflorescence with small scalelike bracts 4
3(2). Lip 3-keeled, > 3.5 cm long; **C:** S, P yellow-green, lip white; **D:** No, CR, SA; **M:** lvs 30 cm long, 15 cm wide, S 5.8–6 cm .... *Vanilla mexicana*
   (Syn. *V. inodora*)
3. Lip with large, fleshy callus on midlobe, < 3.5 cm long; **C:** S, P pale green, lip white; **D:** No, CR, Pma.............. *Vanilla pfaviana*
   The plant of this species is said to be similar to that of *V. mexicana*.
4(2). Flowers 1 or 2 per inflorescence; tube of lip narrow and then abruptly inflated; inflorescence often near base of vine; **C:** S pale green, P greenish cream, lip cream with yellow-brown veins; **D:** CR, Pma (Cb, Pc, Wt); **M:** lvs 15–30 cm long, 5.5–7 cm wide, S 10.5–11.5 cm .... *Vanilla pauciflora*
4. Flowers several to many on each inflorescence; tube of lip gradually inflated from base; inflorescence usually near top of plant 5
5(4). Leaves very large (about 10 cm wide) and fleshy; lip not warty on blade; **C:** pale yellow, darker in throat; **D:** No, CR, Pma, SA (Cb, Pc); **M:** lvs to 20 long, 10 cm wide, S 9 cm ..... *Vanilla pompona*
   Plate 34(1)
   This species is sometimes cultivated as an ornamental but

is considered inferior to *V. planifolia* as a source of flavoring.
5. Leaves only moderately fleshy; lip warty on blade.............. 6
6(5). Margin of lip slightly irregular but not deeply slit, lip not adorned with finger-like appendages; C: S, P pale green, lip greenish cream; D: No, CR, Pma, SA (Cb, Pc); M: lvs to 20 cm long, 5 cm wide, S 5–6 cm..*Vanilla planifolia*
Plate 34(2)

This is the commercial vanilla and widely escaped from cultivation.
6. Margin of lip deeply slit, blade adorned with finger-like appendages; C: S, P greenish yellow, lip orange-yellow; D: No, Pma (Cb); M: lvs 15 cm long, 4 cm wide, S 7–7.5 cm ....*Vanilla insignis*

Wullschlaegelia. Terrestrial saprophytes; roots fleshy, narrowly ellipsoid; without leaves; inflorescence erect, tip nodding, of many flowers, with many Y- or T-shaped hairs on stem and flowers; sepals and petals similar lip simple, column short, with long basal foot, anther sunken in column; pollinia 4, sectile, of thin plates. These are usually self-pollinating, and the flowers open only briefly. The root mass lives between the mineral soil and the leaf mold. The inflorescence appears briefly and quickly develops seed, so the plant is invisible during most of the year. *Wullschlaegelia* plants are inconspicuous at best and may be more widespread in wet forests than the few records indicate.

Key to *Wullschlaegelia*

1. Lip on lower side of flower; spur prominent; hairs pale; branches tapering to sharp points; C: cream; D: No, Pma, So (Pc).......... ....*Wullschlaegelia calcarata*
Plate 33(5)
1. Lip on upper side of flower; spur smaller; hairs dark brown, branches blunt; C: cream; D: No, Pma, So (Cb, Pc).............. .... *Wullschlaegelia aphylla*

Plate 1. Laeliinae, 1 (Chapter 4)

1. *Cattleya dowiana* (*guaria de Turrialba*)
2. *Cattleya patinii*
3. *Cattleya skinneri* (*guaria morada*), the national flower of Costa Rica
4. *Laelia rubescens*
5. *Myrmecophila tibicinis*
6. *Schomburgkia lueddemannii*

Plate 2. Laeliinae, 2 (Chapter 4)

1. *Encyclia cordigera* (Costa Rican form)
2. *Encyclia cordigera* (Panamanian form; *flor de semana santa*)
3. *Encyclia mooreana*
4. *Encyclia alata*
5. *Encyclia prismatocarpa*
6. *Encyclia brassavolae*

Plate 3. Laeliinae, 3 (Chapter 4)

1. *Encyclia ionophlebia*
2. *Encyclia fragrans*
3. *Encyclia vespa*
4. *Homalopetalum pumilio*
5. *Brassavola acaulis*
6. *Brassavola nodosa*

Plate 4. Laeliinae, 4 (Chapter 4)

1. *Epidendrum eburneum*
2. *Epidendrum parkinsonianum*
3. *Epidendrum wercklei*
4. *Epidendrum mirabile*
5. *Epidendrum radicans*
6. *Epidendrum pfavii*

Plate 5. Laeliinae, 5 (Chapter 4)

1. *Epidendrum barbeyanum*
2. *Epidendrum pendens*
3. *Epidendrum laucheanum*
4. *Epidendrum lancilabium*
5. *Epidendrum galeottianum*
6. *Epidendrum rousseauae*

Plate 6. Laeliinae, 6 (Chapter 4)

1. *Epidendrum ciliare*
2. *Epidendrum piliferum*
3. *Oerstedella wallisii*
4. *Oerstedella endresii*
5. *Oerstedella pinnifera*
6. *Oerstedella centradenia*

Plate 7. Laeliinae, 7 (Chapter 4)

1. *Acrorchis roseola*
2. *Barkeria lindleyana*
3. *Dimerandra elegans*
4. *Isochilus major*
5. *Jacquiniella aporophylla*
6. *Ponera striata*

Plate 8. Laeliinae, 8 (Chapter 4)

1. *Scaphyglottis geminata*
2. *Scaphyglottis amparoana*
3. *Scaphyglottis lindeniana*
4. *Nidema ottonis*
5. *Hexisea bidentata*
6. *Reichenbachanthus lankesteri*

Plate 9. Oncidiinae, 1 (Chapter 5)

1. *Oncidium carthagenense*
2. *Oncidium heteranthum*
3. *Oncidium schroederianum*
4. *Oncidium stenotis*
5. *Oncidium fuscatum*
6. *Oncidium obryzatum*

Plate 10. Oncidiinae, 2 (Chapter 5)

1. *Ticoglossum oerstedii*
2. *Osmoglossum egertonii*
3. *Otoglossum chiriquense*
4. *Rossioglossum schlieperianum*
5. *Miltoniopsis roezlii*
6. *Psychopsis krameriana*

Plate 11. Oncidiinae, 3 (Chapter 5)

1. *Trichopilia suavis*
2. *Trichocentrum pfavii*
3. *Aspasia principissa*
4. *Lemboglossum bictoniense*
5. *Ada allenii*
6. *Brassia arcuigera*

Plate 12. Oncidiinae, 4 (Chapter 5)

1. *Systeloglossum panamense*
2. *Goniochilus leochilinus*
3. *Hybochilus inconspicuus*
4. *Leochilus scriptus*
5. *Mesospinidium warscewiczii*
6. *Scelochilus tuerckheimii*

Plate 13. Oncidiinae, 5 (Chapter 5)

1. *Cischweinfia dasyandra*
2. *Rodriguezia lanceolata*
3. *Comparettia falcata*
4. *Ionopsis utricularioides*
5. *Lockhartia amoena*
6. *Macradenia brassavolae*, with its pollinator, *Euglossa hemichlora*

Plate 14. Oncidiinae, 6 (Chapter 5)

1. *Sigmatostalix brownii*
2. *Macroclinium ramonense*
3. *Notylia albida*
4. *Psygmorchis pumilio*
5. *Trizeuxis falcata*
6. *Pachyphyllum hispidulum*

Plate 15. Catasetinae (Chapter 6)

1. *Catasetum maculatum*
2. *Clowesia warscewiczii*
3. *Cycnoches dianae*
4. *Cycnoches warscewiczii*
5. *Dressleria dilecta*
6. *Mormodes skinneri*

Plate 16. Stanhopeinae, 1 (Chapter 6)

1. *Acineta chrysantha*
2. *Coeliopsis hyacinthosma*
3. *Houlletia tigrina*
4. *Paphinia cristata*?
5. *Peristeria elata* (flor de Espíritu Santo, or dove orchid), the national flower of Panama, with its pollinator, *Eufriesea concava*
6. *Peristeria* species

Plate 17. Stanhopeinae, 2 (Chapter 6)

1. *Gongora horichiana*
2. *Gongora tricolor*, with its pollinator, *Euglossa cyanura*
3. *Coryanthes maculata*
4. *Stanhopea cirrhata*
5. *Stanhopea costaricensis*
6. *Stanhopea wardii*

Plate 18. Stanhopeinae, 3 (Chapter 6)

1. *Sievekingia fimbriata*
2. *Horichia dressleri*
3. *Polycycnis ornata*
4. *Polycycnis barbata*
5. *Kegeliella kupperi*
6. *Trevoria glumacea*

Plate 19. Pleurothallidinae, 1 (Chapter 7)

1. *Lepanthes elata*
2. *Lepanthopsis floripecten*
3. *Octomeria valerioi*
4. *Platystele ovalifolia* (*P. jungermannioides*, a related species, is thought to be the world's tiniest orchid.)
5. *Restrepiopsis ujarrensis*
6. *Salpistele brunnea*

Plate 20. Pleurothallidinae, 2 (Chapter 7)

1. *Acostaea costaricensis*
2. *Barbosella anaristella*
3. *Brachionidium folsomii*
4. *Dresslerella pilosissima*
5. *Restrepia subserrata*
6. *Trisetella triaristella*

Plate 21. Pleurothallidinae, 3 (Chapter 7)

1. *Dracula erythrochaete*
2. *Dryadella butcheri*
3. *Masdevallia nicaraguae*
4. *Masdevallia schroederiana*
5. *Masdevallia zahlbruckneri*
6. *Scaphosepalum anchoriferum*

Plate 22. Pleurothallidinae, 4 (Chapter 7)

1. *Ophidion pleurothallopsis*
2. *Myoxanthus hirsuticaulis*
3. *Myoxanthus balaeniceps*
4. *Myoxanthus uncinatus*
5. *Stelis* species
6. *Stelis glossula*

Plate 23. Pleurothallidinae, 5 (Chapter 7)

1. *Pleurothallis crescentilabia*
2. *Pleurothallis cardiothallis*
3. *Pleurothallis palliolata*
4. *Pleurothallis circumplexa*
5. *Pleurothallis lewisae*
6. *Pleurothallis geminicaulina*

Plate 24. Pleurothallidinae, 6 (Chapter 7)

1. *Pleurothallis cogniauxiana*
2. *Pleurothallis tribuloides*
3. *Pleurothallis racemiflora*
4. *Restrepiella ophiocephala*
5. *Trichosalpinx memor*
6. *Zootrophion atropurpureus*

Plate 25. Maxillarieae, 1 (Chapter 8)

1. *Bifrenaria picta*
2. *Lycaste bradeorum*
3. *Lycaste tricolor*
4. *Neomoorea wallisii*
5. *Teuscheria pickiana*
6. *Xylobium elongatum*

Plate 26. Maxillarieae, 2 (Chapter 8)

1. *Chaubardiella pacuarensis*
2. *Pescatorea cerina*
3. *Chondrorhyncha reichenbachiana*
4. *Chondrorhyncha anatona*
5. *Cochleanthes aromatica*
6. *Cochleanthes discolor*

Plate 27. Maxillarieae, 3 (Chapter 8)

1. *Maxillaria amparoana*
2. *Maxillaria rodrigueziana*
3. *Maxillaria reichenheimiana*
4. *Maxillaria wercklei*
5. *Maxillaria pseudoneglecta*
6. *Maxillaria biolleyi*

Plate 28. Maxillarieae, 4 (Chapter 8)

1. *Maxillaria wrightii*
2. *Maxillaria sanguinea*
3. *Maxillaria ampliflora*
4. *Galeottia grandiflora*
5. *Kefersteinia lactea*
6. *Kefersteinia parvilabris*

Plate 29. Maxillarieae, 5 (Chapter 8)

1. *Huntleya burtii*
2. *Huntleya fasciata*
3. *Mormolyca ringens*
4. *Warrea costaricensis*
5. *Telipogon biolleyi*
6. *Telipogon elcimeyae*

Plate 30. Maxillarieae, 6 (Chapter 8)

1. *Dichaea panamensis*
2. *Dichaea trulla*
3. *Ornithocephalus inflexus*
4. *Stellilabium bullpenense*
5. *Trigonidium egertonianum*
6. *Cryptarrhena guatemalensis*

Plate 31. Miscellaneous orchids with pseudobulbs, 1 (Chapter 9)

1. *Arpophyllum giganteum*
2. *Bletia purpurea*
3. *Calanthe calanthoides*
4. *Chysis costaricensis*
5. *Coelia macrostachya*
6. *Spathoglottis plicata*

Plate 32. Miscellaneous orchids with pseudobulbs, 2 (Chapter 9)

1. *Cyrtopodium punctatum*
2. *Eriopsis biloba*
3. *Eulophia alta*
4. *Galeandra batemanii*
5. *Govenia* species
6. *Oeceoclades maculata*

Plate 33. Miscellaneous orchids with pseudobulbs, 3 (Chapter 9);
orchids without pseudobulbs, 1 (Chapter 10)

1. *Polystachya foliosa*
2. *Liparis nervosa*
3. *Malaxis histionantha*
4. *Malaxis tipuloides*
5. *Wullschlaegelia calcarata*
6. *Cleistes rosea*

Plate 34. Miscellaneous orchids without pseudobulbs, 2 (Chapter 10)

1. *Vanilla pompona*
2. *Vanilla planifolia*, in cultivation, with Dora Emilia Mora
3. *Phragmipedium caudatum*
4. *Phragmipedium longifolium*
5. *Selenipedium chica*
6. *Arundina graminifolia*

Plate 35. Miscellaneous orchids without pseudobulbs, 3 (Chapter 10)

1. *Sobralia suaveolens*
2. *Sobralia callosa*
3. *Sobralia atropubescens*
4. *Sobralia macrantha*
5. *Elleanthus lentii*
6. *Elleanthus tonduzii*

Plate 36. Miscellaneous orchids without pseudobulbs, 4 (Chapter 10)

1. *Elleanthus poiformis*
2. *Palmorchis silvicola*
3. *Corymborkis flava*
4. *Tropidia polystachya*
5. *Monophyllorchis maculata*
6. *Habenaria entomantha*

Plate 37. Spiranthinae, 1 (Chapter 10)

1. *Pelexia funckiana*
2. *Pelexia smithii*
3. *Sarcoglottis hunteriana*
4. *Coccineorchis bracteosa*
5. *Stenorrhynchos speciosum*
6. *Stenorrhynchos lanceolatum*

Plate 38. Spiranthinae, 2 (Chapter 10)

1. *Deiregyne hemichrea*
2. *Beloglottis costaricensis*
3. *Cyclopogon elatus*
4. *Eurystyles cotyledon*
5. *Lankesterella* species
6. *Schiedeella llaveana*

Plate 39. Goodyerinae, *Psilochilus*, *Campylocentrum* (Chapter 10)

1. *Aspidogyne* species
2. *Erythrodes killipii*
3. *Platythelys venustula*
4. *Psilochilus physurifolius*
5. *Campylocentrum micranthum*
6. *Campylocentrum fasciola*

Plate 40. Cranichidinae, Prescottiinae (Chapter 10)

1. *Baskervilla* species
2. *Cranichis lankesteri*
3. *Ponthieva maculata*
4. *Pterichis habenarioides*
5. *Solenocentrum costaricense*
6. *Prescottia cordifolia*

Appendix A
Preparing Orchid Materials for Study or Identification

It is difficult to identify plants from even the most detailed descriptions. Photographs are often better than descriptions, and the commoner species can often be identified from a good photograph, but one should always have at least a pressed flower to go with the photograph. One can turn a pressed flower over to see the other side or soften it to see the inside of the lip. In this appendix I briefly treat the different types of specimens and how to prepare them. In any country, be sure to learn the pertinent regulations before collecting any plants or specimens.

How to Prepare Materials

Notes and Data. **Keep records.** However specimens are prepared, they should be **labeled**, and the sooner the better. Specimens lose much of their botanical value when the geographic origin is unknown. After a few days in the field it may be impossible to remember whether a particular *Pleurothallis* was the one picked up the first day or perhaps one found a week later on a different mountain.

Preserving Material. Drying and pickling are the two methods most commonly used by botanists to preserve plant specimens.

DRIED SPECIMENS. The standard in botany is the pressed, dried specimen. It is rarely a thing of beauty, but with reasonable care it will last forever. Though smaller, less fleshy flowers can be pressed between newspaper or tissue paper in a book, botanists regularly carry a large bundle of corrugated cardboard (or plastic or metal corrugates), alternating with felt blotters, all held between wooden frames. The strapped bundle of specimens can be tightened up and allowed to dry over a stove or other heat source; the hot air moving through the corrugates dries the specimens. It is important that the material be pressed flat while drying; otherwise the flowers crumble and are lost. Three-dimensional dried flowers can be kept in cotton or tissue paper in a stiff container, but this is not very practical. The standard plant press is 12 by 18 inches—about the size of a folded sheet of newspaper—large enough to press whole shoots of smaller plants or representative parts of larger ones. Thick pseudobulbs should be split and hollowed before being pressed, or they may mold in the press.

Orchidists usually want their material to be identified, and they want any new and unnamed species to be properly named. Unfortunately, they hate to cut their plants. This has led to a good deal of frustration. 'When are you going to name my new *Epidendrum*?" "Well, when are you going to send me a good specimen?" One can usually make an acceptable botanical specimen without harming the plant. Cutting the inflorescence, or a part of it, does not damage the plant. One should also press the oldest pseudobulb(s) and one or two of the oldest leaves to accompany the flowers. The same system can be used in the field without causing serious harm to the plant. If the plant is left in place even the newest pseudobulb or shoot can usually be cut off without killing the plant. When collecting very small plants, of course, it is easier to press a whole plant, if it is available. Sketches or photographs of the plant can also be useful, especially if one cannot press enough material to show the character of the plant.

PICKLED SPECIMENS. In many ways, flowers preserved in alcohol provide the best material for study. Preservation in liquid gives the botanist a three-dimensional flower with all its details preserved. The color usually fades quickly, so notes on color are useful. There are disadvantages to preserving in liquid: the bottles may leak in one's luggage, and if the bottles are not hermetically sealed they will eventually dry out. Few botanical museums have the staff to regularly replace the liquid, and many specimens are lost by drying.

One can use locally produced alcohol—*aguardiente* or (in Costa Rica) *guaro*—to preserve flowers, but this often makes the material so brittle that it is nearly useless. In the past, mixtures of alcohol and formalin were used, but people have realized that formalin is not only very disagreeable but also dangerous. Ethyl (or grain) alcohol is the most common preservative, and a 60 to 70% solution is about right. If possible, the alcohol solution should also contain about 5% glycerin. Then, even if the container loses most of its liquid the specimens are still usable. If they dry out without glycerin present, they cannot be rehydrated.

PHOTOGRAPHS. Photographs can be very useful, and many people carry a camera in the field. However, it is best to also press a couple of flowers. Even the best photograph cannot show all sides of the flower, and internal details are often critical for identification. Some photographers dissect the flower and spread the parts out with a small ruler to photograph each part. Such photographs show greater detail, but for most people it is easier to press a flower or two. Remember, **label everything.**

Laws Affecting Plant Collectors

Anyone who collects plant material should be aware of laws concerning this activity.

CITES. The Convention for the International Trade in Endangered Species controls international trade in endangered or threatened animals and plants. The convention considers *all* orchids to be threatened by extinction, and the international movement of these plants, or any of their parts, is prohibited without the proper permits. You may not be fined if the customs agent finds a pressed flower of *Epidendrum radicans* in your passport, but under the letter of the law you would be guilty of international trade in endangered species. CITES is now taking a somewhat more lenient approach to "trade" in seeds and in seedlings still in flasks. Under the current laws, orchids can be transported only with the proper export permits from the country of origin and, often, import permits from the country of destination.

Collecting Permits. In some countries, one must obtain a collecting permit *before* collecting any plant materials. Indeed, the authorities may even request a list of exactly what you plan to collect. (This can be a bit frustrating to the botanist who is most interested in new and unnamed plants.) If a collecting permit is required, it is best to have it in hand, or at least requested, before seeking an export permit.

Export Permits. In most cases an export permit from the country of origin, with exactly the correct signature, is required to import orchid materials, including pressed or pickled materials, into another country. In some countries the customs and agriculture people can be quite rigid about this.

Local Institutions. Anyone planning to collect plant material for study or identification would do well to contact local botanists or local orchid societies for help with regulations and permits. In Costa Rica, the staff of the Universidad de Costa Rica, the Museo Nacional, or the Instituto Nacional de Biodiversidad (Inbio) may be able to help. In Panama, the herbarium in the Universidad Nacional or any botanist working at the Smithsonian Tropical Research Institute should be able to give up-to-date advice. These people know the local rules, which may change from year to year, and they may be more sympathetic to your problems than most government bureaucrats.

Generally it is easier for institutions than for individuals to exchange study material. If you cannot get an export permit for your specimens, don't despair; above all, don't throw good specimens in the trash. Give your specimens to one of the local institutions. There may be someone there who can identify the material quickly. Even if not, the material becomes available to visiting specialists. When a botanist undertakes a revision of any group, material is normally borrowed from the important herbaria. Ideally, the herbaria in the countries where the orchids are native should be the most important. Both Costa Rica and Panama now have good herbaria and a good deal of local botanical activity. We can contribute to botanical progress by depositing study material in these herbaria.

Appendix B
Authors of Orchid Names Used in the Guide

Many readers may have little interest in who first published a botanical name or who later transferred the species to a different genus, but botanists expect this information. Authors' names are given in this appendix in the hope of satisfying the botanist without further confusing the hobbyist. When a single name or abbreviation is included after the plant name, the name is as it was originally published. If the species has been transferred to another genus, then the original author's name is given in parentheses, followed by the name of the combining author. Longer names are abbreviated, as is customary. Species of other areas that are mentioned in the guide are also listed in this appendix.

Names marked by asterisks have been used in Costa Rica or Panama, but I have too little information to provide descriptions for their identification. The parenthetical numbers following asterisked *Pleurothallis* species refer to the key groups for that genus in Chapter 7.

Aa paleacea (Kunth) Rchb. f.
Acineta chrysantha Lindl.
A. *erythroxantha Lindl.
A. *gymnostele Schltr.
A. sella-turcica Rchb. f.
A. superba (Kunth) Rchb. f.
Acostaea costaricensis Schltr.
Acrorchis roseola Dressler
Ada allenii (C. Schweinf.) N. H. Williams
A. chlorops (Endres & Rchb. f.) N. H. Williams.

Amparoa costaricensis Schltr.
Arpophyllum giganteum Lindl.
A. medium Rchb. f.
Arundina graminifolia (Don) Hochr.
Aspasia epidendroides Lindl.
A. principissa Rchb. f.
Aspidogyne stictophylla (Schltr.) Garay
A. tuerckheimii (Schltr.) Garay
Barbosella anaristella Kraenzl.
B. circinata Luer

B. orbicularis Luer
B. prorepens Schltr.
Barkeria chinensis (Lindl.) Thien
B. lindleyana Lindl.
B. obovata (Presl) Christenson
B. skinneri (Batem. ex Lindl.) A. Rich. & Gal.
Baskervilla colombiana Garay
B. nicaraguensis Hamer & Garay
Beloglottis bicaudata (Ames) Garay
B. costaricensis (Rchb. f.) Schltr.
B. ecallosa (Ames & C. Schweinf.) Hamer & Garay
B. hameri Garay
B. subpandurata (Ames & C. Schweinf.) Garay
Bifrenaria picta (Schltr.) C. Schweinf.
Bletia campanulata La Llave & Lex.
B. edwardsii Ames
B. purpurata A. Rich. & Gal.
B. purpurea (Lam.) DC.
Bothriochilus macrostachyus (Lindl.) L. O. Williams
Brachionidium cruzae L. O. Williams
B. folsomii Dressler
B. kuhniarum Dressler
B. pusillum Ames & C. Schweinf.
B. valerioi Ames & C. Schweinf.
Brachystele guayanensis (Lindl.) Schltr.
Brassavola acaulis Lindl.
B. cucullata (L.) R. Br.
B. nodosa (L.) Lindl.
B. venosa Lindl.
Brassia arcuigera Rchb. f.
B. caudata (L.) Lindl.
B. gireoudiana Rchb. f. & Warsz.
B. longissima (Rchb. f.) Nash
B. verrucosa Lindl.
Brenesia costaricensis Schltr.

Bulbophyllum aristatum (Rchb. f.) Hemsl.
B. oerstedii (A. Rich. & Gal.) Hamer & Garay
B. vinosum Schltr.
B. wagneri Schltr.
Calanthe calanthoides (A. Rich. & Gal.) Hamer & Garay
Campylocentrum brenesii Schltr.
C. dressleri Dietrich & Díaz
C. fasciola (Lindl.) Cogn.
C. longicalcaratum Ames & C. Schweinf.
C. micranthum (Lindl.) Rolfe
C. pachyrrhizum (Rchb. f.) Rolfe
C. parvulum Schltr.
C. poeppigii (Rchb. f.) Rolfe
C. porrectum (Rchb. f.) Rolfe
C. schiedei (Rchb. f.) Hemsl.
C. schneeanum Foldats
C. tenellum Todzia
C. tyrridion Garay & Dunst.
Catasetum bicolor Klotzsch
C. integerrimum Hook.
C. maculatum Kunth
C. oerstedii Rchb. f.
C. viridiflavum Hook.
Cattleya aurantiaca (Batem.) Lindl.
C. dowiana Batem.
C. deckeri Klotzsch
C. patinii Cogn.
C. skinneri Batem.
Caularthron bilamellatum (Rchb. f.) R. E. Schult.
Chaubardiella chasmatochila (Fowlie) Garay
C. pacuarensis Jenny
C. subquadrata (Schltr.) Garay
Chondrorhyncha albicans Rolfe
C. anatona (Dressler) Senghas
C. bicolor Rolfe
C. crassa Dressler
C. eburnea Dressler

C. lendyana Rchb. f.
C. reichenbachiana Schltr.
Chrysocycnis tigrina (C. Schweinf.) Atwood
Chysis aurea var. maculata Hook.
C. costaricensis Schltr.
C. tricostata Schltr.
Cischweinfia dasyandra (Rchb. f.) Dressler & N. H. Williams
C. pusilla (C. Schweinf.) Dressler & N. H. Williams
Cleistes costaricensis Christenson
C. rosea Lindl.
Clowesia rosea Lindl.
C. russelliana (Hook.) Dodson
C. warscewiczii (Lindl.) Dodson
Coccineorchis bracteosa (Ames & C. Schweinf.) Garay
C. cernua (Lindl.) Garay
C. navarrensis (Ames) Garay
C. standleyi (Ames) Garay
Cochleanthes aromatica (Rchb. f.) R. E. Schult. & Garay
C. discolor (Lindl.) R. E. Schult. & Garay
C. lipscombiae (Rolfe) Garay
C. picta (Rchb. f.) Garay
Coelia macrostachya Lindl.
Coeliopsis hyacinthosma Rchb. f.
Comparettia falcata Poepp. & Endl.
Coryanthes horichiana Jenny
C. hunteriana Schltr.
C. maculata Hook.
C. powellii Schltr.
C. speciosa (Hook.) Hook.
Corymborkis flava (Sw.) Kuntze
C. forcipigera (Rchb. f.) L. O. Williams
Cranichis acuminatissima Ames & C. Schweinf.
C. ciliata (Kunth) Kunth
C. diphylla Sw.
C. lankesteri Ames

C. muscosa Sw.
C. *nigrescens Schltr. (CR, = muscosa?)
C. *pittieri Schltr. (CR, = diphylla?)
C. reticulata Rchb. f.
C. saccata Ames
C. sylvatica A. Rich. & Gal.
C. wageneri Rchb. f.
Crybe rosea Lindl.
Cryptarrhena guatemalensis Schltr.
C. lunata R. Br.
C. quadricornu Kraenzl.
Cryptocentrum calcaratum (Schltr.) Schltr.
C. gracilipes Schltr.
C. gracillimum Ames & C. Schweinf.
C. inaequisepalum C. Schweinf.
C. latifolium Schltr.
C. standleyi Ames
Cryptophoranthus lepidotus L. O. Williams
Cyclopogon cranichoides (Griseb.)
C. elatus (Sw.) Schltr.
C. millei (Schltr.) Schltr.
C. miradorensis Schltr.
C. olivaceus (Rolfe) Schltr.
C. prasophyllum (Rchb. f.) Schltr.
Cycnoches *amparoanum Schltr. (CR)
C. aureum Lindl.
C. chlorochilon Klotzsch
C. dianae Rchb. f.
C. egertonianum Batem.
C. guttulatum Schltr.
C. pachydactylon Schltr.
C. stenodactylum Schltr.
C. *tonduzii Schltr.
C. warscewiczii Rchb. f.
Cyrtopodium punctatum Lindl.
Deiregyne hemichrea (Lindl.) Schltr.
Dichaea acroblephara Schltr.

D. *amparoana* Schltr.
D. *brachypoda* Rchb. f.
D. *costaricensis* Schltr.
D. *cryptarrhena* Kraenzl.
D. *dammeriana* Kraenzl.
D. *eligulata* Folsom
D. *glauca* (Sw.) Lindl.
D. **gracillima* C. Schweinf. (CR)
D. *graminoides* (Sw.) Lindl.
D. *hystricina* Rchb. f.
D. *lankesteri* Ames
D. *morrisii* Fawc. & Rendle
D. *muricata* (Sw.) Lindl.
D. *obovatipetala* Folsom
D. *oxyglossa* Schltr.
D. *panamensis* Lindl.
D. *pendula* (Aubl.) Cogn.
D. *poicillantha* Schltr.
D. *retroflexiligula* Folsom
D. *sarapiquiensis* Folsom
D. *schlechteri* Folsom
D. *squarrosa* Lindl.
D. *standleyi* Ames
D. *tenuifolia* Schltr.
D. *trichocarpa* (Sw.) Lindl.
D. *trulla* Rchb. f.
D. *tuberculilabris* Folsom
D. *tuerckheimii* Schltr.
D. *violacea* Folsom
Dimerandra elegans (Focke) Siegerist
D. *emarginata* (G. F. W. Meyer) Hoehne
D. *latipetala* Siegerist
Discyphus scopulariae (Rchb. f.) Schltr.
Dracula erythrochaete (Rchb. f.) Luer
D. *pusilla* (Rolfe) Luer
D. *ripleyana* Luer
D. *vespertilio* (Rchb. f.) Luer
Dresslerella elvallensis Luer
D. *hispida* (L. O. Williams) Luer
D. *pertusa* (Dressler) Luer

D. *pilosissima* (Schltr.) Luer
D. *powellii* (Ames) Luer
D. *stellaris* Luer & Escobar
Dressleria dilecta (Rchb. f.) Dodson
D. *helleri* Dodson
D. *suavis* (Ames & C. Schweinf.) Dodson
Dryadella butcheri Luer
D. *dressleri* Luer
D. *gnoma* (Luer) Luer
D. *guatemalensis* (Schltr.) Luer
D. *minuscula* Luer & Escobar
D. *odontostele* Luer
D. *sororcula* Luer
Elleanthus albertii Schltr.
E. *aurantiacus* (Lindl.) Rchb. f.
E. *caricoides* Nash
E. *curtii* Schltr.
E. *cynarocephalus* (Rchb. f.) Rchb. f.
E. *fractiflexus* Schltr.
E. *glaucophyllus* Schltr.
E. *graminifolius* (Barb. Rodr.) Løjtnant
E. *hymenophorus* Rchb. f.
E. *isochiloides* Løjtnant
E. *jimenezii* (Schltr.) C. Schweinf.
E. *lancifolius* Presl
E. *laxus* Schltr.
E. *lentii* Barringer
E. *muscicola* Schltr.
E. *poiformis* Schltr.
E. *robustus* (Rchb. f.) Rchb. f.
E. *stolonifer* Barringer
E. *tillandsioides* Barringer
E. *tonduzii* Schltr.
E. *wercklei* Schltr.
E. *xanthocomos* Rchb. f.
Eltroplectris roseoalba (Rchb. f.) Hamer & Garay
Encyclia abbreviata (Schltr.) Dressler
E. *alata* (Batem.) Schltr.

E. amanda (Ames) Dressler & Pollard
E. baculus (Rchb. f.) Dressler & Pollard
E. brassavolae (Rchb. f.) Dressler
E. campylostalix (Rchb. f.) Schltr.
E. ceratistes (Lindl.) Schltr.
E. chacaoensis (Rchb. f.) Dressler & Pollard
E. cochleata (L.) Lemée
E. cordigera (Kunth) Dressler
E. cordigera var. *rosea* (Batem.) H. G. Jones
E. fortunae Dressler
E. fragrans (Sw.) Lemée
E. fragrans subsp. *aemula* (Rchb. f.) Dressler
E. gravida (Lindl.) Schltr.
E. guatemalensis (Klotzsch) Dressler & Pollard
E. ionocentra (Rchb. f.) Mora-Retana & J. García
E. ionophlebia (Rchb. f.) Dressler
E. livida (Lindl.) Dressler
E. luteorosea (A. Rich. & Gal.) Dressler & Pollard
E. mooreana (Rolfe) Schltr.
E. neurosa (Ames) Dressler & Pollard
E. ochracea (Lindl.) Dressler
E. prismatocarpa (Rchb. f.) Dressler
E. pseudopygmaea (A. Finet) Dressler
E. pygmaea (Hook.) Dressler
E. radiata (Lindl.) Dressler
E. selligera (Batem.) Schltr.
E. sima Dressler
E. spondiada (Rchb. f.) Dressler & Pollard
E. tuerckheimii Schltr.
E. vagans (Ames) Dressler
E. varicosa (Batem.) Schltr.
E. vespa (Vell.) Dressler & Pollard

Epidanthus crassus Dressler
Epidendropsis vincentina (Lindl.) Garay
Epidendrum acuñae Dressler
E. adnatum Ames & C. Schweinf.
E. albertii Schltr.
E. alfaroi Ames & C. Schweinf.
E. allenii L. O. Williams
E. amparoanum Schltr.
E. anastasioi Hágsater
E. anceps Jacq.
E. anoglossoides Ames & C. Schweinf.
E. anoglossum Schltr.
E. aporum Garay
E. arcuiflorum Ames & C. Schweinf.
E. atropurpureum Willd.
E. barbae Rchb. f.
E. barbeyanum Kraenzl.
E. baumannianum Schltr.
E. bilobatum Ames
E. bisulcatum Ames
E. blepharistes Barker
E. brachybotrys Ackerman & Montalvo
E. bracteosum Ames & C. Schweinf.
E. brenesii Schltr.
E. candelabrum Hágsater
E. carolii Schltr.
E. carpophorum Barb. Rodr.
E. ciliare L.
E. circinatum Ames
E. cocleense Ames, Hubb. & C. Schweinf.
E. concavilabium C. Schweinf.
E. confertum Ames & C. Schweinf.
E. congestoides Ames & C. Schweinf.
E. congestum Rolfe
E. cordiforme C. Schweinf.
E. coriifolium Lindl.
E. coronatum Ruiz & Pavon

E. criniferum Rchb. f.
E. cristobalense Ames
E. cryptanthum L. O. Williams
E. curvicolumna Ames, Hubb. & C. Schweinf.
E. curvisepalum Hágsater
E. dentiferum Ames & C. Schweinf.
E. dentilobum Ames, Hubb. & C. Schweinf.
E. difforme Jacq.
E. dolabrilobum Ames, Hubb. & C. Schweinf.
E. dolichostachyum Ames & C. Schweinf.
E. dosbocasense Hágsater
E. eburneum Rchb. f.
E. ellipsophyllum L. O. Williams
E. epidendroides (Garay) Mora-Retana & J. García
E. *estrellense Ames (CR)
E. exiguum Ames & C. Schweinf.
E. exile Ames
E. firmum Rchb. f.
E. flexicaule Schltr.
E. flexuosissimum C. Schweinf.
E. fuscopurpureum Schltr.
E. galeottianum A. Rich. & Gal.
E. glumibracteum Rchb. f.
E. goniorhachis Schltr.
E. gregorioi Hágsater
E. guanacastense Ames & C. Schweinf.
E. hammellii Hágsater
E. hawkesii Heller
E. hellerianum A. D. Hawkes
E. hunterianum Schltr.
E. ibaguense Kunth
E. imatophyllum Lindl.
E. incisum Vell.
E. incomptum Rchb. f.
E. insolatum Barringer
E. insulanum Schltr.
E. isomerum Schltr.

E. isthmii Schltr.
E. kerichilum Hágsater
E. lacustre Lindl.
E. lagenocolumna Hágsater
E. lancilabium Schltr.
E. lankesteri Ames
E. laterale Rolfe
E. latifolium Garay & Sweet
E. laucheanum Bonhof
E. lockhartioides Schltr.
E. luerorum Hágsater
E. lutheri Hágsater
E. macroclinium Hágsater
E. macrostachyum Lindl.
E. majale Schltr.
E. mantis-religiosae Hágsater
E. *microcardium Schltr. (CR)
E. microdendron Rchb. f.
E. microphyllum Lindl.
E. mirabile Ames & C. Schweinf.
E. miserrimum Rchb. f.
E. modestiflorum Schltr.
E. mora-retanae Hágsater
E. moyobambae Kraenzl.
E. muscicola Schltr.
E. myodes Rchb. f.
E. nervosiflorum Ames & C. Schweinf.
E. nitens Rchb. f.
E. nocturnum Jacq.
E. notabile Schltr.
E. nubium Rchb. f.
E. nutantirachis Ames & C. Schweinf.
E. obesum Ames
E. obliquifolium Ames, Hubb. & C. Schweinf.
E. octomerioides Schltr.
E. odontochilum Hágsater
E. oerstedii Rchb. f.
E. oxyglossum Schltr.
E. pachyceras Hágsater
E. pachyrhachis Ames
E. pallens Rchb. f.

E. *palmense* Ames
E. *panamense* Schltr.
E. *paniculatum* Ruiz & Pavon
E. *paranthicum* Rchb. f.
E. *parkinsonianum* Hook.
E. *paucifolium* Schltr.
E. *pendens* L. O. Williams
E. *pentotis* Rchb. f.
E. *peperomia* Rchb. f.
E. *pfavii* Rolfe
E. *phragmites* Heller & L. O. Williams
E. *phyllocharis* Rchb. f.
E. *physodes* Rchb. f.
E. *piliferum* Rchb. f.
E. *platystigma* Rchb. f.
E. *pleurothalloides* Hágsater
E. *polyanthum* Lindl.
E. *polyanthum* var. *myodes* (Rchb. f.) Ames, Hubb. & C. Schweinf.
E. *polychlamys* Schltr.
E. *porpax* Rchb. f.
E. *powellii* Schltr.
E. *probiflorum* Schltr.
E. *pseudepidendrum* Rchb. f.
E. *pseudoramosum* Schltr.
E. *pudicum* Ames
E. *purpurascens* Focke
E. *puteum* Standl. & L. O. Williams
E. *radicans* Lindl.
E. *rafael-lucasii* Hágsater
E. *ramonianum* Schltr.
E. *ramosissimum* Ames & C. Schweinf.
E. *ramosum* Jacq.
E. *raniferum* Lindl.
E. *repens* Cogn.
E. *resectum* Rchb. f.
E. *rigidiflorum* Schltr.
E. *rigidum* Jacq.
E. *rousseauae* Schltr.
E. *rugosum* Ames
E. *sanchoi* Ames

E. *sanchoi* var. *exasperatum* Ames & C. Schweinf.
E. *sancti-ramoni* Kraenzl.
E. *santaclarense* Ames
E. *schlechterianum* Ames
E. *schomburgkii* Lindl.
E. *sculptum* Rchb. f.
E. *selaginella* Schltr.
E. *simulacrum* Ames
E. *smaragdinum* Lindl.
E. *sobralioides* Ames & Correll
E. *stamfordianum* Batem.
E. *stangeanum* Rchb. f.
E. *stevensii* Hágsater
E. *storkii* Ames
E. *strobiliferum* Rchb. f.
E. *subnutans* Ames & C. Schweinf.
E. *summerhayesii* Hágsater
E. *talamancanum* (Atwood) Mora-Retana & J. García
E. *trachythece* Schltr.
E. *trialatum* Hágsater
E. *triangulabium* Ames & C. Schweinf.
E. *trianthum* Schltr.
E. *turialvae* Rchb. f.
E. *veraguasense* Hágsater
E. *vincentinum* Lindl.
E. *volutum* Lindl. & Paxton
E. *warscewiczii* Rchb. f.
E. *wercklei* Schltr.
Epilyna jimenezii Schltr.
Eriopsis biloba Lindl.
Erythrodes calophylla (Rchb. f.) Ames
E. *killipii* Ames
E. **lehmannii* (Schltr.) Ames
E. *lunifera* Ames
E. **nigrescens* (Schltr.) Ames
E. *purpurea* Ames
E. *vescifera* (Rchb. f.) Ames
Eulophia alta (L.) Fawc. & Rendle
Eurystyles auriculata Schltr.
E. *cotyledon* Wawra

E. standleyi Ames
Fregea amabilis Rchb. f.
Funkiella stolonifera (Ames & Correll) Garay
Galeandra batemanii Rolfe
G. baueri Lindl.
G. beyrichii Rchb. f.
G. juncea Lindl.
G. styllomisantha (Vell.) Hoehne
Galeottia grandiflora A. Rich.
Galeottiella minutiflora (A. Rich. & Gal.) Szlach.
G. nutantiflora (Schltr.) Szlach.
Gomphichis costaricensis (Schltr.) Ames, Hubb. & C. Schweinf.
Gongora amparoana Schltr.
G. armeniaca (Lindl.) Rchb. f.
G. armeniaca var. *bicornuta* C. Schweinf. & P. H. Allen
G. atropurpurea Hook.
G. charontis Rchb. f.
G. claviodora Dressler
G. gibba Dressler
G. horichiana Fowlie
G. quinquenervis Ruiz & Pavon
G. tricolor (Lindl.) Rchb. f.
G. truncata Lindl.
G. unicolor Schltr.
Goniochilus leochilinus (Rchb. f.) Chase
*Goodyera *bradeorum* Schltr. (CR)
G. erosa (Ames & C. Schweinf.) Ames, Hubb. & C. Schweinf.
G. major Ames & Correll
*G. *micrantha* Schltr. (CR)
*G. *modesta* Schltr. (CR)
G. ovatilabia Schltr.
*G. *turrialbae* Schltr. (CR)
Govenia capitata Lindl.
G. ciliilabia Ames & C. Schweinf.
G. liliacea Lindl.
G. superba (La Llave & Lex.) Lindl.
G. utriculata Lindl.

Habenaria alata Hook.
*H. *amparoana* Schltr. (CR)
H. avicula Schltr.
H. aviculoides Ames & C. Schweinf.
H. bicornis Lindl.
H. bractescens Lindl.
*H. *brenesii* Schltr. (CR)
H. clypeata Lindl.
H. distans Griseb.
*H. *endresiana* Schltr. (CR)
H. entomantha (La Llave & Lex.) Lindl.
H. floribunda Lindl.
*H. *gymnadenioides* Schltr. (CR)
*H. *irazuensis* Schltr. (CR)
*H. *jimenezii* Schltr. (CR)
H. lankesteri Ames
H. leprieuri Rchb. f.
H. mesodactyla Griseb.
H. monorrhiza (Sw.) Rchb. f.
H. novemfida Lindl.
H. odontopetala Rchb. f.
H. pauciflora (Lindl.) Rchb. f.
H. quinqueseta (Michx.) Sw.
H. repens Nutt.
H. trifida Kunth
H. verecunda Schltr.
*H. *wercklei* Schltr. (CR)
Hapalorchis cheirostyloides Schltr.
H. lineatus (Lindl.) Schltr.
H. pumilus (C. Schweinf.) Garay
Harrisella porrecta (Rchb. f.) Fawc. & Rendle
Helleriella nicaraguensis A. D. Hawkes
Hexisea bidentata Lindl.
H. imbricata (Lindl.) Rchb. f.
H. sigmoidea Ames & C. Schweinf.
Homalopetalum pumilio (Rchb. f.) Schltr.
Horichia dressleri Jenny
Houlletia landsbergii Linden & Rchb. f.

H. odoratissima Linden ex Lindl. & Paxton
H. tigrina Lindl. & Paxton
Huntleya burtii Rchb. f.
H. fasciata Fowlie
Hybochilus inconspicuus (Kraenzl.) Schltr.
Ionopsis costaricensis Schltr.
I. satyrioides (Sw.) Rchb. f.
I. utricularioides (Sw.) Lindl.
Isochilus amparoanus Schltr.
I. carnosiflorus Lindl.
I. latibracteatus A. Rich. & Gal.
I. linearis (Jacq.) R. Br.
I. major Cham. & Schltdl.
Jacquiniella aporophylla (L. O. Williams) Dressler
J. cobanensis (Ames & Schltr.) Dressler
J. equitantifolia (Ames) Dressler
J. globosa (Jacq.) Schltr.
J. pedunculata Dressler
J. standleyi (Ames) Dressler
J. teres (Rchb. f.) Hamer & Garay
J. teretifolia (Sw.) Britton & Wilson
Kefersteinia alba Schltr.
K. auriculata Dressler
K. costaricensis Schltr.
K. deflexipetala Fowlie
K. lactea (Rchb. f.) Schltr.
K. maculosa Dressler
K. microcharis Schltr.
K. mystacina (Rchb. f.) Rchb. f.
K. parvilabris Schltr.
K. wercklei Schltr.
Kegeliella atropilosa L. O. Williams & Heller
K. houtteana (Rchb. f.) Mansf.
K. kupperi Mansf.
Koellensteinia kellneriana Rchb. f.
K. lilijae Foldats
*Kreodanthus *secundus* (Ames) Garay
Lacaena bicolor Lindl.
L. spectabilis (Klotzsch) Rchb. f.
Laelia rubescens Lindl.
Lanium microphyllum Lindl. ex Benth.
Lankesterella orthantha (Kraenzl.) Garay
Lemboglossum bictoniense (Batem.) Christenson
L. cordatum (Lindl.) Halbinger
L. hortensiae (Rodríguez) Halbinger
L. maculatum (La Llave & Lex.) Halbinger
L. stellatum (Lindl.) Halbinger
Leochilus labiatus (Sw.) Kuntze
L. scriptus (Scheidw.) Rchb. f.
L. tricuspidatus (Rchb. f.) Kraenzl.
*Lepanthes *acoridilabia* Ames & C. Schweinf.
*L. *acostaei* Schltr.
*L. *acuminata* Schltr.
*L. *ankistra* Luer & Dressler
*L. *antilocapra* Luer & Dressler
*L. *arachnion* Luer & Dressler
*L. *atwoodii* Luer
*L. *barbae* Schltr.
*L. *blephariglossa* Schltr.
*L. *blepharistes* Rchb. f.
*L. *bradei* Schltr.
*L. *brenesii* Schltr.
*L. *cascajalensis* Ames
*L. *chameleon* Ames
*L. *chiriquensis* Schltr.
*L. *ciliisepala* Schltr.
*L. *collaris* Luer
*L. *comet-halleyi* Luer
*L. *confusa* Ames & C. Schweinf.
*L. *costaricensis* Schltr.
*L. *crossota* Luer
*L. *decipiens* Ames & C. Schweinf.
*L. *droseroides* Luer
*L. *eciliata* Schltr.
*L. *edwardsii* Ames

L. *elata* Rchb. f.
L. *endresii* Luer
L. *erinacea* Rchb. f.
L. *estrellensis* Ames
L. *exasperata* Ames & C. Schweinf.
L. *eximia* Ames
L. *exposita* Luer
L. *glicensteinii* Luer
L. *grandiflora* Ames & C. Schweinf.
L. *guanacastensis* Ames & C. Schweinf.
L. *horichii* Luer
L. *horrida* Rchb. f.
L. *inaequiloba* Ames & C. Schweinf.
L. *incantata* Luer
L. *inescata* Luer
L. *ingramii* Luer
L. *inornata* Schltr.
L. *insectiflora* C. Schweinf.
L. *jennyi* Luer
L. *jimenezii* Schltr.
L. *jugum* Luer
L. *lancifolia* Schltr.
L. *latisepala* Ames & C. Schweinf.
L. *lindleyana* Oerst. & Rchb. f.
L. *macalpinii* Luer
L. *maxonii* Schltr.
L. *mentosa* Luer
L. *microglottis* Schltr.
L. *microtica* Luer & Escobar
L. *minutilabia* Ames & C. Schweinf.
L. *monteverdensis* Luer & Escobar
L. *mulderae* Luer
L. *myiophora* Luer
L. *mystax* Luer & Escobar
L. *odontolabis* Luer
L. *otopetala* Luer
L. *ova-rajae* Luer
L. *pantomima* Luer & Dressler
L. *pexa* Luer
L. *posthon* Luer
L. *pygmaea* Luer
L. *ramonensis* Schltr.
L. *regularis* Luer
L. *rotundifolia* L. O. Williams
L. *samacensis* Ames
L. *sannio* Luer
L. *standleyi* Ames
L. *stenophylla* Schltr.
L. *subdimidiata* Ames & C. Schweinf.
L. *tetroptera* Luer
L. *tipulifera* Rchb. f.
L. *tonduziana* Schltr.
L. *tridens* Ames
L. *truncata* Luer & Dressler
L. *turialvae* Rchb. f.
L. *volsella* Luer & Escobar
L. *wendlandii* Rchb. f.
L. *wercklei* Schltr.
Lepanthopsis comet-halleyi Luer
L. floripecten (Rchb. f.) Ames
L. obliquipetala (Ames & C. Schweinf.) Luer
Leucohyle subulata (Sw.) Schltr.
Ligeophila clavigera (Rchb. f.) Garay
Liparis elata Lindl.
L. fantastica Ames & C. Schweinf.
L. fratrum Schltr.
L. nervosa (Thunb.) Lindl.
L. vexillifera (La Llave & Lex.) Cogn.
Lockhartia acuta (Lindl.) Rchb. f.
L. amoena Endres & Rchb. f.
L. chocoensis Kraenzl.
L. *dipleura* Schltr. (CR, = amoena?)
L. hercodonta Kraenzl.
L. integra Ames & C. Schweinf.
L. micrantha Rchb. f.
L. obtusata L. O. Williams

L. *odontochila* Kraenzl. (CR, = *micrantha*?)
L. *oerstedii* Rchb. f.
L. *pittieri* Schltr.
Lycaste *bradeorum* Schltr.
L. *brevispatha* Lindl.
L. *campbellii* C. Schweinf.
L. *dowiana* Endres & Rchb. f.
L. *leucantha* Lindl.
L. *macrophylla* Lindl.
L. *powellii* Schltr.
L. *schilleriana* Rchb. f.
L. *tricolor* (Klotzsch) Rchb. f.
L. *xytriophora* Linden & Rchb. f.
Macradenia *brassavolae* Rchb. f.
Macroclinium *bicolor* (Lindl.) Dodson
M. *cordesii* (L. O. Williams) Dodson
M. *glicensteinii* Atwood
M. *juncta* (Dressler) Dodson
M. *lineare* (Ames & C. Schweinf.) Dodson
M. *paniculatum* (Ames & C. Schweinf.) Dodson
M. *ramonense* (Schltr.) Dodson
M. *simplex* Dressler
Malaxis *adolphi* (Schltr.) Ames
M. *aurea* Ames
M. *blephariglottis* (Schltr.) Ames
M. *brachyrrhynchos* (Rchb. f.) Ames
M. *carnosa* (Kunth) C. Schweinf.
M. *crispifolia* (Rchb. f.) Kuntze
M. *excavata* (Lindl.) Kuntze
M. *fastigiata* (Lindl.) Kuntze
M. *histionantha* (Link, Klotzsch & Otto) Garay & Dunst.
M. *lagotis* (Rchb. f.) Kuntze (CR)
M. *majanthemifolia* Cham. & Schltdl.
M. *nana* C. Schweinf.
M. *pandurata* (Schltr.) Ames
M. *parthonii* Morren

M. *pittieri* (Schltr.) Ames
M. *simillima* (Rchb. f.) Kuntze
M. *soulei* L. O. Williams
M. *tipuloides* (Lindl.) Kuntze
M. *tonduzii* (Schltr.) Ames
M. *wendlandii* (Rchb. f.) L. O. Williams
M. *woodsonii* L. O. Williams
Masdevallia *attenuata* Rchb. f.
M. *calura* Rchb. f.
M. *chasei* Luer
M. *chontalensis* Rchb. f.
M. *collina* L. O. Williams
M. *cupularis* Rchb. f.
M. *demissa* Rchb. f.
M. *erinacea* Rchb. f.
M. *flaveola* Rchb. f.
M. *floribunda* Lindl.
M. *fulvescens* Rolfe
M. *lata* Rchb. f.
M. *laucheana* Woolward
M. *livingstoneana* Rchb. f.
M. *marginella* Roezl & Rchb. f.
M. *molossoides* Kraenzl.
M. *nicaraguae* Luer
M. *nidifica* Rchb. f.
M. *pelecaniceps* Luer
M. *picturata* Rchb. f.
M. *pleurothalloides* Luer
M. *pygmaea* Kraenzl.
M. *rafaeliana* Luer
M. *reichenbachiana* Endres & Rchb. f.
M. *rolfeana* Kraenzl.
M. *scabrilinguis* Luer
M. *schizopetala* Kraenzl.
M. *schroederiana* Veitch
M. *striatella* Rchb. f.
M. *thienii* Luer
M. *tokachiorum* Luer
M. *tonduzii* Woolward
M. *tubuliflora* Ames
M. *utriculata* Luer
M. *walteri* Luer

M. *zahlbruckneri* Kraenzl.
Maxillaria *aciantha* Rchb. f.
M. *acostaei* Schltr.
M. *acutifolia* Lindl.
M. *adendrobium* (Rchb. f.) Dressler
M. *alba* (Hook.) Lindl.
M. *albertii Schltr. (CR)
M. *alfaroi* Ames & C. Schweinf.
M. *allenii* L. O. Williams
M. *amparoana* Schltr.
M. *ampliflora* C. Schweinf.
M. *anceps* Ames & C. Schweinf.
M. *angustisegmenta* Ames & C. Schweinf.
M. *angustissima* Ames, Hubb. & C. Schweinf.
M. *appendiculoides* C. Schweinf.
M. *arachnitiflora* Ames & C. Schweinf.
M. *attenuata* Ames & C. Schweinf.
M. *aurea* (Poepp. & Endl.) L. O. Williams
M. *bicallosa* (Rchb. f.) Garay
M. *biolleyi* (Schltr.) L. O. Williams
M. *brachybulbon* Schltr.
M. *bracteata* (Schltr.) Ames & Correll
M. *bradeorum* (Schltr.) L. O. Williams
M. *brenesii* Schltr.
M. *brevilabia* Ames & Correll
M. *brevipes* Schltr.
M. *brunnea* Rchb. f.
M. *caespitifica* Rchb. f.
M. *camaridii* Rchb. f.
M. *campanulata* C. Schweinf.
M. *chartacifolia* Ames & C. Schweinf.
M. *cobanensis* Schltr.
M. *concavilabia* Ames & Correll
M. *conduplicata* (Ames & C. Schweinf.) L. O. Williams
M. *confusa* Ames & C. Schweinf.

M. *costaricensis* Schltr.
M. *crassifolia* (Lindl.) Rchb. f.
M. *cryptobulbon* Carnevali & Atwood
M. *ctenostachya* Rchb. f.
M. *cucullata* Lindl.
M. *curtipes* Hook. f.
M. *dendrobioides* (Schltr.) L. O. Williams
M. *densa* Lindl.
M. *discolor* (Lodd.) Rchb. f.
M. *diuturna* Ames & C. Schweinf.
M. *elatior* Rchb. f.
M. *endresii* Rchb. f.
M. *exaltata* (Kraenzl.) C. Schweinf.
M. *falcata* Ames & C. Schweinf.
M. *flava* Ames & Correll
M. *foliosa* Ames & C. Schweinf.
M. *friedrichsthalii* Rchb. f.
M. *fulgens* (Rchb. f.) L. O. Williams
M. *hedwigae* Hamer & Dodson
M. *hematoglossa* A. Rich. & Gal.
M. *hennisiana* Schltr.
M. *horichii* Senghas
M. *houtteana* Rchb. f.
M. *inaudita* Rchb. f.
M. *insolita* Dressler
M. *linearifolia* Ames & C. Schweinf.
M. *longipetiolata* Ames & C. Schweinf.
M. *lueri* Dodson
M. *luteoalba* Lindl.
M. *maleolens* Schltr.
M. *meridensis* Lindl.
M. *microphyton* Schltr.
M. *minor* (Schltr.) L. O. Williams
M. *nasuta* Rchb. f.
M. *neglecta* (Schltr.) L. O. Williams
M. *nicaraguensis* (Hamer & Garay) Atwood
M. *oreocharis* Schltr.

M. pachyacron Schltr.
M. paleata (Rchb. f.) Ames & Correll
M. parviflora (Poepp. & Endl.) Garay
M. piestopus Schltr.
M. pittieri L. O. Williams
M. planicola C. Schweinf.
M. ponerantha Rchb. f.
M. powellii Schltr.
M. pseudoneglecta Atwood
M. pubilabia Schltr.
M. *quadrata Ames & Correll (CR)
M. ramonensis Schltr.
M. reichenheimiana Rchb. f.
M. repens L. O. Williams
M. ringens Rchb. f.
M. rodrigueziana Atwood & Mora-Retana
M. rousseauae Schltr.
M. rubrilabia Schltr.
M. rufescens Lindl.
M. sanguinea Rolfe
M. scorpioidea Kraenzl.
M. semiorbicularis Ames & C. Schweinf.
M. serrulata Ames & Correll
M. sigmoidea (C. Schweinf.) Ames & Correll
M. splendens Poepp. & Endl.
M. strumata Ames & Correll
M. suaveolens Barringer
M. subulifolia Schltr.
M. tenuifolia Lindl.
M. tigrina C. Schweinf.
M. tonduzii (Schltr.) Ames & Correll
M. trilobata Ames & C. Schweinf.
M. umbratilis L. O. Williams
M. uncata Lindl.
M. vagans Ames & C. Schweinf.
M. vaginalis Rchb. f.
M. valenzuealana (A. Rich.) Nash
M. valerioi Ames & C. Schweinf.
M. variabilis Lindl.
M. vittariifolia L. O. Williams
M. wercklei (Schltr.) L. O. Williams
M. wrightii (Schltr.) Ames & Correll
Mendoncella grandiflora (A. Rich.) A. D. Hawkes
Mesadenella petenensis (L. O. Williams) Garay
M. tonduzii (Schltr.) Pabst & Garay
Mesadenus polyanthus (Rchb. f.) Schltr.
Mesospinidium endresii (Kraenzl.) Garay
M. horichii Bock
M. panamense Garay
M. warscewiczii Rchb. f.
Miltonia clowesii (Lindl.) Beer
M. endresii Nichols
M. schroederiana (Rchb. f.) Garay & Stacy
M. warscewiczii Rchb. f.
Miltoniopsis roezlii (Rchb. f.) Godefroy-Lebeuf
M. warscewiczii (Rchb. f.) Garay & Dunst.
Monophyllorchis maculata Garay
M. microstyloides (Rchb. f.) Garay
Mormodes atropurpurea Lindl.
M. cartonii Hook.
M. colossa Rchb. f.
M. flavida Klotzsch
M. fractiflexa Rchb. f.
M. hookeri Lem.
M. horichii Fowlie
M. ignea Lindl. & Paxton
M. lancilabris Pabst
M. lobulata Schltr.
M. powellii Schltr.
M. punctata Rolfe
M. skinneri Rchb. f.

Mormolyca ringens (Lindl.) Schltr.
Myoxanthus aspasicensis (Rchb. f.) Luer
M. balaeniceps (Luer & Dressler) Luer
M. colothrix (Luer) Luer
M. hirsuticaulis (Ames & C. Schweinf.) Luer
M. lappiformis (Heller & L. O. Williams) Luer
M. octomeriae (Schltr.) Luer
M. pan (Luer) Luer
M. scandens (Ames) Luer
M. sempergemmatus (Luer) Luer
M. stonei (Luer) Luer
M. trachychlamys (Schltr.) Luer
M. uncinatus (Fawc.) Luer
Myrmecophila brysiana Lem.
M. tibicinis (Batem.) Rolfe
Neomoorea irrorata Rolfe
N. wallisii (Rchb. f.) Schltr.
Neourbania adendrobium Fawc. & Rendle
Neowilliamsia cuneata Dressler
N. tenuisulcata Dressler
Nidema boothii Schltr.
N. ottonis Britton & Millsp.
Notylia albida Klotzsch
N. barkeri Lindl.
N. *brenesii Schltr. (CR)
N. *lankesteri Ames (CR)
N. latilabia Ames & C. Schweinf.
N. panamensis Ames
N. pentachne Rchb. f.
N. *trisepala Lindl. (No, CR)
N. *turialbae Schltr.
Octomeria apiculata (Lindl.) Garay & Sweet
O. costaricensis Schltr.
O. graminifolia (L.) R. Br.
O. hondurensis Ames
O. surinamensis Focke
O. valerioi Ames & C. Schweinf.
Oeceoclades maculatum (Lindl.) Lindl.

Oerstedella* aberrans (Schltr.) Hamer (CR)
O. *acrochordonia (Schltr.) Hágsater (CR)
O. caligaria (Rchb. f.) Hágsater
O. centradenia Rchb. f.
O. centropetala (Rchb. f.) Rchb. f.
O. crescentiloba (Ames) Hágsater
O. endresii (Rchb. f.) Hágsater
O. exasperata (Rchb.f.) Hágsater
O. fuscina Dressler
O. intermixta (Ames & C. Schweinf.) Hágsater
O. lactea (Dressler) Hágsater
O. ornata Dressler
O. pajitense (C. Schweinf.) Hágsater
O. pansamalae (Schltr.) Hágsater
O. pentadactyla (Rchb. f.) Hágsater
O. pinnifera (C. Schweinf.) Hágsater
O. pseudoschummaniana (Fowlie) Hágsater
O. pseudowallisii (Schltr.) Hágsater
O. pumila (Rolfe) Hágsater
O. schummaniana (Schltr.) Hágsater
O. tetraceros (Rchb. f.) Hágsater
O. wallisii (Rchb. f.) Hágsater
Oncidium *advena Rchb. f. (CR, Pma)
O. altissimum Sw.
O. ampliatum Lindl.
O. *angustisepalum Kraenzl. (CR)
O. ansiferum Rchb. f.
O. anthocrene Rchb. f.
O. ascendens Lindl.
O. baueri Lindl.
O. bracteatum Warsz. & Rchb. f.
O. cabagrae Schltr.
O. cariniferum (Rchb. f.) Beer
O. carthagenense (Jacq.) Sw.
O. cebolleta (Jacq.) Sw.

O. *cheirophorum* Rchb. f.
O. **costaricense* Schltr. (CR)
O. *crista-galli* Rchb. f.
O. **dichromaticum* Rchb. f. (CR)
O. *endocharis* Rchb. f.
O. *ensatum* Lindl.
O. *exalatum* Hágsater
O. *exauriculatum* Jiménez
O. *fuscatum* Rchb. f.
O. *globuliferum* Kunth
O. *graminifolium* Lindl.
O. *guttatum* (L.) Rchb. f.
O. *guttulatum* Rchb. f.
O. *heteranthum* Poepp. & Endl.
O. *incurvum* Barker ex Lindl.
O. *isthmii* Schltr.
O. *klotzschianum* Rchb. f.
O. *luridum* Lindl.
O. *luteum* Rolfe
O. *maculatum* (Lindl.) Lindl.
O. *nudum* Batem.
O. *obryzatum* Rchb. f.
O. *ochmatochilum* Rchb. f.
O. *ornithorrhynchum* Kunth
O. *paleatum* Schltr.
O. *panamense* Schltr.
O. *panduriforme* Ames & C. Schweinf.
O. *parviflorum* L. O. Williams
O. **peliogramma* Linden & Rchb. f. (CR, Pma)
O. **pergameneum* Lindl. (No, CR)
O. *pittieri* Schltr.
O. *planilabre* Lindl.
O. *powellii* Schltr.
O. *scansor* Rchb. f.
O. *schroederianum* (Rchb. f.) Garay & Stacy
O. *sphacelatum* Lindl.
O. *stenoglossum* (Schltr.) Dressler & N. H. Williams
O. *stenotis* Rchb. f.
O. *stipitatum* Lindl.
O. *storkii* Ames & C. Schweinf.
O. *teres* Ames & C. Schweinf.
O. **tetraskelidion* Kraenzl. (CR, = obryzatum?)
O. **turialbae* Schltr. (CR)
O. *warscewiczii* Rchb. f.
O. **wentworthianum* Lindl. (No, CR)
Ophidion pleurothallopsis (Kraenzl.) Luer
Ornithocephalus bicornis Lindl.
O. *cochleariformis* C. Schweinf.
O. *inflexus* Lindl.
O. *lankesteri* Ames
O. *powellii* Schltr.
O. **valerioi* Ames & C. Schweinf. (CR)
Osmoglossum anceps Schltr.
O. *convallarioides* Schltr.
O. *egertonii* (Lindl.) Schltr.
O. *pulchellum* (Lindl.) Schltr.
Otoglossum chiriquense (Rchb. f.) Garay & Dunst.
Pachyphyllum costaricense (Ames & C. Schweinf.) L. O. Williams
P. *crystallinum* Lindl.
P. *hispidulum* (Rchb. f.) Garay & Dunst.
P. *pastii* Rchb. f.
Palmorchis nitida Dressler
P. *powellii* (Ames) C. Schweinf. & Correll
P. *silvicola* L. O. Williams
P. *trilobulata* L. O. Williams
Paphinia clausula Dressler
P. *cristata* Lindl.
P. *cristata* var. *modiglianiana* Rchb. f.
Pelexia adnata (Sw.) Spreng.
P. *callifera* (C. Schweinf.) Garay
P. *congesta* Ames & C. Schweinf.
P. *funckiana* (A. Rich. & Gal.) Schltr.
P. *laxa* (Poepp. & Endl.) Lindl.
P. *obliqua* (J. J. Sm.) Garay
P. *olivacea* Rolfe
P. *smithii* (Rchb. f.) Garay

Peristeria cochlearis Garay
P. elata Hook.
Pescatorea cerina Rchb. f.
P. dayana Rchb. f.
Phragmipedium caudatum (Lindl.) Rolfe
P. longifolium (Rchb. f. & Warsz.) Rolfe
P. warscewiczianum Rchb. f.
Phymatidium panamense Dressler
Platyglottis coriacea L. O. Williams
Platystele brenneri Luer
P. calymma Luer
P. caudatisepala (C. Schweinf.) Garay
P. compacta (Ames) Ames
P. dressleri Luer
P. jungermannioides (Schltr.) Garay
P. lancilabris (Rchb. f.) Schltr.
P. microtatantha (Schltr.) Garay
P. minimiflora (Schltr.) Garay
P. ortiziana Luer
P. ovalifolia (Focke) Garay & Dunst.
P. ovatilabia (Ames & C. Schweinf.) Garay
P. oxyglossa (Ames & C. Schweinf.) Garay
P. perpusilla (Rchb. f.) Garay
P. propinqua (Ames) Garay
P. resimula Luer & Hirtz
P. stenostachya (Rchb. f.) Garay
P. taylori Luer
Platythelys maculata (Hook.) Garay
P. vaginata (Hook.) Garay
P. venustula (Ames) Garay
Plectrophora alata (Rolfe) Garay
Pleurothallis abbreviata Schltr.
P. aberrans Luer

P. abjecta Ames
P. acicularis Ames & C. Schweinf.
P. *acostaei Schltr. (CR, Pma, 1)
P. acrisepala Ames & C. Schweinf.
P. alexandrae Schltr.
P. alexii Heller
P. allenii L. O. Williams
P. alpina Ames
P. amparoana Schltr.
P. *angusta Ames & C. Schweinf. (CR, 1)
P. annectans Luer
P. antonensis L. O. Williams
P. archicolonae Luer
P. areldii Luer
P. arietina Ames
P. aristata Hook.
P. aryter Luer
P. aurita C. Schweinf.
P. barboselloides Schltr.
P. barbulata Lindl.
P. brighamii S. Wats.
P. broadwayi Ames
P. butcheri L. O. Williams
P. *cachensis Ames (CR, 2)
P. cactantha Luer
P. caligularis Luer
P. calyptrosepala L. O. Williams
P. calyptrostele Schltr.
P. campicola Luer
P. canae Ames
P. caniceps Luer
P. cardiochila L. O. Williams
P. cardiothallis Rchb. f.
P. carnosilabia Heller & A. D. Hawkes
P. carpinterae Schltr.
P. casualis Ames
P. circumplexa Lindl.
P. citrophila Luer
P. cobanensis Schltr.
P. cobraeformis L. O. Williams
P. cogniauxiana Schltr.
P. condylata Luer

P. connata Luer
P. convallaria Schltr.
P. corniculata Lindl.
P. costaricensis Schltr.
P. *crassilabia Ames & C. Schweinf. (CR, 2)
P. crescentilabia Ames
P. crocodiliceps Rchb. f.
P. cucumeris Luer
P. cuspidata Luer
P. decipiens Ames & C. Schweinf.
P. dentipetala Ames
P. deregularis (Barb. Rodr.) Luer
P. divaricans Schltr.
P. dolichopus Schltr.
P. dorotheae Luer
P. dracontea Luer
P. dressleri Luer
P. ellipsophylla L. O. Williams
P. endotrachys Rchb. f.
P. erinacea Rchb. f.
P. eumecocaulon Schltr.
P. excavata Schltr.
P. exesilabia Heller & A. D. Hawkes
P. falcatiloba Ames
P. fantastica Ames
P. floribunda Poepp. & Endl.
P. fortunae Luer
P. fractiflexa Ames & C. Schweinf.
P. fuegii Rchb. f.
P. fulgens Rchb. f.
P. gacayana Schltr.
P. gelida Lindl.
P. geminicaulina Ames
P. geminiflora Ames
P. ghiesbreghtiana A. Rich. & Gal.
P. glandulosa Ames
P. grandis Rolfe
P. grobyi Lindl.
P. guanacastensis Ames & C. Schweinf.
P. guttata Luer
P. harpago Luer

P. hastata Ames
P. helleri A. D. Hawkes
P. hemileuca Luer
P. herpestes Luer
P. homalantha Schltr.
P. homalanthoides Schltr.
P. imago Luer
P. immersa Rchb. f.
P. imraei Lindl.
P. instar Luer
P. isthmica Luer
P. janetiae Luer
P. johnsonii Ames
P. juxtaposita Luer
P. lanceana Lodd.
P. lanceola (Sw.) Spreng.
P. lateritia Rchb. f.
P. lentiginosa Lehm. & Kraenzl.
P. lepidota L. O. Williams
P. leucantha Schltr.
P. *leucopyramis Rchb. f. (CR)
P. lewisae Ames
P. listerophora Schltr.
P. longipedicellata Ames & C. Schweinf.
P. loranthophylla Rchb. f.
P. luctuosa Rchb. f.
P. macrantha L. O. Williams
P. mammillata Luer
P. microphylla A. Rich. & Gal.
P. *minor (Rendle) L. O. Williams (CR, 2)
P. minuta Ames & C. Schweinf.
P. *monocardia Rchb. f. (Pma, 1)
P. mystax Luer
P. *naraniensis Rchb. f. (CR, 2)
P. nemorum Schltr.
P. nervosa Braid
P. nitida Luer
P. obovata Lindl.
P. oscitans Ames
P. pacayana Schltr.
P. pachyglossa Lindl.
P. pallida Luer

P. *palliolata* Ames
P. *pantasmi* Rchb. f.
P. *peculiaris* Luer
P. *peperomioides* Ames
P. *periodica* Ames
P. *pfavii* Rchb. f.
P. *phyllocardia* Rchb. f.
P. *phyllocardioides* Schltr.
P. *picta* Hook.
P. **platycardia* Rchb. f. (CR, 1)
P. **platysepala* Schltr. (Pma, 1)
P. *polygonoides* Griseb.
P. *polysticta* Luer
P. *pompalis* Ames
P. *powellii* Schltr.
P. *praegrandis* Ames
P. *pruinosa* Lindl.
P. *pubescens* Lindl.
P. *pyrsodes* Rchb. f.
P. *quadrifida* (La Llave & Lex.) Lindl.
P. *quinqueseta* Ames
P. *racemiflora* Lodd.
P. *radula* Luer
P. *ramonensis* Schltr.
P. **rectipetala* Ames & C. Schweinf. (CR, 1)
P. *rhodoglossa* Schltr.
P. *rowleei* Ames
P. *rubella* Luer
P. *ruscifolia* (Jacq.) R. Br.
P. *saccata* Ames
P. *saccatilabia* C. Schweinf.
P. *samacensis* Ames
P. *sanchoi* Ames
P. *scitula* Luer
P. *sclerophylla* Lindl.
P. *segoviensis* Rchb. f.
P. *segregatifolia* Ames & C. Schweinf.
P. *sertularioides* Spreng.
P. *setigera* Lindl.
P. *setosa* C. Schweinf.
P. *sicaria* Lindl.

P. **simplex* Ames & C. Schweinf. (CR, 2)
P. *simulans* L. O. Williams
P. *stevensii* Luer
P. *strumosa* Ames
P. *telamon* Luer
P. *testaefolia* (Sw.) Lindl.
P. *thymochila* Luer
P. *titan* Luer
P. *tonduzii* Schltr.
P. *tribuloides* (Sw.) Lindl.
P. *tripterantha* Rchb. f.
P. *tuerckheimii* Schltr.
P. *turrialbae* Luer
P. *undulata* Poepp. & Endl.
P. *uniflora* Lindl.
P. *ventricosa* Lindl.
P. *veraguacensis* Luer
P. *verecunda* Schltr.
P. *vittariifolia* Schltr.
P. *volcanica* Luer
P. *wercklei* Schltr.
Polycycnis barbata (Lindl.) Rchb. f.
P. *gratiosa* Endres & Rchb. f.
P. *lehmannii* Rolfe
P. *muscifera* (Lindl. & Paxton) Rchb. f.
P. *ornata* Garay
P. *tortuosa* Dressler
Polystachya cerea Lindl.
P. **cingulata* Rchb. f. (CR)
P. *concreta* (Jacq.) Garay & Sweet
P. *flavescens* (Bl.) J. J. Sm.
P. *foliosa* (Hook.) Rchb. f.
P. *luteola* (Sw.) Hook.
P. *masayensis* Rchb. f.
P. *minuta* (Aubl.) Cordem.
Ponera striata Lindl.
Ponthieva brenesii Schltr.
P. *ephippium* Rchb. f.
P. *formosa* Schltr.
P. *maculata* Lindl.
P. *racemosa* (Walt.) Mohr

P. tuerckheimii Schltr.
Prescottia cordifolia Rchb. f.
P. oligantha (Sw.) Lindl.
P. stachyodes (Sw.) Lindl.
Pseudocentrum hoffmannii (Rchb. f.) Rchb. f.
Psilochilus carinatus Garay
P. macrophyllus (Lindl.) Ames
P. modestus Barb. Rodr.
P. physurifolius (Rchb. f.) Løjtnant
Psychopsis krameriana (Rchb. f.) H. G. Jones
Psygmorchis glossomystax (Rchb. f.) Dodson & Dressler
P. gnoma (Kraenzl.) Dodson & Dressler
P. pumilio (Rchb. f.) Dodson & Dressler
P. pusilla (L.) Dodson & Dressler
Pterichis costaricensis Schltr.
P. habenarioides (Lehm. & Kraenzl.) Schltr.
P. leo L. D. Gómez & J. Gómez
Reichenbachanthus cuniculatus (Schltr.) Pabst
R. lankesteri (Ames) Mora-Retana & García
R. reflexus (Lindl.) Brade
Restrepia lankesteri Ames & C. Schweinf.
R. muscifera (Rchb. f.) Lindl.
R. subserrata Schltr.
R. xanthophthalma Rchb. f.
Restrepiella ophiocephala (Lindl.) Garay & Dunst.
Restrepiopsis reichenbachiana (Endres) Luer
R. tubulosa (Lindl.) Luer
R. ujarrensis (Rchb. f.) Luer
Rodriguezia compacta Schltr.
R. lanceolata Ruiz & Pavon
R. secunda Kunth
Rossioglossum schlieperianum (Rchb. f.) Garay & Kennedy

Salpistele brunnea Dressler
S. dressleri Luer
S. lutea Dressler
S. parvula Luer
Sarcoglottis hunteriana Schltr.
S. neglecta Christenson
S. valida Ames
S. woodsonii (L. O. Williams) Garay
Scaphosepalum anchoriferum (Rchb. f.) Rolfe
S. clavellatum Luer
S. microdactylum Rolfe
S. viviparum Luer
Scaphyglottis acostaei (Schltr.) C. Schweinf.
S. amethystina Schltr.
S. amparoana (Schltr.) Dressler
S. arctata (Dressler) B. R. Adams
S. behrii Hemsl.
S. bifida C. Schweinf.
S. bilineata Schltr.
S. boliviensis (Rolfe) B. R. Adams
S. chlorantha B. R. Adams
S. confusa (Schltr.) Ames & Correll
S. corallorhiza (Ames) Ames, Hubb. & C. Schweinf.
S. crurigera (Lindl.) Ames & Correll
S. cuneata Schltr.
S. densa (Schltr.) B. R. Adams
S. fusiformis R. E. Schult.
S. geminata Dressler & Mora-Retana
S. gigantea Dressler
S. gracilis Schltr.
S. huebneri Schltr.
S. jimenezii Schltr.
S. laevilabia Ames
S. leucantha Rchb. f.
S. limonensis B. R. Adams
S. lindeniana (A. Rich. & Gal.) L. O. Williams
S. longicaulis S. Wats.

S. mesocopis Hemsl.
S. micrantha (Lindl.) Ames & Correll
S. minuta (A. Rich. & Gal.) Garay
S. minutiflora Ames & Correll
S. panamensis B. R. Adams
S. prolifera Cogn.
S. pulchella (Schltr.) L. O. Williams
S. punctulata (Rchb. f.) C. Schweinf.
S. robusta B. R. Adams
S. sessiliflora B. R. Adams
S. sigmoidea (Ames & C. Schweinf.) B. R. Adams
S. spathulata C. Schweinf.
S. stellata Lindl.
S. subulata Schltr.
S. tenella L. O. Williams
Scelochilus tuerckheimii Schltr.
Schiedeella fauci-sanguinea Dod
S. llaveana (Lindl.) Schltr.
S. trilineata (Lindl.) Balogh
S. valerioi (Ames & C. Schweinf.) Szlach.
S. wercklei (Schltr.) Garay
Schlimia jasminodora Planchon & Linden ex Lindl. & Paxton
Schomburgkia lueddemannii Prill.
S. undulata Lindl.
Selenipedium chica Rchb. f.
Sepalosaccus strumatus (Endres & Rchb. f.) Garay
Sievekingia butcheri Dressler
S. fimbriata Rchb. f.
S. suavis Rchb. f.
Sigmatostalix abortiva L. O. Williams
S. brownii Garay
S. guatemalensis Schltr.
S. hymenantha Schltr.
S. macrobulbon Kraenzl.
S. picta Rchb. f.
S. picturatissima Kraenzl.
S. racemifera L. O. Williams

S. unguiculata C. Schweinf.
Sobralia allenii L. O. Williams
S. amabilis (Rchb. f.) L. O. Williams
S. atropubescens Ames & C. Schweinf.
*S. *bletiae* Rchb. f. (Pma)
S. bouchei Ames & C. Schweinf.
S. bradeorum Schltr.
S. callosa L. O. Williams
S. corazoi Lank. & Ames
S. decora Batem.
S. decora var. *aerata* P. H. Allen & L. O. Williams
S. fenzliana Rchb. f.
S. fimbriata Poepp. & Endl.
S. fragrans Lindl.
S. helleri A. D. Hawkes
*S. *labiata* Warsz. & Rchb. f. (Pma)
S. lepida Rchb. f.
S. lindleyana Rchb. f.
S. luteola Rolfe
S. macra Schltr.
S. macrantha Lindl.
S. macrophylla Rchb. f.
S. mucronata Ames & C. Schweinf.
S. neglecta Schltr.
S. pfavii Schltr.
S. pleiantha Schltr.
S. powellii Schltr.
S. suaveolens Rchb. f.
S. undatocarinata C. Schweinf.
S. valida Rolfe
S. warscewiczii Rchb. f.
S. wercklei (Schltr.) L. O. Williams
Solenocentrum costaricense Schltr.
Spathoglottis plicata Bl.
Sphyrastylis cryptantha (C. Schweinf. & P. H. Allen) Garay
Spiranthes amesiana Schltr.
S. graminea Lindl.
S. pubicaulis L. O. Williams
S. torta (Thunb.) Garay & Sweet

Stanhopea avicula Dressler
S. cirrhata Lindl.
S. costaricensis Rchb. f.
S. ecornuta Lem.
S. gibbosa Rchb. f.
S. oculata (Lodd.) Lindl.
S. panamensis N. H. Williams & Whitten
S. platyceras Rchb. f.
S. pulla Rchb. f.
S. wardii Lodd. ex Lindl.
S. warscewicziana Klotzsch
Stelis *aemula Schltr.
S. *allenii L. O. Williams
S. *aprica Lindl.
S. *argentata Lindl.
S. *barbata Rolfe
S. *bidentata Schltr.
S. *carnosiflora Ames & C. Schweinf.
S. *catharinensis Lindl.
S. *chihobensis Ames
S. *ciliaris Lindl.
S. *cleistogama Schltr.
S. *conmixta Schltr.
S. *cooperi Schltr.
S. *costaricensis Rchb. f.
S. *crescentiicola Schltr.
S. *crystallina Ames
S. *cucullata Ames
S. *cuspidata Ames
S. *cyclopetala Ames
S. *despectans Schltr.
S. *effusa Schltr.
S. *fimbriata R. K. Baker
S. *glossula Rchb. f.
S. *gracilifolia C. Schweinf.
S. *gracilis Ames
S. *guatemalensis Schltr.
S. *inaequalis Ames
S. *jimenezii Schltr.
S. *lankesteri Ames
S. *latipetala Ames
S. *leucopogon Rchb. f.

S. *longipetiolata Ames
S. *maxima Lindl.
S. *maxonii Schltr.
S. *micragrostis Schltr.
S. *microchila Schltr.
S. *minutiflora Hoffsgg.
S. *morganii Dodson & Garay
S. *nubis Ames
S. *ovatilabia Schltr.
S. *panamensis Schltr.
S. *pardipes Rchb. f.
S. *parvula Lindl.
S. *pendulispica Ames
S. *persimilis Ames
S. *planipetala Ames
S. *poasensis Ames
S. *powellii Schltr.
S. *propinqua Ames
S. *purpurascens A. Rich. & Gal.
S. *rowleei Ames
S. *sanchoi Ames
S. *skutchii Ames
S. *spathulata Poepp. & Endl.
S. *standleyi Ames
S. *storkii Ames
S. *superbiens Lindl.
S. *thecoglossa Rchb. f.
S. *tonduziana Schltr.
S. *transversalis Ames
S. *triangulabia Ames
S. *tricuspis Schltr.
S. *tridentata Lindl.
S. *vestita Ames
S. *vulcanica Schltr.
S. *wercklei Schltr.
S. *williamsii Ames
Stellilabium boylei Atwood
S. bullpenense Atwood
S. campbellorum Atwood
S. distantiflorum Ames
S. lankesteri (Ames) L. O. Williams
S. minutiflorum (Kraenzl.) L. O. Williams
S. monteverdense Atwood

S. standleyi (Ames) L. O. Williams
Stenorrhynchos lanceolatum
  (Aubl.) Spreng.
S. speciosum (Jacq.) Spreng.
Systeloglossum acuminatum Ames
  & C. Schweinf.
S. costaricense Schltr.
S. panamense Dressler & N. H.
  Williams
Telipogon ampliflorus C. Schweinf.
T. ardeltianus Braas
T. ballesteroi Dodson & Escobar
T. biolleyi Schltr.
T. caroliae Dodson & Escobar
T. cascajalensis Dodson & Escobar
T. christobalensis Kraenzl.
T. costaricensis Schltr.
T. *dendriticus Rchb. f. (Pma).
T. elcimeyae Braas & Horich
T. endresianum Kraenzl
T. glicensteinii Dodson & Escobar
T. gracilipes Schltr.
T. guila Dodson & Escobar
T. leila-alexandrae Braas
T. monticola L. O. Williams
T. parvulus C. Schweinf.
T. pfavii Schltr.
T. portilloi Dodson & Escobar
T. *radiatus Rchb. f. (Pma)
T. retanarum Dodson & Escobar
T. setosus Ames
T. storkii Ames & C. Schweinf.
T. vampirus Braas & Horich
Teuscheria horichiana Jenny &
  Braem
T. pickiana (Schltr.) Garay
Ticoglossum krameri (Rchb. f.)
  Halbinger
T. oerstedii (Rchb. f.) Halbinger
Trevoria glumacea Garay
T. zahlbruckneriana (Schltr.)
  Garay
Trichocentrum brenesii Schltr.
T. caloceras Rchb. f.

T. candidum Lindl.
T. capistratum Rchb. f.
T. pfavii Rchb. f.
Trichopilia leucoxantha L. O. Williams
T. maculata Rchb. f.
T. marginata Henfr.
T. *punctata Rolfe (CR)
T. suavis Lindl.
T. tortilis Lindl.
T. turialbae Rchb. f.
Trichosalpinx arbuscula Lindl.
T. blaisdellii (S. Wats.) Luer
T. carinilabia (Luer) Luer
T. cedralensis (Luer) Luer
T. ciliaris (Lindl.) Luer
T. dura (Lindl.) Luer
T. foliata (Griseb.) Luer
T. intricata (Lindl.) Luer
T. membraniflora (C. Schweinf.)
  Luer
T. memor (Rchb. f.) Luer
T. moschata (Rchb. f.) Luer
T. navarrensis (Ames) Mora-
  Retana & García
T. orbicularis (Lindl.) Luer
T. operculata Luer
T. pergrata (Ames) Luer
T. rotundata (C. Schweinf.) Luer
T. tantilla Luer
T. tropida Luer
Trigonidium egertonianum Lindl.
T. insigne Rchb. f. ex Benth. &
  Hook.
T. lankesteri Ames
T. riopalenquense Dodson
Triphora gentianoides (Sw.) Ames
  & C. Schweinf.
T. mexicana (S. Wats.) Schltr.
T. nitida Schltr.
T. ravenii (L. O. Williams) Garay
T. wagneri Schltr.
Trisetella dressleri (Luer) Luer
T. tenuissima (C. Schweinf.) Luer

T. triaristella (Rchb. f.) Luer
T. triglochin (Rchb. f.) Luer
Trizeuxis falcata Lindl.
Tropidia polystachya (Sw.) Ames
Uleiorchis ulaei (Cogn.) Handro
Vanilla inodora Schiede
V. insignis Ames
V. mexicana Miller
V. odorata Presl
V. pauciflora Dressler
V. pfaviana Rchb. f.
V. planifolia C. Jackson
V. pompona Schiede
Warmingia margaritacea Johansen
Warrea costaricensis Schltr.
Warreopsis parviflora (L. O. Williams) Garay
Wullschlaegelia aphylla (Sw.) Rchb. f.
W. calcarata Benth.
Xylobium colleyi (Batem. & Lindl.) Rolfe
X. elongatum (Lindl.) Hemsl.
X. foveatum (Lindl.) Nichols.
X. powellii Schltr.
X. squalens Lindl.
Zootrophion atropurpureus (Lindl.) Luer
Z. endresianus (Kraenzl.) Luer
Z. gracilentus (Rchb. f.) Luer
Z. hypodiscus (Rchb. f.) Luer
Z. moorei (Rolfe) Luer
Z. vulturiceps (Luer) Luer

Glossary

Acuminate Ending in a long, narrow point, this distinctly narrower than the blade (see Fig. 6M).

Acute Sharp, ending in an acute angle (see Fig. 6K).

Anther The part of the flower that produces and contains pollen.

Apex Tip or distal end.

Apicule A short, sharp point at the tip of any structure; apiculate, having an apicule (see Fig. 6L).

Appressed Lying on the ground or substrate; pressed close to.

Arched Curved, with the convex side upward.

Attenuate Gradually tapering, with the sides shallowly concave (see Fig. 6O).

Axil The angle between a leaf (or bract) and a stem; a bud is normally formed there; axillary, positioned in an axil.

Basal Near the base or point of attachment.

Bilobed Divided into two lobes.

Blade The flat portion of any foliar organ, as in leaf or lip.

Bowed Curved, with the concave side upward.

Bract A scale- or sheath-like structure corresponding to a leaf but without a blade.

Callus (pl. calli) A crest or fleshy outgrowth of the lip (see Fig. 2).

Caudicle A slender, mealy, or elastic extension of the pollinium, or a mealy portion at one end of the pollinium; this structure is part of the pollen mass and is produced within the anther.

Chin A chinlike projection at the base of a flower, consisting of the column foot with the bases of the lateral sepals; mentum (see Fig. 2).

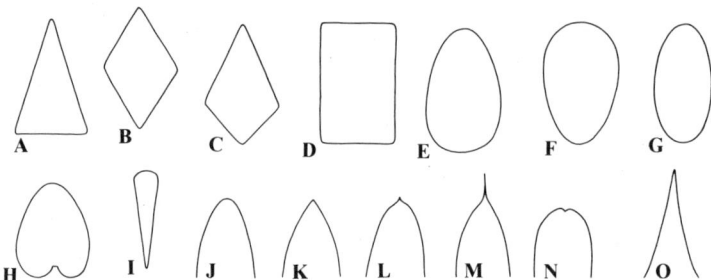

Figure 6. Outlines showing some of the shapes described by botanical terms. The lower side is the base in all cases. A–G, terms describing outlines. H and I, terms describing either the base or the outline. J–O, terms describing apices. A, triangular. B, rhombic. C, trullate. D, quadrate. E, ovate. F, obovate. G, elliptic. H, cordate. I, cuneate. J, obtuse. K, acute. L, apiculate. M, acuminate. N, retuse. O, attenuate.

Clinandrium The anther bed or portion of the column on which the anther is borne; also refers to a collar or hood of columnar tissue surrounding or covering the anther.

Column The central structure in an orchid flower, comprising the style united with the filaments of one or more anthers (see Fig. 1).

Column foot A ventral extension at the base of the column; the lip is normally attached at its tip (see Fig. 2).

Column wing A winglike or armlike appendage of the column, usually lateral (see Fig. 1).

Conduplicate Of leaflike organs, with a single median fold, each half being flat.

Connivent Refers to structures that are parallel and close together without being united.

Convolute Refers to leaves that are rolled during their development.

Coralloid Thick and irregularly branched, coral-like.

Cordate Heart-shaped, with the base at the broad, notched end; may be used to refer either to a deeply notched base or to the shape of the whole leaf (see Fig. 6H). *Heart* is used here in the sense of the symmetrical valentine heart.

Corm A short, thick, underground stem, usually of several internodes, as in *Gladiolus* and *Bletia*.

Crenate Having small, rounded teeth on the margin.

Cuneate Wedge-shaped, tapering evenly to the base (see Fig. 6I).

Dentate Toothed, with toothlike projections along the margin.

Distal Toward the tip or away from the base. One commonly uses *above* in this sense, but this becomes unclear when the plant in question is hanging downward. Using *basal* and *distal* avoids any possible confusion.

Dorsal Refers to the upper side of the flower; in orchid flowers the side away from the lip, even when the lip is uppermost.

Duplicate Refers to leaves with a single median fold during their development.

Ellipsoid Spindle-shaped; narrow and tapering at the ends, three-dimensional.

Elliptic Spindle-shaped, two-dimensional (see Fig. 6G).

Entire Undivided, not fringed, notched, or lobed.

Ephemeral Short-lived; of flowers, lasting only a few hours.

Epichile The terminal portion of a complex lip, as in *Stanhopea*.

Epiphyte Any plant that grows upon another plant.

Erose A minutely jagged edge, as though chewed.

Falcate Sickle-shaped; a flat organ that is curved in the plane in which it is flattened.

Genus (pl. genera) A taxonomic category above species; the generic name forms the first part of the binomial species name.

Globose Spherical, ball-like.

Herbaceous Not woody; applies especially to herbs whose shoots die back each year. *Soft-herbaceous* refers to thin leaves that are convolute in development but not plicate (typical of terrestrial orchids whose leaves die back each year).

Herbarium A botanical museum; a collection of pressed and dried plant specimens.

Hypochile The basal part of a complex lip, as in *Stanhopea*.

Inflorescence A flower cluster or (if solitary) a flower.

Internode The section of a stem between two nodes.

Isthmus A narrow portion of a lip or petal.

Laminar Thin and flattened, bladelike.

Lance A contraction of *lanceolate* used in combination with other shape terms.

Lanceolate Lance-shaped, widest near the base and tapering to a narrow tip. In the past this term was also used to mean 'narrowly elliptic,' so some authors avoid the term altogether, using instead *narrowly ovate*.

Lateral To either side of a vertical line drawn in the center of a bilaterally symmetrical structure.

Leaves Normal foliage leaves of any plant, excluding bracts and other leaflike structures.

Ligule A small, straplike appendage; in the orchids it refers to the appendage borne under the stigma of *Dichaea*.

Lip One of the three petals that is usually larger and different in shape from the other two; the median petal (see Fig. 1).

Lithophyte A plant that grows on rocks.

Lobule A small lobe or projection.

Lunate Crescent-shaped, half-moon-shaped; *lunulate* is the diminutive form.

Median Referring to a middle portion or middle position.

Mesochile The middle portion of a complex lip, as in *Stanhopea*.

Midlobe A middle lobe, referring especially to orchid lips, which commonly have three lobes, with the midlobe usually larger (see Fig. 1).

Monopodial A growth habit in which the stem may continue to grow indefinitely, as opposed to sympodial.

Mycorrhiza A symbiotic relationship between vascular plant roots and fungi.

Myrmecophyte Any plant associated with ants; the plant may grow in an ant nest or it may provide a structure inhabited by ants.

Nectary Any nectar-producing structure or gland.

Node The point on a stem at which a leaf or bract is attached.

Oblique Asymmetrical, with the sides unequal.

Oblong Longer than wide, with the sides nearly or quite parallel.

Obovate Egg-shaped, with the wider end at the tip, two-dimensional (see Fig. 6F).

Obtuse Blunt, rounded at the free end.

Ovary The part of a flower that develops into the fruit.

Ovate Egg-shaped in outline with the wide end basal, two-dimensional (see Fig. 6E).

Ovoid Egg-shaped, three-dimensional.

Papillose Refers to a surface with small, nipple-like bumps.

Pedicel The stem that supports an individual flower, usually jointed at the base, above the floral bract.

Peduncle The stem that supports a solitary flower or an inflorescence.

Petal In orchids two of the inner perianth parts; the other is the lip (see Fig. 1).

Petiole The narrow, stemlike basal portion of a leaf.

Pleated Refers to leaves having several or many major longitudinal veins and folded or creased at each one; plicate.

Pollinium (pl. pollinia) A more or less compact and coherent mass of pollen; usually the contents of an anther cell, or of one-half of an anther cell.

Procumbent Lying down on the soil or substrate, not erect.

Pseudobulb A thickened stem, usually aerial.

Quadrate Four-angled, square or rectangular (see Fig. 6D).

Raceme An unbranched inflorescence in which the pedicellate flowers normally open from the base upward.

Rachis The axis of an inflorescence to which the pedicels are attached, above the peduncle.

Reed-stem Having stems that are long and relatively slender, commonly with leaves scattered along the stem.

Reflexed Turned back or curled back.

Resupinate The flower having the lip on the lower side.

Retrorse Turned backward or downward toward the base.

Retuse Having a notch in a broad apex (see Fig. 6N).

Rhizome A horizontal stem, usually on or in the substrate, made up of the bases of successive shoots.

Rhombic Diamond-shaped (see Fig. 6B).

Rosette A densely clustered spiral of leaves, usually borne near the ground.

Rostellar remnant The portion of the rostellum that remains on the column after the pollinia and associated structures have been removed; especially with reference to the *Spiranthes* complex.

Rostellum A portion of the stigma that aids in gluing the pollinia to the pollinating agent; the tissue that separates the anther from the fertile stigma; it is sometimes beaklike.

Rugose With raised lines or wrinkles.

Saccate Saclike, deeply concave.

Scabrous Rough to the touch.

Scurfy Refers to a surface with small, loosely attached scales.

Sectile The condition in which soft, granular pollinia are subdivided into small packets usually connected by elastic threads, as in *Habenaria*.

Secund Having the flowers in an inflorescence all face to one side, usually by twisting.

Sepal In orchids, any of the outer three segments of the flower (see Fig. 1).

Sheath A leaflike structure that enfolds a stem, pseudobulb, or young inflorescence.

Shoot A growth, especially a new growth arising from the base of an older one.

Sinus The gap between two lobes.

Spathe A folded, leaflike sheath that protects an inflorescence bud. When the inflorescence grows, it breaks through the tip of the spathe.

Spatulate Spoon-shaped, having a narrow base and an abruptly wider tip.

Species A population or group of populations that share a common pool of genetic and structural features; the basic unit in biological classification.

Spur A slender tubular or saclike projection from a flower part, usually a nectary, commonly formed by the base of the lip (see Fig. 2).

Staminode, staminodium A sterile stamen or anther; in the orchids it may form a column wing.

Stem The axis of a shoot, especially the axis of the vegetative shoot. Technically, the rhizome and the peduncle and rachis of the inflorescence are all stems, but in description the term *stem* does not normally include structures for which other special terms are available.

Stigma The sticky receptive part of the pistil or column; it produces a viscid, sugary material that receives the pollinia and permits the pollen grains to germinate.

Sub Prefix meaning about or almost.

Sympodial A modular habit of growth in which each shoot has limited growth, new shoots usually arising from the bases of older ones; most orchids have this growth habit.

Synsepal A compound organ formed by the union of the two lateral sepals.

Taxon (pl. taxa) A taxonomic group of any rank (family, genus, species, etc.).

Taxonomy The science of classification and naming.

Terminal At the tip or apex.

Throat The tubular portion of the lip, as in *Cattleya*.

Trullate Trowel-shaped, referring to a bricklayer's trowel (see Fig. 6C).

Truncate Having the apex nearly square, as though cut off.

Umbel A cluster of flowers whose pedicels are densely clustered so that the cluster of flowers is more or less flat-topped or rounded; umbellate.

Ventral On the lower side; in orchids refers to the lip side of the flower, regardless of position.

Viscidium (pl. viscidia) A viscid part of the rostellum that is clearly defined and removed with the pollinia as a unit; it serves to attach the pollinia to an insect or other agent.

Illustration Credits

The author gratefully acknowledges the use of illustrations from the following individuals.

Calaway H. Dodson: *Oerstedella pinnifera, Otoglossum chiriquense, Stanhopea costaricensis, S. wardii*
Kerry Ann Dressler: All color photographs not listed for other contributors, except *Coccineorchis bracteosa, Pleurothallis geminicaulina,* and *Ponera striata* (supplied by R. L. Dressler)
E. W. Greenwood: *Arpophyllum giganteum, Beloglottis costaricensis, Deiregyne hemichrea, Schiedeella llaveana*
Eric Hágsater: *Coelia macrostachya*
Helen A. Kennedy: *Epidendrum radicans*
Lynn S. Kimsey: *Sobralia callosa*
Carlyle A. Luer: *Tropidia polystachya*
Kiat W. Tan: *Monophyllorchis maculata*
Kerry S. Walter: *Cochleanthes aromatica, Encyclia cordigera* (Costa Rican)
Norris H. Williams: *Cischweinfia dasyandra, Systeloglossum panamense*

Index

Color plates of the orchids follow page 326. Abbreviations and symbols used in the keys are on pages 37 and 38.

Aa paleacea, 278, 289
Acineta, 30, 141, 143, 144–145
   chrysantha, 145, Pl. 16(1)
   erythroxantha, 331
   gymnostele, 331
   sella-turcica, 145
   superba, 144
Acostaea costaricensis, 30, 159, 162, Pl. 20(1)
Acrorchis roseola, 47, 51, Pl. 7(1)
Ada, 106, 108
   allenii, 108, Pl. 11(5)
   chlorops, 108
Aglae, 28
Amparoa costaricensis, 106, 108
Arpophyllum, 44, 46, 265, 266
   alpinum, 266
   giganteum, 266, Pl. 31(1)
   medium, 266
Arrocillo, 69
Arundina graminifolia, 282, 290, Pl. 34(6)
Aspasia, 7, 105, 108–109
   epidendroides, 109
   principissa, 108, Pl. 11(3)
Aspidogyne, 278, 284, 290
   sp., 290, Pl. 39(1)

   stictophylla, 290
   tuerckheimii, 290

Barbosella, 160, 162–163
   anaristella, 163, Pl. 20(2)
   circinata, 162
   orbicularis, 157, 162
   prorepens, 163
Barkeria, 47, 48, 51
   chinensis, 51
   lindleyana, 51, Pl. 7(2)
   obovata, 51
   skinneri, 51
Baskervilla, 39, 277, 288, 290–291
   sp., 291, Pl. 40(1)
   colombiana, 291
   nicaraguensis, 291
Beloglottis, 278, 285, 291–292
   bicaudata, 292
   costaricensis, 292, Pl. 38(2)
   ecallosa, 291
   hameri, 292
   subpandurata, 291
Bifrenaria picta, 218, 223, Pl. 25(1)
Bletia, 19, 264, 266–267
   campanulata, 267
   edwardsii, 267

   purpurata, 266
   purpurea, 6, 266, Pl. 31(2)
Bothriochilus macrostachyus, 269
Brachionidium, 10, 158, 163
   cruzae, 163
   folsomii, 163, Pl. 20(3)
   kuhniarum, 163
   pusillum, 163
   valerioi, 163
Brachystele guayanensis, 278, 285, 292
Brassavola, 46, 49, 51–52
   acaulis, 41, 52, Pl. 3(5)
   cucullata, 52
   nodosa, 6, 52, Pl. 3(6)
   venosa, 52
Brassia, 7, 106, 109
   arcuigera, 109, Pl. 11(6)
   caudata, 109
   gireoudiana, 109
   longissima, 109
   verrucosa, 109
Brenesia costaricensis, 189, 196
Bulbophyllum, 28, 265, 267–268
   aristatum, 267
   oerstedii, 267
   pachyrachis, 267

Bulbophyllum (cont.)
  vinosum, 268
  wagneri, 267

Calanthe calanthoides, 264, 268, Pl. 31(3)
Campylocentrum, 20, 43, 279, 282, 292–294
  brenesii, 293
  dressleri, 294
  fasciola, 294, Pl. 39(6)
  longicalcaratum, 293
  micranthum, 293, Pl. 39(5)
  pachyrrhizum, 294
  parvulum, 293
  poeppigii, 293
  porrectum, 294
  schiedei, 293
  schneeanum, 293
  tenellum, 293
  tyrridion, 294
Catasetinae, 133–141
Catasetum, 25, 26, 29, 30, 31, 133, 135, 136
  bicolor, 136
  maculatum, 6, 136, Pl. 15(1)
  oerstedii, 136
  viridiflavum, 136
Cattleya, 28, 50, 52–54
  aurantiaca, 53, 54
  deckeri, 54
  dowiana, 53, Pl. 1(1)
  patinii, 7, 53–54, Pl. 1(2)
  skinneri, 7, 53, 54, Pl. 1(3)
Caularthron bilamellatum, 49, 54
Centropetalum costaricensis, 110
Chaubardiella, 30, 217, 222, 223
  chasmatochila, 223
  pacuarensis, 223, Pl. 26(1)
  subquadrata, 223
Chondrorhyncha, 29, 217, 222, 223–224
  albicans, 224
  anatona, 224, 225, Pl. 26(4)
  bicolor, 224
  crassa, 224
  eburnea, 225
  lendyana, 224
  reichenbachiana, 224, Pl. 26(3)
Chrysocycnis tigrina, 249
Chysis, 263, 268
  aurea var. *maculata*, 268

  costaricensis, 268, Pl. 31(4)
  tricostata, 268
Cischweinfia, 104, 109–110
  dasyandra, 110, Pl. 13(1)
  pusilla, 109
CITES, 53, 329
Classification, 33–36
Cleistes, 28, 282, 294–295
  costaricensis, 294
  rosea, 295, Pl. 33(6)
Clowesia, 133, 135, 136–137
  rosea, 137
  russelliana, 137
  warscewiczii, 30, 135, 136, Pl. 15(2)
Coccineorchis, 278, 286, 295
  bracteosa, 295, Pl. 37(4)
  cernua, 295
  navarrensis, 295
  standleyi, 295
Cochleanthes, 7, 28, 217, 222, 225
  aromatica, 225, Pl. 26(5)
  discolor, 225, Pl. 26(6)
  lipscombiae, 28, 225
  picta, 225
Cockleshell orchids, 57
Coelia macrostachya, 265, 269, Pl. 31(5)
Coeliopsis hyacinthosma, 142, 145, Pl. 16(2)
Comparettia falcata, 103, 110, Pl. 13(3)
Coryanthes, 29, 30, 143, 145–146
  horichiana, 146
  hunteriana, 146
  maculata, 146, Pl. 17(3)
  powellii, 146
  speciosa, 145
Corymborkis, 279, 281, 295–296
  flava, 296, Pl. 36(3)
  forcipigera, 296
Costa Rica
  climate of, 3–5
  geography of, 1–3
  orchid study in, 13–14
  people and culture of, 11–13
  vegetation of, 5–11
Cranichidinae, 277
Cranichis, 277, 289, 296–297
  acuminatissima, 297
  ciliata, 296
  diphylla, 297

  lankesteri, 296, Pl. 40(2)
  muscosa, 297
  nigrescens, 333
  pittieri, 333
  reticulata, 296
  saccata, 297
  sylvatica, 297
  wageneri, 296
Crybe rosea, 266
Cryptarrhena, 219, 226
  guatemalensis, 226, Pl. 30(6)
  lunata, 226
  quadricornu, 226
Cryptocentrum, 221, 226–227
  calcaratum, 226
  gracilipes, 227
  gracillimum, 227
  inaequisepalum, 227
  latifolium, 226
  standleyi, 226
Cryptophoranthus lepidotus, 216
Cyclopogon, 278, 287, 297–298
  cranichoides, 298
  elatus, 298, Pl. 38(3)
  millei, 297
  miradorensis, 298
  olivaceus, 298
  prasophyllum, 297
Cycnoches, 30, 31, 135, 137–138
  amparoanum, 333
  aureum, 138
  chlorochilon, 137
  dianae, 138, Pl. 15(3)
  egertonianum, 138
  guttulatum, 138
  pachydactylon, 138
  stenodactylon, 138
  warscewiczii, 137, Pl. 15(4)
Cypripedioideae, 277
Cyrtopodium punctatum, 25, 26, 44, 263, 269, Pl. 32(1)

Deiregyne hemichrea, 278, 284, 298, Pl. 38(1)
Dichaea, 29, 221, 227–230
  acroblephara, 230
  amparoana, 230
  brachypoda, 229
  costaricensis, 229
  cryptarrhena, 229
  dammeriana, 228
  eligulata, 229

Dichaea (cont.)
glauca, 230
gracillima, 334
graminoides, 230
hystricina, 227
lankesteri, 230
morrisii, 227, 229
muricata, 227
obovatipetala, 228
oxyglossa, 228
panamensis, 230, Pl. 30(1)
pendula, 228
poicillantha, 229
retroflexiligula, 229
sarapiquiensis, 228
schlechteri, 229
squarrosa, 228
standleyi, 230
tenuifolia, 227
trichocarpa, 228
trulla, 229, Pl. 30(2)
tuberculilabris, 229
tuerckheimii, 227
violacea, 229
Dimerandra, 48, 54–55
elegans, 54, Pl. 7(3)
emarginata, 55
latipetala, 54
Discyphus scopulariae, 278, 286, 298
Dove orchid, 150–151
Dracula, 160, 164
erythrochaete, 164, Pl. 21(1)
pusilla, 164
ripleyana, 164
vespertilio, 164
Dresslerella, 161, 164–165
elvallensis, 165
hispida, 165
pertusa, 165
pilosissima, 165, Pl. 20(4)
powellii, 165
stellaris, 165
Dressleria, 30, 133, 135, 138–139
sp., 139
dilecta, 139, Pl. 15(5)
helleri, 139
suavis, 139
Dryadella, 160, 165–166
butcheri, 166, Pl. 21(2)
dressleri, 166
gnoma, 166
guatemalensis, 166
minuscula, 166

odontostele, 166
sororcula, 166

Ecology of orchids, 24–32
Elleanthus, 278, 281, 298–301
albertii, 300
aurantiacus, 299
caricoides, 299, 300
curtii, 300
cynarocephalus, 299
fractiflexus, 301
glaucophyllus, 300
graminifolius, 301
hymenophorus, 300
near *isochiloides*, 301
jimenezii, 300
lancifolius, 300
laxus, 300
lentii, 299, Pl. 35(5)
linifolius, 301
muscicola, 300
poiformis, 301, Pl. 36(1)
robustus, 300
stolonifer, 301
tillandsioides, 301
tonduzii, 299, Pl. 35(6)
wercklei, 299
xanthocomos, 299
Eloyella, 313
Eltroplectris roseoalba, 278, 288, 301
Encyclia, 7, 22, 31, 55–60
abbreviata, 60
aemula, 59
alata, 56, Pl. 2(4)
amanda, 56
baculus, 60
brassavolae, 58, Pl. 2(6)
campylostalix, 58
ceratistes, 57
chacaoensis, 60
chimborazoensis, 59
cochleata, 60
cordigera, 55, Pl. 2(1, 2)
cordigera var. *rosea*, 55
fortunae, 58
fragrans, 59, Pl. 3(2)
fragrans subsp. *aemula*, 59
gravida, 56
guatemalensis, 56
ionocentra, 58
ionophlebia, 60, Pl. 3(1)
livida, 58
luteorosea, 55
mooreana, 56, Pl. 2(3)

neurosa, 60
ochracea, 57
polybulbon, 55
prismatocarpa, 58, Pl. 2(5)
pseudopygmaea, 57
pygmaea, 57
radiata, 60
selligera, 56
sima, 59
spondiada, 59
subgenus *Dinema*, 55
subgenus *Encyclia*, 49, 55, 57
subgenus *Osmophyta*, 50, 55, 57
tuerckheimii, 56
vagans, 60
varicosa, 58
vespa, 59, Pl. 3(3)
Epidanthus, 83
crassus, 83
Epidendropsis vincentina, 69
Epidendrum, 17, 28, 39, 40, 45, 61–84
acunae, 80
adnatum, 66
albertii, 66
alfaroi, 84
allenii, 67
amparoanum, 74
anastasioi, 76
anceps, 68
anoglossoides, 81
anoglossum, 84
aporum, 78
arcuiflorum, 65, 68
atropurpureum, 55
barbae, 77
barbeyanum, 74, Pl. 5(1)
baumannianum, 67–68
bilobatum, 78
bisulcatum, 77
blepharistes, 63
brachybotrys, 66
bracteosum, 63
brenesii, 77
candelabrum, 75
carolii, 69
carpophorum, 73
ciliare, 64, Pl. 6(1)
circinatum, 78
cocleense, 79
concavilabium, 78
confertum, 80
congestoides, 74
congestum, 74

Epidendrum (cont.)
  cordiforme, 81
  coriifolium, 78
  coronatum, 69, 72
  criniferum, 65
  cristobalense, 69
  cryptanthum, 77
  cuneatum, 84
  curvicolumna, 76
  curvisepalum, 80
  dentiferum, 75
  dentilobum, 71
  difforme, 73
  dolabrilobum, 63
  dolichostachyum, 69
  dosbocasense, 82
  eburneum, 73, Pl. 4(1)
  ellipsophyllum, 63
  epidendroides, 84
  estrellense, 336
  exiguum, 80, 82
  exile, 80
  firmum, 75
  flexicaule, 82
  flexuosissimum, 70
  fuscopurpureum, 78
  galeottianum, 68, Pl. 5(5)
  glumibracteum, 65
  goniorhachis, 83
  gregorioi, 76
  guanacastense, 79
  hammelii, 75
  hawkesii, 67
  hellerianum, 68, 73
  hunterianum, 76
  ibaguense, 336
  imatophyllum, 67
  incisum, 68
  incomptum, 77
  insolatum, 83
  insulanum, 79
  isomerum, 79
  isthmii, 71
  kerichilum, 75, 76
  lacustre, 65
  lagenocolumna, 75
  lancilabium, 79, Pl. 5(4)
  lankesteri, 66
  laterale, 63
  latifolium, 73
  laucheanum, 69, Pl. 5(3)
  lockhartioides, 78
  luerorum, 78
  lutheri, 77
  macroclinium, 70
  macrostachyum, 76
  majale, 75
  mantis-religiosae, 75
  microcardium, 336
  microdendron, 82
  microphyllum, 72
  mirabile, 80, Pl. 4(4)
  miserrimum, 70, 83
  modestiflorum, 82
  mora-retanae, 81
  moyobambae, 69, 72
  muscicola, 83
  myodes, 69
  nervosiflorum, 84
  nitens, 79
  nocturnum, 73
  notabile, 73
  nubium, 77
  nutantirachis, 82
  obesum, 65
  obliquifolium, 66
  octomerioides, 64, 72
  odontochilum, 75
  oerstedii, 64
  oxyglossum, 80
  pachyceras, 76
  pachyrhachis, 68, 72
  pallens, 70
  palmense, 78
  panamense, 67
  paniculatum, 70
  paranthicum, 83
  parkinsonianum, 63, Pl. 4(2)
  paucifolium, 70, 72
  pendens, 74, Pl. 5(2)
  pentotis, 60
  peperomia, 74
  pfavii, 65, Pl. 4(6)
  phragmites, 66
  phyllocharis, 67, 73
  physodes, 70
  piliferum, 71, Pl. 6(2)
  platystigma, 81
  pleurothalloides, 78
  polyanthum, 69
  polyanthum var. *myodes*, 69
  polychlamys, 77
  porpax, 74
  powellii, 68
  probiflorum, 69
  pseudepidendrum, 68
  pseudoramosum, 82
  pudicum, 74
  purpurascens, 65
  puteum, 71, 77
  radicans, 28, 67, Pl. 4(5)
  rafael-lucasii, 81
  ramonianum, 81
  ramosissimum, 79
  ramosum, 82
  raniferum, 65
  repens, 80
  resectum, 71
  rigidiflorum, 76
  rigidum, 79
  rousseauae, 41, 63, Pl. 5(6)
  rugosum, 81
  sanchoi, 81
  sanchoi var. *exasperatum*, 80
  sancti-ramoni, 83
  santaclarense, 81
  schlechterianum, 74
  schomburgkii, 68
  sculptum, 80
  selaginella, 70, 83
  simulacrum, 76
  smaragdinum, 68, 72
  sobralioides, 72
  stamfordianum, 6, 41, 63, 64
  stangeanum, 64
  stevensii, 82
  storkii, 74
  strobiliferum, 82
  subnutans, 71
  summerhayesii, 78
  talamancanum, 83
  tenuisulcatum, 84
  trachythece, 81
  trialatum, 76
  triangulabium, 72
  trianthum, 65, 73
  turialvae, 70
  veraguasense, 82
  vincentinum, 69
  volutum, 64
  warscewiczii, 72
  wercklei, 84, Pl. 4(3)
Epilyna jimenezii, 300
Eriopsis biloba, 265, 269, Pl. 32(2)
Erythrodes, 278, 283, 301–302
  calophylla, 302
  killipii, 302, Pl. 39(2)
  lehmannii, 337
  lunifera, 302
  nigrescens, 337
  purpurea, 302
  vescifera, 302
Eufriesea, 29

Index

Euglossa, 29, 31
Eulaema, 29
Eulophia alta, 26, 216, 264, 269, Pl. 32(3)
Eurystyles, 278, 284, 302–303
  auriculata, 303
  cotyledon, 303, Pl. 38(4)
  standleyi, 303
Exaerete, 28

Fernandezia, 10, 99, 100, 110
Flor de Espíritu Santo, 150–151
Fregea amabilis, 321
Funkiella stolonifera, 278, 287, 303

Galeandra, 263, 270
  batemanii, 270, Pl. 32(4)
  baueri, 270
  beyrichii, 270
  juncea, 270
  styllomisantha, 270
Galeottia grandiflora, 219, 230, Pl. 28(4)
Galeottiella, 278, 285, 303
  minutiflora, 304
  nutantiflora, 304
Gomphichis costaricensis, 278, 289, 304
Gongora, 30, 143, 146–148
  amparoana, 146
  armeniaca, 148
  armeniaca subsp. *armeniaca*, 148
  armeniaca subsp. *cornuta*, 148
  armeniaca var. *bicornuta*, 147
  atropurpurea, 147
  charontis, 147
  claviodora, 147
  fulva, 147
  gibba, 147
  horichiana, 147, Pl. 17(1)
  quinquenervis, 147
  tricolor, 147, Pl. 17(2)
  truncata, 147
  unicolor, 147
Goniochilus leochilinus, 106, 110, Pl. 12(2)
Goodyera, 278, 283, 304–305
  bradeorum, 338
  erosa, 304
  major, 304

  micrantha, 338
  modesta, 338
  ovatilabia, 304
  turrialbae, 338
Goodyerinae, 278, 283
Govenia, 26, 263, 270–271
  sp., Pl. 32(5)
  capitata, 271
  ciliilabia, 271
  liliacea, 271
  superba, 271
  utriculata, 271
Guaria de Turrialba, 53
Guaria morada, 53

Habenaria, 18, 23, 283, 305–307
  alata, 305
  amparoana, 338
  avicula, 306
  aviculoides, 306
  bicornis, 306
  bractescens, 306
  brenesii, 338
  clypeata, 307
  distans, 306
  endresiana, 338
  entomantha, 306, Pl. 36(6)
  floribunda, 305
  gymnadenioides, 338
  irazuensis, 338
  jimenezii, 338
  lankesteri, 305
  leprieuri, 306
  mesodactyla, 306
  monorrhiza, 307
  novemfida, 306
  odontopetala, 305
  pauciflora, 307
  quinqueseta, 306
  repens, 307
  trifida, 307
  verecunda, 306
  wercklei, 338
Hapalorchis, 278, 288, 307
  cheirostyloides, 307
  lineatus, 307
  pumilus, 307
Harrisella porrecta, 204
Helleriella nicaraguensis, 47, 84–85
Hexisea, 48, 85
  bidentata, 85, Pl. 8(5)
  imbricata, 85
  sigmoidea, 95
Holy Ghost orchid, 150–151

Homalopetalum pumilio, 50, 85, Pl. 3(4)
Horichia dressleri, 144, 148, Pl. 18(2)
Houlletia, 144, 148
  landsbergii, 148
  odoratissima, 148
  tigrina, 148, Pl. 16(3)
Huntleya, 217, 221, 230–231
  burtii, 231, Pl. 29(2)
  fasciata, 231, Pl. 29(1)
Hybochilus inconspicuus, 105, 110, Pl. 12(3)

Ionopsis, 101, 105, 111
  satyrioides, 111
  utricularioides, 111, Pl. 13(4)
Isochilus, 47, 85–86
  amparoanus, 86
  carnosiflorus, 86
  latibracteatus, 86
  linearis, 86
  major, 86, Pl. 7(4)

Jacquiniella, 46, 86–87
  aporophylla, 87, Pl. 7(5)
  cobanensis, 87
  equitantifolia, 87
  globosa, 87
  pedunculata, 87
  standleyi, 87
  teres, 87
  teretifolia, 87

Kefersteinia, 7, 30, 222, 231–233
  alba, 232
  auriculata, 232
  costaricensis, 233
  deflexipetala, 232
  lactea, 231, Pl. 28(5)
  maculosa, 233
  microcharis, 232
  mystacina, 231
  parvilabris, 232, Pl. 28(6)
  wercklei, 233
Kegeliella, 7, 30, 142, 148–149
  atropilosa, 149
  houtteana, 149
  kupperi, 149, Pl. 18(5)
Key, general, 41–44
Koellensteinia, 219, 233
  kellneriana, 233
  lilijae, 233

Kreodanthus, 278, 283, 307–308
  secundus, 308

Lacaena, 143, 149
  bicolor, 149
  spectabilis, 149
Laelia rubescens, 50, 87, 88, Pl. 1(4)
Laeliinae, 45
Lanium, 88, 92
  microphyllum, 72
Lankesterella, 278, 285, 308
  sp., 308, Pl. 38(5)
  orthantha, 308
Lemboglossum, 106, 111–112
  bictoniense, 111, Pl. 11(4)
  cordatum, 112
  hortensiae, 112
  maculatum, 111
  stellatum, 111
Leochilus, 106, 112
  labiatus, 112
  scriptus, 112, Pl. 12(4)
  tricuspidatus, 112
Lepanthes, 156, 157, 166–167
  elata, Pl. 19(1)
  species list, 339–340
Lepanthopsis, 157, 167
  comet-halleyi, 167
  floripecten, 167, Pl. 19(2)
  obliquipetala, 167, 191
Leucohyle subulata, 102, 112
Ligeophila clavigera, 278, 284, 308
Liparis, 265, 271–272
  elata, 271
  fantastica, 271
  fratrum, 272
  nervosa, 263, 271, Pl. 33(2)
  vexillifera, 272
Lockhartia, 20, 29, 99, 100, 113–114
  acuta, 113
  amoena, 113, Pl. 13(5)
  chocoensis, 114
  dipleura, 340
  hercodonta, 113
  integra, 114
  micrantha, 113
  obtusata, 113
  odontochila, 341
  oerstedii, 113
  pittieri, 114
Lycaste, 20, 29, 218, 234–235

  bradeorum, 234, Pl. 25(2)
  brevispatha, 234
  campbellii, 234
  dowiana, 235
  leucantha, 235
  macrophylla, 235
  powellii, 235
  schilleriana, 235
  tricolor, 234, Pl. 25(3)
  xytriophora, 235
Lyroglossa pubicaulis, 278, 287, 308

Macradenia brassavolae, 29, 30, 104, 114, Pl. 13(6)
Macroclinium, 29, 99, 101, 114–115
  bicolor, 114
  cordesii, 115
  glicensteinii, 115
  junctum, 115
  lineare, 114
  paniculatum, 115
  ramonense, 114, Pl. 14(2)
  simplex, 114
Malaxis, 262, 265, 272–275, 282
  adolphi, 274
  aurea, 273
  blephariglottis, 41, 272, 273
  brachyrrhynchos, 275
  carnosa, 274
  crispifolia, 274
  excavata, 274
  fastigiata, 275
  histionantha, 274, Pl. 33(3)
  lagotis, 341
  majanthemifolia, 273
  nana, 274
  pandurata, 274
  parthonii, 274
  pittieri, 273
  simillima, 274
  soulei, 273
  tonduzii, 274
  tipuloides, 41, 273, Pl. 33(4)
  wendlandii, 273
  woodsonii, 274
Masdevallia, 28, 160, 167–177
  attenuata, 170, 171
  calura, 170, 171
  chasei, 169, 171
  chontalensis, 169, 171
  collina, 169, 172

  cupularis, 169, 170, 172
  demissa, 170, 172
  erinacea, 169, 172
  flaveola, 169, 172
  floribunda, 169, 172
  fulvescens, 171, 173
  lata, 170, 173
  laucheana, 170, 173
  livingstoneana, 168, 173
  marginella, 170, 173
  molossoides, 169, 173
  nicaraguae, 168, 174, Pl. 21(3)
  nidifica, 169, 174
  pelecaniceps, 168, 174
  picturata, 169, 174
  pleurothalloides, 168, 174
  pygmaea, 168, 174
  rafaeliana, 169, 175
  reichenbachiana, 171, 175
  rolfeana, 170, 175
  scabrilinguis, 170, 175
  schizopetala, 170, 175
  schroederiana, 171, 175, Pl. 21(4)
  striatella, 168, 169, 176
  thienii, 171, 176
  tokachiorum, 169, 176
  tonduzii, 170, 176
  tubuliflora, 168, 176
  utriculata, 168, 176
  walteri, 170, 177
  zahlbruckneri, 169, 177, Pl. 21(5)
Maxillaria, 7, 20, 23, 40, 99, 220, 221, 222, 235–252
  aciantha, 242
  acostaei, 240, 241
  acutifolia, 239
  adendrobium, 251
  alba, 244
  albertii, 342
  alfaroi, 246, 250
  allenii, 251
  amparoana, 241, Pl. 27(1)
  ampliflora, 236, 247, Pl. 28(3)
  anceps, 244
  angustisegmenta, 238, 240
  angustissima, 237
  appendiculoides, 251
  arachnitiflora, 237
  attenuata, 237
  aurea, 250
  bicallosa, 252
  biolleyi, 250

Maxillaria (cont.)
  brachybulbon, 236, 239
  bracteata, 246, 248
  bradeorum, 247
  brenesii, 238, 242
  brevilabia, 245
  brevipes, 240
  brunnea, 241
  caespitifica, 244
  camaridii, 243
  campanulata, 247
  chartacifolia, 252
  cobanensis, 239
  concavilabia, 245
  conduplicata, 250
  confusa, 241
  costaricensis, 244
  crassifolia, 236
  cryptobulbon, 240
  ctenostachya, 248
  cucullata, 240
  curtipes, 244
  dendrobioides, 251
  densa, 243
  discolor, 238
  diuturna, 244, 247, 248
  elatior, 248
  endresii, 238
  exaltata, 248, 250
  falcata, 250
  flava, 249
  foliosa, 246
  friedrichsthalii, 242
  fulgens, 246
  hedwigae, 239
  hematoglossa, 240
  hennissiana, 238
  horichii, 245
  houtteana, 244
  inaudita, 250
  insolita, 243
  linearifolia, 251
  longipetiolata, 240
  lueri, 242, 248
  luteoalba, 238, 240
  maleolens, 238
  meridensis, 248, 250
  microphyton, 249
  minor, 249
  nasuta, 239
  neglecta, 243, 245
  nicaraguensis, 251
  oreocharis, 247
  pachyacron, 237
  paleata, 247
  parviflora, 245
  piestopus, 239
  pittieri, 246
  planicola, 247
  ponerantha, 246
  powellii, 241
  pseudoneglecta, 245, Pl. 27(5)
  pubilabia, 238, 242
  quadrata, 343
  ramonensis, 241
  reichenheimiana, 237, Pl. 27(3)
  repens, 245
  ringens, 238, 241
  rodrigueziana, 237, Pl. 27(2)
  rousseauae, 238, 242
  rubrilabia, 240
  rufescens, 239
  sanguinea, 243, Pl. 28(2)
  scorpioidea, 242
  semiorbicularis, 249
  serrulata, 248
  sigmoidea, 246
  splendens, 237, 241
  strumata, 243
  suaveolens, 246, 249
  subulifolia, 242
  tenuifolia, 243
  tigrina, 249
  trilobata, 250
  tonduzii, 250
  umbratilis, 247
  uncata, 242
  vagans, 246, 248
  vaginalis, 248, 251
  valenzuelana, 252
  valerioi, 251
  variabilis, 244
  vittariifolia, 239
  wercklei, 249, Pl. 27(4)
  wrightii, 246, Pl. 28(1)
Mendoncella grandiflora, 230
Mesadenella, 278, 286, 309
  petenensis, 309
  tonduzii, 309
Mesadenus polyanthus, 278, 287, 309
Mesospinidium, 105, 115–116
  endresii, 116
  horichii, 116
  panamense, 116
  warscewizcii, 116, Pl. 12(5)
Miltonia, 99, 100
  clowesii, 120
  endresii, 116
  schroederiana, 120
  warscewiczii, 120
Miltoniopsis, 107, 116
  roezlii, 6, 116, Pl. 10(5)
  warscewiczii, 116
Monophyllorchis maculata, 280, 309, Pl. 36(5)
  microstyloides, 309
Mormodes, 31, 32, 134, 139–141
  atropurpurea, 140
  cartonii, 140
  colossa, 141
  flavida, 141
  fractiflexa, 141
  hookeri, 140
  horichii, 141
  ignea, 141
  lancilabris, 141
  lobulata, 140
  powellii, 141
  punctata, 140
  skinneri, 140, Pl. 15(6)
Mormolyca ringens, 222, 252, Pl. 29(3)
Myoxanthus, 156, 158, 161, 177–179
  aspacicensis, 178
  balaeniceps, 179, Pl. 22(3)
  colothrix, 178
  hirsuticaulis, 177, Pl. 22(2)
  lappiformis, 179
  octomeriae, 178
  pan, 179
  scandens, 178
  sempergemmatus, 178
  speciosus, 178
  stonei, 179
  trachychlamys, 178, 179
  uncinatus, 178, Pl. 22(4)
Myrmecophila, 49, 88
  brysiana, 88
  thomsoniana, 88
  tibicinis, 88, Pl. 1(5)

Names of orchids, 39
Neomoorea, 218, 252
  irrorata, 252
  wallisii, 252, Pl. 25(4)
Neourbania adendrobium, 251
Neowilliamsia, 39, 84
  cuneata, 39, 84
  tenuisulcata, 84
Nidema, 50, 88
  boothii, 88
  ottonis, 88, Pl. 8(4)

Notylia, 29, 30, 104, 116–117
  albida, 117, Pl. 14(3)
  barkeri, 117
  brenesii, 344
  lankesteri, 344
  latilabia, 117
  panamensis, 117
  pentachne, 117
  trisepala, 344
  turialbae, 344

Octomeria, 161, 179–180
  apiculata, 179
  costaricensis, 180
  graminifolia, 179
  hondurensis, 179
  surinamensis, 180
  valerioi, 180, Pl. 19(3)
Odontoglossum, 99
Oeceoclades maculata, 26, 265, 275, Pl. 32(6)
Oerstedella, 47, 89–91
  aberrans, 344
  acrochardonia, 344
  caligaria, 90
  centradenia, 90, Pl. 6(6)
  centropetala, 90
  crescentiloba, 90
  endresii, 91, Pl. 6(4)
  exasperata, 89
  fuscina, 89
  intermixta, 89
  lactea, 91
  ornata, 89
  pajitensis, 91
  pansamalae, 90
  pentadactyla, 90
  pinnifera, 90, Pl. 6(5)
  pseudoschummaniana, 6, 91
  pseudowallisii, 91
  pumila, 91
  schummaniana, 91
  tetraceros, 89
  wallisii, 91, Pl. 6(3)
Oncidiinae, 99
Oncidium, 23, 27, 31, 40, 100, 101, 102, 107, 117–124
  advena, 344
  altissimum, 118
  ampliatum, 6, 119
  angustisepalum, 344
  ansiferum, 123
  anthocrene, 121
  ascendens, 118
  baueri, 123
  bracteatum, 123
  cabagrae, 124
  cariniferum, 120
  carthagenense, 118, Pl. 9(1)
  cebolleta, 118
  cheirophorum, 119
  costaricense, 344
  crista-galli, 121
  dichromaticum, 344
  endocharis, 120
  ensatum, 122
  exalatum, 119
  exauriculatum, 119
  fuscatum, 120, Pl. 9(5)
  globuliferum, 122
  graminifolium, 120
  guttatum, 118
  guttulatum, 123
  heteranthum, 119, Pl. 9(2)
  incurvum, 122
  isthmii, 124
  klotzschianum, 123
  luridum, 118
  luteum, 119
  maculatum, 121
  nudum, 118
  obryzatum, 122, Pl. 9(6)
  ochmatochilum, 122
  ornithorrhynchum, 119
  paleatum, 121
  panamense, 31, 124
  panduriforme, 123
  parviflorum, 123
  peliogramma, 344
  pergameneum, 345
  pittieri, 119
  planilabre, 123
  powellii, 121
  scansor, 122
  schroederianum, 120, Pl. 9(3)
  sphacelatum, 124
  stenoglossum, 120
  stenotis, 121, Pl. 9(4)
  stipitatum, 118
  storkii, 121
  teres, 118
  tetraskelidion, 344
  turialbae, 344
  warscewiczii, 121
  wentworthianum, 344
Ophidion pleurothallopsis, 159, 180, Pl. 22(1)
Orchids
  classification of, 33–36
  ecology of, 24–32
  names and synonyms of, 39
  pollination of, 26–32
  structure of, 12–24
  study of, in Costa Rica and Panama, 13–14
Ornithocephalus, 127, 220, 252–253
  bicornis, 253
  cochleariformis, 253
  inflexus, 253, Pl. 30(3)
  lankesteri, 253
  powellii, 253
  valerioi, 345
Osmoglossum, 104, 124–125
  anceps, 124
  convallarioides, 125
  egertonii, 124, Pl. 10(2)
  pulchellum, 125
Otoglossum chiriquense, 107, 125, Pl. 10(3)

Pachyphyllum, 10, 100, 125
  costaricense, 110, 125
  crystallinum, 125
  hispidulum, 125, Pl. 36(2)
  pastii, 125
Palmorchis, 31, 281, 310
  nitida, 310
  powellii, 310
  silvicola, 310, Pl. 36(2)
  trilobulata, 310
Palumbina, 124
Panama
  climate of, 3–5
  geography of, 1–3
  orchid study in, 13–14
  people and cuture of, 11–13
  vegetation of, 5–11
Paphinia, 142, 149–150
  "*clausula*," 150
  cristata?, 150, Pl. 16(4)
  cristata var. *modiglianiana*, 150
Pelexia, 278, 286, 310–311
  adnata, 311
  callifera, 311
  congesta, 311
  funckiana, 311, Pl. 37(1)
  laxa, 311
  obliqua, 311
  olivacea, 311
  smithii, 311, Pl. 37(2)
Peristeria, 30, 142, 150–151
  sp., 150, Pl. 16(6)

Index

Peristeria (cont.)
  cochlearis, 150
  elata, 30, 150, Pl. 16(5)
Pescatorea, 217, 221, 253
  cerina, 253, Pl. 26(2)
  dayana, 253
Phragmipedium, 28, 277, 279, 312
  caudatum, 312, Pl. 34(3)
  longifolium, 312, Pl. 34(4)
  warscewiczianum, 312
Phymatidium panamense, 220, 312
Platyglottis, 47, 85, 92
Platyglottis coriacea, 47, 92
Platystele, 158, 180–182
  brenneri, 182
  calymma, 181
  caudatisepala, 182
  compacta, 182
  dressleri, 181
  jungermannioides, 180
  lancilabris, 182
  microtatantha, 182
  minimiflora, 180
  ortiziana, 181
  ovalifolia, 181, Pl. 19(4)
  ovatilabia, 181
  oxyglossa, 182
  perpusilla, 181
  propinqua, 182
  resimula, 182
  stenostachya, 181
  taylori, 182
Platythelys, 278, 284, 313
  maculata, 313
  vaginata, 313
  venustula, 313, Pl. 39(3)
Plectrophora alata, 103, 126
Pleurothallidinae, 156
Pleurothallis, 20, 40, 156, 157, 158, 160, 161, 183–208
  abbreviata, 205
  aberrans, 197
  abjecta, 199
  Acianthera group, 183, 189
  acicularis, 200
  acostaei, 346
  acrisepala, 201
  alexandrae, 178
  alexii, 200
  allenii, 186
  alpina, 197
  amparoana, 192, 202
  angusta, 346
  annectans, 187
  antonensis, 189
  archicolonae, 186
  areldii, 206
  arietina, 184
  aristata, 204
  aryter, 200, 204
  aurita, 185
  barboselloides, 201
  barbulata, 199
  brighamii, 201
  broadwayi, 213
  butcheri, 192
  cachensis, 346
  cactantha, 199
  caligularis, 196
  calyptrosepala, 207
  calyptrostele, 205
  campicola, 196
  canae, 202
  caniceps, 184
  cardiochila, 188
  cardiothallis, 184, 188, Pl. 23(2)
  carnosilabia, 196
  carpinterae, 193
  casualis, 199
  circumplexa, 197, Pl. 23(4)
  citrophila, 198
  cobanensis, 207
  cobraeformis, 186
  cogniauxiana, 190, Pl. 24(1)
  condylata, 200
  connata, 167, 191
  convallaria, 193
  corniculata, 201
  costaricensis, 207
  crassilabia, 347
  crescentilabia, 185, Pl. 23(1)
  crocodiliceps, 184
  cucumeris, 196
  cuspidata, 204, 205
  decipiens, 198
  dentipetala, 191
  deregularis, 191
  divaricans, 196
  dolichopus, 192
  dorotheae, 186
  dracontea, 194
  dressleri, 208
  ellipsophylla, 198
  endotrachys, 192, 199
  erinacea, 192, 198
  eumecocaulon, 184
  excavata, 186
  exesilabia, 204
  falcatiloba, 203
  fantastica, 184
  floribunda, 191, 195
  foliata, 213
  fortunae, 194
  fractiflexa, 202
  fuegii, 206
  fuegii var. *echinata*, 206
  fulgens, 200
  gacayana, 197
  gelida, 191
  geminicaulina, 198, Pl. 23(6)
  geminiflora, 205, 207
  ghiesbreghtiana, 194
  glandulosa, 199
  grandis, 192
  grobyi, 206, 207
  guanacastensis, 200
  guttata, 205
  harpago, 184
  hastata, 199
  helleri, 195
  hemileuca, 187
  herpestes, 204
  homalantha, 189
  homalanthoides, 188
  imago, 186, 187
  immersa, 202
  imraei, 190, 195
  instar, 184
  isthmica, 187
  janetiae, 205
  johnsonii, 189, 196
  juxtaposita, 190
  lanceana, 194
  lanceola, 207
  lateritia, 207
  lentiginosa, 185, 196
  lepidota, 196
  leucantha, 186
  leucopyramis, 347
  lewisae, 208, Pl. 23(5)
  listerophora, 196
  longipedicellata, 191
  loranthophylla, 194
  luctuosa, 195
  macrantha, 193
  mammillata, 193
  microphylla, 207
  minor, 346
  minuta, 207
  monocardia, 346
  mystax, 192
  naraniensis, 347
  nemorum, 189
  nervosa, 187

Pleurothallis (cont.)
  nitida, 187
  obovata, 185
  oscitans, 192
  pacayana, 197
  pachyglossa, 194
  pallida, 195
  palliolata, 188, Pl. 23(3)
  pantasmi, 197
  peculiaris, 187
  peperomioides, 208
  periodica, 201
  pfavii, 199
  phyllocardia, 189
  phyllocardioides, 188
  picta, 206
  platycardia, 346
  platysepala, 346
  polygonoides, 201, 202, 206
  polysticta, 195
  pompalis, 193, 202
  powellii, 191
  praegrandis, 190
  pruinosa, 195
  pubescens, 198
  pyrsodes, 200
  quadrifida, 194
  quinqueseta, 204
  racemiflora, 194, Pl. 24(3)
  radula, 188
  ramonensis, 194
  rectipetala, 346
  rhodoglossa, 189
  rowleei, 193
  rubella, 203
  ruscifolia, 183, 184, 185
  saccata, 190
  saccatilabia, 185
  samacensis, 203
  sanchoi, 187
  scitula, 189
  sclerophylla, 192
  segoviensis, 203, 206
  segregatifolia, 207
  sertularioides, 206
  setigera, 204
  setosa, 203
  sicaria, 197
  simplex, 346
  simulans, 186
  stevensii, 190, 191
  strumosa, 203
  telamon, 187
  testaefolia, 199, 208
  thymochila, 194
  titan, 188
  tonduzii, 184
  tribuloides, 159, 205, Pl. 24(2)
  tripterantha, 205
  tuerckheimii, 193
  turrialbae, 199
  undulata, 188
  uniflora, 202
  ventricosa, 185
  veraguacensis, 189
  verecunda, 198
  vitariifolia, 200
  volcanica, 191, 195
  wercklei, 202
Pollination, 26–32
Polycycnis, 144, 151
  barbata, 151, Pl. 18(4)
  gratiosa, 151
  lehmannii, 151
  muscifera, 151
  ornata, 151, Pl. 18(3)
  tortuosa, 151
Polystachya, 44, 49, 265, 275–276
  cerea, 276
  cingulata, 346
  concreta, 276
  flavescens, 276
  foliosa, 276, Pl. 33(1)
  luteola, 276
  masayensis, 276
  minuta, 276
Ponera striata, 47, 92, Pl. 7(6)
Ponthieva, 277, 289, 313–314
  brenesii, 314
  ephippium, 314
  formosa, 314
  maculata, 314, Pl. 40(3)
  racemosa, 314
  tuerckheimii, 314
Prescottia, 278, 289, 314–315
  cordifolia, 315, Pl. 40(6)
  oligantha, 314
  stachyodes, 315
Prescottiinae, 278
Prosthechea, 57
Pseudocentrum hoffmannii, 277, 288, 315
Psilochilus, 31, 282, 315–316
  macrophyllus, 316
  modestus, 315
  physurifolius, 315, Pl. 39(4)
Psychilus, 55
Psychopsis krameriana, 102, 107, 126, Pl. 10(6)
Psygmorchis, 100, 101, 126
  glossomystax, 126
  pumilio, 126, Pl. 14(4)
  pusilla, 126
Pterichis, 277, 289, 316
  costaricensis, 316
  habenarioides, 316, Pl. 40(4)
  leo, 316

Quince de septiembre, 51

Reichenbachanthus, 46, 48, 92–93
  cuniculatus, 92
  lankesteri, 92, Pl. 8(6)
  reflexus, 93
Restrepia, 161, 209
  lankesteri, 209
  muscifera, 209
  subserrata, 209, Pl. 20(5)
  xanthophthalma, 209
Restrepiella ophiocephala, 162, 209, Pl. 24(4)
Restrepiopsis, 162, 209–210
  reichenbachiana, 209
  tubulosa, 210
  ujarrensis, 210, Pl. 19(5)
Rodriguezia, 103, 126–127
  compacta, 127
  lanceolata, 127, Pl. 13(2)
  secunda, 127
Rosario, 69
Rossioglossum schlieperianum, 107, 127, Pl. 10(4)
Rudolfiella. See *Bifrenaria*

Salpistele, 159, 210
  brunnea, 210, Pl. 19(6)
  dressleri, 210
  lutea, 210
  parvula, 210
Sarcoglottis, 278, 286, 316–317
  hunteriana, 317, Pl. 37(3)
  neglecta, 317
  valida, 311
  woodsonii, 316
Scaphosepalum, 159, 210–211
  anchoriferum, 211, Pl. 21(6)
  clavellatum, 211
  microdactylum, 211
  viviparum, 211
Scaphyglottis, 17, 48, 50, 85, 93–98, 156
  acostaei, 93
  amethystina, 94

Index

Scaphyglottis (cont.)
  amparoana, 95, Pl. 8(2)
  arctata, 95
  behrii, 96
  bifida, 93
  bilineata, 97
  boliviensis, 95
  chlorantha, 94
  confusa, 94
  corallorhiza, 94, 95
  crurigera, 93
  cuneata, 97
  densa, 94
  fusiformis, 96
  geminata, 96, Pl. 8(1)
  gigantea, 95, 97
  gracilis, 95, 97
  huebneri, 95
  jimenezii, 94
  laevilabia, 95
  leucantha, 97
  limonensis, 98
  lindeniana, 93, Pl. 8(3)
  longicaulis, 97
  mesocopis, 94
  micrantha, 93
  minuta, 94
  minutiflora, 97
  panamensis, 97
  prolifera, 97
  pulchella, 96
  punctulata, 97
  robusta, 97
  sessiliflora, 96
  sigmoidea, 95
  spathulata, 96
  stellata, 94
  subulata, 94
  tenella, 96
Scelochilus tuerckheimii, 103, 127, Pl. 12(6)
Schiedeella, 278, 286, 317
  fauci-sanguinea, 317
  llaveana, 317, Pl. 38(6)
  trilineata, 317
  valerioi, 317
  wercklei, 317
Schlimia jasminodora, 144, 152
Schomburgkia, 50, 98
  lueddemannii, 98, Pl. 1(6)
  undulata, 6, 98
Scleria, 10
Selenipedium chica, 277, 279, 318, Pl. 34(5)
Semana santa, 55
Sepalosaccus strumatus, 243

Sievekingia, 143, 152
  butcheri, 152
  fimbriata, 152, Pl. 18(1)
  suavis, 152
Sigmatostalix, 27, 103, 127–128
  abortiva, 128
  brownii, 128, Pl. 14(1)
  guatemalensis, 128
  hymenantha, 128
  macrobulbon, 128
  picta, 128
  picturatissima, 128
  racemifera, 128
  unguiculata, 128
Sobralia, 31, 278, 281, 318–322
  allenii, 319
  amabilis, 321
  atropubescens, 321, Pl. 35(3)
  bletiae, 350
  bouchei, 321
  bradeorum, 321
  callosa, 319, Pl. 35(2)
  corazoi, 320
  decora, 322
  decora var. aerata, 321
  fenzliana, 322
  fimbriata, 320
  fragrans, 318
  helleri, 319
  labiata, 350
  lepida, 321
  lindleyana, 320
  luteola, 320
  macra, 319
  macrantha, 321, Pl. 35(4)
  macrophylla, 320
  mucronata, 320
  neglecta, 322
  pfavii, 319
  pleiantha, 320
  powellii, 320
  suaveolens, 319, Pl. 35(1)
  undatocarinata, 319
  valida, 320
  warscewiczii, 321
  wercklei, 321
Sobraliinae, 278
Solenocentrum costaricense, 277, 288, 322, Pl. 40(5)
Spathoglottis plicata, 264, 276, Pl. 31(6)
Sphyrastylis cryptantha, 220, 253–254

Spiranthes, 278, 287, 322
  amesiana, 322
  graminea, 322
  torta, 322
Spiranthinae, 278
Stanhopea, 30, 143, 152–154
  avicula, 154
  cirrhata, 154, Pl. 17(4)
  costaricensis, 154, Pl. 17(5)
  ecornuta, 154
  gibbosa, 154
  oculata, 153
  panamensis, 153
  platyceras, 153
  pulla, 154
  wardii, 153, Pl. 17(6)
  warscewicziana, 153
Stanhopeinae, 134, 135, 141–155
Stelis, 158, 211–212, Pl. 22(5, 6)
  species list, 351
Stellilabium, 221, 254–255
  boylei, 255
  bullpenense, 254, Pl. 30(4)
  campbellorum, 254
  distantiflorum, 255
  minutiflorum, 255
  monteverdense, 254
  lankesteri, 255
  standleyi, 255
Stenorrhynchos, 278, 287, 323
  lanceolatum, 323, Pl. 37(6)
  speciosum, 323, Pl. 37(5)
Structure of orchids, 12–24
Synonyms, 39
Systeloglossum, 23, 104, 129
  acuminatum, 129
  costaricense, 129
  panamense, 129, Pl. 12(1)

Telipogon, 10, 28, 43, 220, 255–258
  ampliflorus, 256
  ardeltianus, 256
  ballesteroi, 258
  biolleyi, 257, Pl. 29(5)
  caroliae, 256
  cascajalensis, 258
  christobalensis, 257
  costaricensis, 256
  dendriticus, 352
  elcimeyae, 256, Pl. 29(6)
  endresianum, 257
  glicensteinii, 257
  gracilipes, 256

Telipogon (cont.)
  guila, 257
  leila-alexandrae, 256
  monticola, 258
  parvulus, 257
  pfavii, 258
  portilloi, 256
  radiatus, 352
  retanarum, 257
  setosus, 255
  storkii, 258
  storkii subsp. *magnificus*, 258
  vampirus, 255
Teuscheria, 218, 258–259
  horichiana, 259
  pickiana, 259, Pl. 25(5)
Ticoglossum, 107, 129
  krameri, 129
  oerstedii, 129, Pl. 10(1)
Torito, 153
Torito sin cacho, 154
Tortuga, 119
Trevoria, 144, 154–155
  glumacea, 155, Pl. 18(6)
  zahlbruckneriana, 155
Trichocentrum, 29, 30, 102, 130–131
  brenesii, 130
  caloceras, 130, 131
  candidum, 130
  capistratum, 130
  panamense, 130
  pfavii, 130, Pl. 11(2)
Trichoceros, 28
Trichopilia, 29, 104, 131–132
  leucoxantha, 131
  maculata, 131
  marginata, 132
  punctata, 352
  suavis, 131, Pl. 11(1)
  tortilis, 131
  turialbae, 132
Trichosalpinx, 42, 157, 212–214

  arbuscula, 213
  blaisdellii, 214
  carinilabia, 213
  cedralensis, 212
  ciliaris, 214
  dura, 213
  intricata, 212
  membraniflora, 213
  memor, 214, Pl. 24(5)
  moschata, 213
  navarrensis, 212
  operculata, 213
  orbicularis, 213
  pergrata, 213
  rotundata, 213
  tantilla, 213
  tropida, 213
Trigonidium, 222, 259
  egertonianum, 259, Pl. 30(5)
  insigne, 259
  lankesteri, 259
  riopalenquense, 259
Triphora, 31, 282, 323–324
  gentianoides, 323
  mexicana, 324
  nitida, 324
  ravenii, 324
  trianthophora, 324
  wagneri, 323
Trisetella, 160, 214–215
  dressleri, 215
  tenuissima, 215
  triaristella, 214, Pl. 20(6)
  triglochin, 215
Tropidia polystachya, 279, 281, 324, Pl. 36(4)
Tropidieae, 278
Trizeuxis falcata, 101, 132, Pl. 14(5)
Turtle orchid, 119

Uleiorchis ulaei, 26, 278, 280, 324

Vanda, 18
Vanilla, 280, 324–326
  inodora, 325
  insignis, 326
  mexicana, 325
  odorata, 325
  pauciflora, 325
  pfaviana, 325
  planifolia, 326, Pl. 34(2)
  pompona, 325, Pl. 34(1)
Variation, 40–41

Warmingia margaritacea, 104, 132
Warrea costaricensis, 219, 259–260, Pl. 29(4)
Warreopsis parviflora, 219, 260
Wullschlaegelia, 26, 278, 280, 326
  aphylla, 326
  calcarata, 326, Pl. 33(5)

Xylobium, 219, 260–261
  colleyi, 260
  elongatum, 260, Pl. 25(6)
  foveatum, 261
  powellii, 260
  squalens, 261

Zamia pseudoparasitica, 7
Zootrophion, 161, 215–216
  atropurpureus, 215, Pl. 24(6)
  endresianus, 216
  gracilentus, 216
  hypodiscus, 216
  moorei, 215
  vulturiceps, 215
Zygopetalum umbonatum, 233